Neurology and Neurobiology

EDITORS
Victoria Chan-Palay and Sanford L. Palay
The Harvard Medical School

ADVISORY BOARD

Günter Baumgartner
University Hospital, Zürich

Gösta Jonnson
Karolinska Institute

Bruce McEwen
Rockefeller University

Masao Ito
Tokyo University

The Physiology
of Excitable Cells

Professor Susumu Hagiwara

The Physiology of Excitable Cells

Proceedings of a Symposium in Honor of
Professor Susumu Hagiwara
Held in Santa Monica, California
November 6–8, 1982

Editors

Alan D. Grinnell

Department of Physiology
University of California
Los Angeles

William J. Moody, Jr.

Department of Zoology
University of Washington
Seattle

ALAN R. LISS, INC., NEW YORK

Address all Inquiries to the Publisher
Alan R. Liss, Inc., 150 Fifth Avenue, New York, NY 10011

Copyright © 1983 Alan R. Liss, Inc.

Printed in the United States of America.

Library of Congress Cataloging in Publication Data

Main entry under title:

The physiology of excitable cells.

(Neurology and neurobiology ; v. 5)
Includes index.
1. Cell physiology—Congresses. 2. Excitation (Physiology)—Congresses. 3. Ion channels—Congresses. 4. Hagiwara, S. (Susumu), 1922- —Congresses. I. Hagiwara, S. (Susumu), 1922- . II. Grinnell, Alan, 1936- . III. Moody, William J. (William James), 1950- . [DNLM: 1. Neurons—Physiology—Congresses. 2. Ion channels—Physiology—Congresses. W1 NE337B v.5 / WL 102.5 P5786 1982]
QH631.P48   1983   591.87'6   83-12063
ISBN 0-8451-2704-7

Contents

SECTION V. SENSORY AND CNS PHYSIOLOGY

Contributors

Frances M. Ashcroft, University Laboratory of Physiology, Oxford University, Oxford OX1 3PT, United Kingdom [97]

Christiane Baud, Department of Physiology, and Jerry Lewis Neuromuscular Research Center, UCLA School of Medicine, Los Angeles, CA 90024 [257]

Malcolm S. Brodwick, Department of Physiology and Biophysics, University of Texas Medical Branch, Galveston, TX 77550 [149]

H. Mack Brown, Department of Physiology, University of Utah, Salt Lake City, UT 84108 [327]

Theodore H. Bullock, Department of Neurosciences, School of Medicine, University of California, San Diego, La Jolla, CA 92093 [587]

John Chad, Department of Biology, University of California, Los Angeles, CA 90024 [25]

Shiko Chichibu, Department of Physiology, Kinki University School of Medicine, Sayama, Osaka 589, Japan [515]

Sergio Ciani, Jerry Lewis Neuromuscular Research Center, UCLA School of Medicine, Los Angeles, CA 90024 [73, 83]

A.J. D'Alonzo, Jerry Lewis Neuromuscular Research Center, UCLA School of Medicine, Los Angeles, CA 90024 [435]

Joachim W. Deitmer, Abteilung für Biologie, Ruhr-Universität, D-4630 Bochum, Federal Republic of Germany [51]

Michael Delay, Department of Physiology, University of California, Los Angeles, CA 90024 [343]

Douglas C. Eaton, Department of Physiology and Biophysics, University of Texas Medical Branch, Galveston, TX 77550 [149]

Roger Eckert, Department of Biology, University of California, Los Angeles, CA 90024 [25]

Charles Edwards, Department of Biological Sciences, State University of New York, Albany, NY 12222 [497]

George Eisenman, Department of Physiology, University of California, Los Angeles, CA 90024 [191]

Douglas Ewald, Department of Biology, University of California, Los Angeles, CA 90024 [25]

Jun Fukuda, Department of Physiology, Faculty of Medicine, University of Tokyo, Bunkyo-ku, Tokyo 113, Japan [305]

Kaare M. Gautvik, Institute of Physiology, University of Oslo, Oslo 1, Norway [357]

Alan D. Grinnell, Jerry Lewis Neuromuscular Research Center, UCLA School of Medicine, Los Angeles, CA 90024 [xv,435]

Raphael Gruener, Department of Physiology, University of Arizona College of Medicine, Tucson, AZ 85724 [139]

Richard Gunning, Jerry Lewis Neuromuscular Research Center, UCLA School of Medicine, Los Angeles, CA 90024 [73]

The number in brackets is the opening page number of the contributor's article.

Susumu Hagiwara, Jerry Lewis Neuromuscular Research Center, UCLA School of Medicine, Los Angeles, CA 90024 **[13, 83]**

Egil Haug, Institute of Physiology, University of Oslo, Oslo 1, Norway **[357]**

Judith Heiny, Department of Physiology, University of California, Los Angeles, CA 90024 **[343]**

Maryanna Henkart, Immunology Branch, National Cancer Institute, Bethesda, MD 20205 **[371]**

A.A. Herrera, Department of Biology, University of Southern California, Los Angeles, CA 90007 **[435]**

Tomoo Hirano, Laboratory of Neurobiology, Institute of Brain Research, University of Tokyo, Tokyo, Japan **[279]**

Richard Horn, Department of Physiology, University of California, Los Angeles, CA 90024 **[181]**

Susanne L. Huttner, Department of Biology, University of California, Los Angeles, CA 90024 **[461]**

Kazuo Ikeda, Department of Neurosciences, City of Hope Research Institute, Duarte, CA 91010 **[411]**

Laurinda A. Jaffe, Department of Physiology, University of Connecticut Health Center, Farmington, CT 06032 **[211]**

Douglas Junge, School of Dentistry and Department of Physiology, University of California, Los Angeles, CA 90024 **[109]**

Raymond T. Kado, Laboratoire de Neurobiologie Cellulaire, Centre National de la Recherche Scientifique, 91190 Gif sur Yvette, France **[247]**

Ann E. Kammer, Division of Biology, Kansas State University, Manhattan, KS 66506 **[393]**

Raj Kapur, Department of Biology, University of California, Los Angeles, CA 90024 **[461]**

Satoru Kato, Department of Neurophysiology, Neuroinformation Research Institute, University of Kanazawa School of Medicine, 13-1 Takara-machi, Kanazawa 920, Japan **[535]**

Yasuji Katsuki, National Institute for Physiological Sciences, Myodaiji-machi, Okazaki, Japan **[505]**

Yoshiaki Kidokoro, Molecular Neurobiology Laboratory, The Salk Institute, San Diego, CA 92138 **[127]**

Frank Kirschbaum, Zoologisches Institut, Universität zu Köln, Weyertal 119, Köln Lindenthal, Federal Republic of Germany **[451]**

J.H. Koenig, Department of Neurosciences, City of Hope Research Institute, Duarte, CA 91010 **[411]**

Hiroyuki Koike, Department of Neurophysiology, Tokyo Metropolitan Institute for Neurosciences, 2-6 Musashidai, Fuchu City, Tokyo 183, Japan **[523]**

Yukio Komatsu, Department of Physiology, University of Nagoya School of Medicine, 65 Tsurumai, Showaku, Nagoya, Japan **[557]**

Sally Krasne, Department of Physiology, University of California, Los Angeles, CA 90024 **[83]**

Jeffry B. Lansman, Department of Physiology, and Jerry Lewis Neuromuscular Research Center, UCLA School of Medicine, Los Angeles, CA 90024 **[233]**

James Lechleiter, Department of Physiology, University of Arizona College of Medicine, Tucson, AZ 85724 **[139]**

Michael S. Letinsky, Department of Physiology, University of California, Los Angeles, CA 90024 **[423]**

Diana C. Linden, Department of Biology, Occidental College, Los Angeles, CA 90041 **[423]**

Pier L. Marchiafava, Istituto di Neurofisiologia del CNR, Via S. Zeno, 51, 56100-Pisa, Italy **[549]**

Robert W. Meech, Department of Physiology, University of Bristol, Bristol BS8 1TD, United Kingdom **[65]**

Shun-ichi Miyazaki, Department of Physiology, Jichi School of Medicine, Tochigi-ken 329-04, Japan **[83, 219]**

William J. Moody, Jr., Department of Zoology, University of Washington, Seattle, WA 98104 **[xv]**

Kathleen Morrison-Graham, Department of Biology, University of California, Los Angeles, CA 90024 **[461]**

Shigehiro Nakajima, Department of Biological Sciences, Purdue University, West Lafayette, IN 47907 **[7]**

Koroku Negishi, Department of Neurophysiology, Neuroinformation Research Institute, University of Kanazawa School of Medicine, 13-1 Takaramachi, Kanazawa 920, Japan **[535]**

Richard Nuccitelli, Department of Zoology, University of California, Davis, CA 95616 **[475]**

B.M. Nudell, Jerry Lewis Neuromuscular Research Center, UCLA School of Medicine, Los Angeles, CA 90024 **[435]**

Harunori Ohmori, Department of Neurophysiology, Institute of Brain Research, Tokyo University School of Medicine, Hongo 7-3-1, Bunkyo-ku, Tokyo, Japan **[13]**

Paul H. O'Lague, Department of Biology, University of California, Los Angeles, CA 90024 **[461]**

Yutaka Oomura, Department of Physiology, Faculty of Medicine, Kyushu University 60, Fukuoka 812, Japan **[569]**

Seiji Ozawa, Department of Physiology, Jichi School of Medicine, Tochigi-ken 329-04, Japan **[357]**

Joseph B. Patlak, Department of Physiology and Biophysics, University of Vermont, Burlington, VT 05405 **[165]**

P.A. Pawson, Jerry Lewis Neuromuscular Research Center, UCLA School of Medicine, Los Angeles, CA 90024 **[435]**

T.D. Plant, Physiologisches Institut, Universität des Saarlandes, D-6650 Homburg/Saar, Federal Republic of Germany **[39]**

Mary B. Rheuben, Department of Anatomy, Michigan State University, East Lansing, MI 48824 **[393]**

Bernard Ribalet, Department of Physiology, University of California, Los Angeles, CA 90024 **[343]**

Toshiie Sakata, First Department of Medicine, Kyushu University 60, Fukuoka 812, Japan **[569]**

Olav Sand, Department of Biology, University of Oslo, Blindern Oslo 3, Norway **[357]**

John Sandblom, Department of Medical Biophysics, University of Uppsala, Uppsala, Sweden **[191]**

Nobuaki Shimizu, Department of Physiology, Faculty of Medicine, Kyushu University 60, Fukuoka 812, Japan **[569]**

Kjersti Sletholt, Department of Physiology, Norwegian College of Veterinary Medicine, Oslo Dep, Norway **[357]**

Nicholas B. Standen, Department of Physiology, University of Leicester, Leicester LE1 7RH, United Kingdom **[39, 181]**

Peter R. Stanfield, Department of Physiology, Leicester University, Leicester LE1 7RH, United Kingdom **[97]**

Enrica Strettoi, Istituto di Neurofisiologia del CNR, Via S. Zeno, 51, 56100-Pisa, Italy **[549]**

Thomas Szabo, Neurophysiologie Sensorielle, Laboratoire de Physiologie Nerveuse, 91190 Gif sur Yvette, France **[451]**

Kunitaro Takahashi, Laboratory of Neurobiology, Institute of Brain Research, University of Tokyo, Tokyo, Japan **[279]**

Tsunenobu Teranishi, Department of Neurophysiology, Neuroinformation Research Institute, University of Kanazawa School of Medicine, 13-1 Takara-machi, Kanazawa 920, Japan **[535]**

R.C. Thomas, Department of Physiology, University of Bristol, Bristol BS8 1TD, United Kingdom **[65]**

Keisuke Toyama, Department of Physiology, Kyoto Prefectural School of Medicine, Kawaramachi, Hirokoji, Kamigyoku, Kyoto, Japan **[557]**

L.O. Trussell, Jerry Lewis Neuromuscular Research Center, UCLA School of Medicine, Los Angeles, CA 90024 **[435]**

Julio Vergara, Department of Physiology, University of California, Los Angeles, CA 90024 **[343]**

T.A. Ward, Department of Physiology, University of Leicester, Leicester LE1 7RH, United Kingdom **[39]**

Akira Watanabe, Department of Cell Physiology, National Institute for Physiological Sciences, Myodaiji-cho, Okazaki 444, Japan **[295]**

Reto Weiler, Zoologisches Institut der Universität, München, Federal Republic of Germany **[549]**

Keiji Yanagisawa, Department of Physiology, Tsurumi University School of Dental Medicine, 2-1-3 Tsurumi, Tsurumi-ku, Yokohama, Japan **[505]**

Shigeru Yoshida, Department of Physiology, Nagasaki University School of Medicine, Nagasaki 852, Japan **[267]**

Tohru Yoshioka, Department of Physiology, Yokohama City University School of Medicine, 2-33 Urafune-cho, Minami-ku, Yokohama, Japan **[505]**

Guido A. Zampighi, Department of Anatomy, and Jerry Lewis Neuromuscular Research Center, UCLA School of Medicine, Los Angeles, CA 90024 **[317]**

Preface

This volume is the outgrowth of a symposium held November 6–8, 1982 in Santa Monica, California in honor of Susumu Hagiwara on his 60th birthday—an age when professors normally retire in Japan. (In fact, at the symposium banquet, Mr. Eiiche Narishige presented him with the traditional red hat and vest, like those in which a newborn baby is wrapped, symbolizing the recycling and fresh start of a person's life on his 60th birthday.) It seemed a good excuse to hold a meeting honoring this man who has contributed in so many important ways to our understanding of the properties of excitable cells.

Hagi, as he is affectionately known to all of us, was born in Hokkaido, Japan, the son of a schoolmaster who was also an enthusiastic collector of butterflies and primitive art and artifacts—fascinations that the son has carried on and perfected. Hagi received his M.D. from Tokyo University in 1946, and his Ph.D., also from Tokyo University, in 1951. His Ph.D. research, under the direction of T. Wakabayashi, concerned the nervous control of insect muscle and mechanisms of information coding in impulse trains. He held a faculty position in Tokyo University from 1946 to 1948, was a research fellow in the Japanese Pharmacological Institute from 1948 to 1950, and became Associate Professor of Physiology at Tokyo Medical and Dental University in 1950, rising to Professor of Physiology and Chairman of the Second Department of Physiology (Professor Katsuki being Chairman of the First) in 1959.

Between 1954 and 1957, with a three-year travel award from the Japanese Ministry of Education, Hagi worked for periods of several months each in five prominent laboratories around the world: with Y. Zotterman in Stockholm, T.H. Bullock at UCLA, Steve Kuffler at the Wilbur Institute in Baltimore, David Lloyd at the Rockefeller Institute in New York, and I. Tasaki at the N.I.H. In 1961, he moved to the United States permanently, first as Professor of Zoology at UCLA, then as Professor of Physiology and Head of the Marine Neurobiology Facility at the University of California, San Diego. In 1969 he returned to UCLA as Professor of Physiology, and was named the Eleanor I. Leslie Professor of Neuroscience in 1978.

Few areas of neurophysiology have escaped Hagiwara's imprint. His pioneering contributions have ranged from explanations of physical chemical mechanisms of membrane channel function to the neural correlates of whole animal behavior. His interests and important contributions span the entire range of behavioral, integrative, cellular, and molecular neurobiology. His work has been strikingly innovative and of such high caliber that he finds himself an authority in every one of these areas. Time and again, he has introduced new procedures or preparations which have become major foci of work for others. Unquestionably, however, the problems that have most occupied his attention are those of how the properties of cell membranes and of the ionic permeation channels in them account for observed cell behavior. It is appropriate that, almost simultaneous with this volume, there will appear a monograph by Hagiwara, **Membrane Potential Dependent Ion Channels in Cell Membranes: Phylogenetic and Developmental Approaches,** Volume 3 in the Distinguished Lecture Series of the Society of General Physiologists.

This emphasis is reflected in the papers in this volume. All are by one-time students, postdoctoral associates, or colleagues of Hagiwara; a large majority are concerned with the properties of ion permeation channels. We have somewhat arbitrarily subdivided them into five sections. The first two are devoted to the properties of ion channels, especially Ca^{++} channels and the inwardly rectifying K^+ channel, both of which are particularly closely identified with Hagi's lab. Section I deals predominantly with nerve and muscle cell membranes and Section II with oocyte and egg cell membranes, and the ontogeny of excitability. Section III deals with the physiological roles of ion channels and intracellular ions. Section IV contains papers on synaptic mechanisms, development, and trophic interaction between excitable cells. Finally, in Section V, a number of papers relate highlights of recent studies on sensory transduction, information processing, and plasticity in the CNS.

Clearly, a subject as enormous as the "physiology of excitable cells" cannot be covered thoroughly in a volume of this size, nor have we attempted to do so. These papers, being in most cases an outgrowth of recent or distant work strongly influenced by Hagiwara, inevitably reflect the interests and approaches he has fostered. On the other hand, the range of topics represented does span much of the field of neurobiology, providing, we hope, something of the flavor of contemporary investigation throughout the field.

We have been extremely fortunate, both in organizing the symposium and in preparing this volume, to have the invaluable assistance of many colleagues, especially Carol Gallion, Donald Simpson, Gretchen Wooden, Martha Bosma, Pat Ulrich, Frances Knight, and Mike Barish. The tireless efforts and good humor of Pat Ulrich and Sandra Nath Singh have been essential to the completion of this volume.

We are also deeply grateful to the financial donors who made such an international symposium possible. A major share of the support came from the UHI Corporation, Los Angeles. We are particularly indebted to Mr. S. Katayama, Chairman of the Board, and Mr. D.W. Weill, President, for their generosity, and to Dr. Robert Watanabe for his role in arranging the gift. We also are grateful for important support from Dr. and Mrs. M. Hirose of Santa Monica, Pacesetter Systems, Inc., the Schering Corp., A.G. Heinze Co., Max Erb Co., and Roboz Surgical Instrument Company. We are also deeply indebted to Vice Chancellor Albert Barber, Dean Sherman Mellinkoff, and Associate Dean Frederick Rasmussen of the UCLA School of Medicine for their constant and vital support for the symposium and the volume.

Alan D. Grinnell
William J. Moody, Jr.

SECTION I
PROPERTIES OF ION CHANNELS

Top: W.F.H.M. Mommaerts; C. Gallion; W.J. Moody, Jr.
Bottom: A. Watanabe; E. Narashige; K. Kusano; Y. Oomura

The Physiology of Excitable Cells, pages 3–6
© 1983 Alan R. Liss, Inc., 150 Fifth Avenue, New York, NY 10011

INTRODUCTION

Perhaps more than any other area of neurobiology, it is the properties and functions of voltage-dependent ion channels which have occupied the attention of Professor Hagiwara and his coworkers during the last twenty-five years. Hagi's studies of the Ca^{2+} channel and the inwardly rectifying K^+ channel have been particularly influential in the field of membrane biophysics (see Hagiwara and Byerly, 1981, and Hagiwara, 1983 for recent reviews), but anion permeation mechanisms have also been the subject of several important studies (e.g., Hagiwara, Gruener, Hayashi, Sakata, Grinnell 1968; Hagiwara, Toyama, Hayashi 1971; Hagiwara, Takahashi, 1974).

In 1964 Hagiwara and his colleagues began elucidating the detailed properties of the Ca^{2+} channel in barnacle muscle fibers. The paper by NAKAJIMA, one of Hagi's co-workers during those experiments, discusses the influence of the early Ca^{2+} work on the field of neurobiology. More sophisticated techniques for the study of channel properties have become available recently, and in the paper by OHMORI & HAGIWARA two of these methods--the whole cell suction pipet voltage-clamp and the patch clamp--are used to study the properties of single Ca^{2+} channels in GH_3 cells. The mechanism of inactivation of Ca^{2+} currents by intracellular Ca^{2+} is detailed in the papers by ECKERT et al. and PLANT et al. As these papers discuss, the first observations which suggested this mechanism of inactivation were those of Hagiwara and Naka (1964) and Hagiwara and Nakajima (1966) in barnacle muscle fibers. The Ca^{2+} channel appears to be phylogenetically quite old compared to the Na^+ channel of the neuron; DEITMER emphasizes its role in motility in the protozoon *Stylonychia*, and discusses the possibility of three types of Ca^{2+} channels in this ciliate. Attempts at rigorous analyses of the properties of the Ca^{2+} current have often been thwarted by the inability to block completely the interfering outward currents present at positive membrane potentials. Even under the rather extreme conditions of replacement of internal K^+ with the supposedly impermeant ion Cs^+, these currents persist (see Byerly and Hagiwara, 1982). MEECH & THOMAS provide a possible explanation for this dilemma in the existence of H^+ current activated by depolarization in molluscan neurons. Prior to these experiments, the idea of a voltage-dependent H^+ channel had not been seriously considered by many people, but the possibility that H^+ flux through this pathway participates in maintaining the cell's alkaline intracellular pH is incentive for a careful characterization of this channel in the future.

Another ion channel whose biophysical properties have been extensively studied is the inwardly rectifying K^+ channel. This conductance is particularly prominent in vertebrate skeletal muscle fibers and certain oocytes. In the past several years, Hagi and his collaborators have focused their attention on oocytes, since their simple geometry makes voltage clamp analysis of those currents relatively straightforward. In fact, one of the earlier oocyte studies associated with the laboratory was performed during a voyage of the research vessel Alpha Helix to the Great Barrier Reef. The aim was to obtain relatively large oocytes which would permit biophysical studies of the anomolous rectifier channel that were not possible in other preparations. A long series of papers on the anomalous rectifier channel in starfish and tunicate eggs coupled with those on striated muscle where it was first discovered have made it one of the most extensively studied ion channels. The papers by GUNNING & CIANI, and KRASNE et al. discuss theoretical models of this channel, based in large part on the oocyte experiments. ASHCROFT & STANFIELD take up the problem of the effect of the permeant ion on the kinetics of the inward rectifier in skeletal muscle, a question which is also considered in a slightly different manner for the delayed K^+ channel by JUNGE.

Two papers in this section are concerned with the acetylcholine receptor channel. KIDOKORO analyzes the properties of these channels in adrenal chromaffin cells using both noise analysis and single channel recording techniques and discusses two aspects of their function: 1) The idea that Ca^{2+} entering the cell through these channels contributes to secretion; and 2) The problems inherent in statistical fluctuations in the number of open channels in a high input resistance cell in which the opening of a single channel can depolarize the membrane to threshold. LECHLEITER & GRUENER consider the effect of changes in membrane fluidity on the kinetics of ACh receptor channels in myocytes. The interaction of channel molecules with their lipid environment is an area which has not received sufficient attention in the past.

Three strategies for studying the operation of channels at the molecular level are represented in the final four papers of this section. EATON & BRODWICK use chemical modification of the Na^+ channel to elucidate the amino acid residues which reside in that portion of the Na^+ channel responsible for inactivation. The next two papers discuss kinetic schemes for single channel opening and inactivation in terms of the statistical analysis of single channel records. PATLAK considers the limitations of conditional probability measurements, especially in terms of deciding between models in which Na^+ channel inactivation is or is not coupled to channel opening. HORN & STANDEN present a general discussion of the various kinetic states of channels and the behavior of the transitions between these states. They then consider the question of whether the Na^+ channel has one or more than one kinetically distinct open state. In the final paper of this section, EISENMAN & SANDBLOM report data on the effect of TEA binding to the gramicidin channel. The effect of this compound on a simple peptide channel is especially interesting in light of its well-known blocking action on the delayed K^+ channel in nerve, an effect first reported in squid axon by Tasaki and Hagiwara in 1957.

It is an appropriate tribute to the breadth and depth of Professor Hagiwara's work on ion channels that these papers by his students and colleagues represent a comprehensive overview of the present state of research in the area of channel biophysics.

References

Byerly L, Hagiwara S (1982). Calcium currents in internally-perfused nerve cell bodies of *Limnea stagnalis*. J Physiol 322:503.

Hagiwara S (1983). Membrane potential dependent ion channels in cell membranes: phylogenetic and developmental approaches. Soc for Gen Physiologists, Vol 3.

Hagiwara S, Byerly L (1981). Ca channel. Ann Rev Neurosci 4:69.

Hagiwara S, Gruener R, Hayashi H, Sakata H, Grinnell A (1968). Effect of external and internal pH changes on K and Cl conductances in the muscle fiber membrane of a giant-barnacle. J Gen Physiol 52:773.

Hagiwara S, Naka K (1964). The initiation of spike potential in barnacle muscle fibers under low intracellular Ca^{++}. J Gen Physiol 48:141.

Hagiwara S, Nakajima S (1966). Effects of the intracellular Ca ion concentration upon the excitability of the muscle fiber membrane of a barnacle. J Gen Physiol 49:807.

Hagiwara S, Toyama K, Hayashi H (1971). Mechanisms of anion and cation permeations in the resting membrane of a barnacle muscle fiber. J Gen Physiol 57:408.

Hagiwara S, Takahashi K (1974). Mechanism of anion permeation through the muscle fibre membrane of an elasmobranch fish, *Taeniura lymma*. J Physiol 238:109.

The Physiology of Excitable Cells, pages 7-11

HAGIWARA AND THE CALCIUM CHANNEL: HISTORICAL PERSPECTIVE

Shigehiro Nakajima

Department of Biological Sciences
Purdue University
West Lafayette, IN   47907

From 1964 to 1967 Hagiwara and his co-workers published five papers (Hagiwara and Naka 1964; Hagiwara et
al. 1964; Hagiwara and Nakajima 1966a and b; Hagiwara and
Takahashi 1967) on calcium spikes in the barnacle giant
muscle fibers. As is the case with many important scientific contributions, these papers by Hagiwara were not the
first to describe the occurrence of calcium spikes. Previously, Fatt and Katz 1953 and Oomura et al. 1961 had described the possible occurrence of calcium spikes in crustacean muscle fibers and molluscan neurons. Nevertheless,
the five papers of Hagiwara made several extemely important
points. The papers (a) established beyond reasonable doubt
the presence of a calcium spike in crustacean muscle fibers,
(b) presented evidence that a calcium spike occurs in
vertebrate heart muscle (see also Niedergerke and Orkand
1966) (c) provided pharmacological criteria to distinguish
calcium spikes from sodium spikes through the use of manganese, cobalt, and tetrodotoxin, and (d) showed that the
intracellular calcium concentration, $[Ca]_{in}$, controls
the calcium channels.

The importance of this work can be appreciated more
fully by considering the impact that it has had on subsequent progress. The establishment of the calcium spike and
the pharmacological tools for studying it have had a direct
influence on the subsequent discovery of calcium spikes in
many other systems: nerve cell bodies (Geduldig, Junge
1968; Geduldig, Gruener 1970), axons (Baker et al. 1971),
and presynaptic nerve terminals (Katz, Miledi 1969; Linas
et al. 1972), sensory receptors (Zipser, Bennett 1973;

Clusin, Bennett 1977), glands (Matthews, Sakamoto 1975;
Kidokoro 1975; Kater 1977), oocytes (Miyazaki et al. 1972),
Paramecium (Naitoh et al. 1972), and epithelial cells
(Herrera 1979). The calcium spike in vertebrate smooth muscle was established early (Nonomura et al. 1966), and the
calcium spike in vertebrate heart muscle has also been studied extensively (Reuter 1967). While it is difficult to
assess how much the work of Hagiwara influenced the development of organic calcium blockers (Kohlhardt et al. 1972)
which will soon be used for the treatment of angina pectoris, it would clearly have been impossible to consider the
development of channel blockers without knowing of the
existence of the channels.

More recently, the effect of $[Ca]_{in}$ on the gating of
the calcium channel has been studied extensively (for review
see Hagiwara, Byerly 1981, and papers by Eckert and Plant et
al. in this volume), and it has been proposed that increased
$[Ca]_{in}$ may be responsible for the inactivation of calcium
currents (Brehm, Eckert 1978; Tillotson 1979; Ashcroft,
Stanfield 1981). Perhaps the work of Hagiwara had an indirect effect on the discovery of calcium-induced potassium
channel activation (Meech 1974).

In addition to the pivotal role played by the work of
Hagiwara in the development of knowledge related directly
to the calcium spike, the work of Hagiwara probably has had
an indirect influence on the establishment of the idea that
calcium ion is indeed a regulator or a messenger for many
important cellular functions. The development of this idea
has come from observations of the role of calcium in processes such as: muscle contraction (Ebashi, Endo 1968), the
activation of oocyte development (Gilkey et al. 1978), cell
death (Schanne et al. 1979), motility of nonmuscle cells
(Hitchcock 1977), and the regulation of electrical coupling
between cells via gap junctions (Rose, Loewenstein 1975).
The identificaton of the Ca-binding proteins, troponin
which confers Ca sensitivity on the control of contraction
in striated muscle (Ebashi et al. 1968) and calmodulin, a
ubiquitous Ca-dependent modulator of many enzymatic activities (Kakiuchi et al. 1970; Cheung 1970) has provided a new
level of understanding of the mechanisms by which Ca regulates these cellular functions. Thus, the last decade has
become a golden age for calcium, and it is clear that the
work of Hagiwara has been central in the development of
this remarkable progress. The future will show in greater

detail how these calcium-mediated functions are related to
the function of calcium-dependent regulatory proteins and
how calcium channels are influenced by modifiers.

References

Ashcroft FM, Stanfield PR (1981). Calcium dependence of
the inactivation of calcium currents in skeletal muscle
fibers of an insect. Science 213:224.

Baker PF, Hodgkin AL, Ridgway EB (1971). Depolarization
and calcium entry in squid giant axons. J Physiol (Lond)
218:709.

Brehm P, Eckert R (1978). Calcium entry leads to inactivation of calcium channel in Paramecium. Science 202:1203.

Cheung WY (1970). Cyclic 3', 5'-Nucleotide phosphodiesterase. Demonstration of an activator. Biochem Biophys
Res Commun 38:533.

Clusin WT, Bennett MVL (1977). Calcium-activated conductance in skate electroreceptors. J Gen Physiol 69:121.

Ebashi S, Endo M (1968). Calcium ion and muscle contraction.
Prog Biophys Mol Biol 18:125.

Ebashi S. Kodama A, Ebashi F (1968). Troponin I. Preparation and physiological function. J Biochem 64:465.

Fatt P, Katz B (1953). The electrical properties of crustacean muscle fibers. J Physiol (Lond) 120:171.

Geduldig D, Greuner R (1970). Voltage clamp of the Aplysia
giant neurone: Early sodium and calcium currents. J
Physiol (Lond) 211:217.

Geduldig D, Junge D (1968). Sodium and calcium components
of action potentials in the Aplysia giant neurone. J
Physiol (Lond) 199:347.

Gilkey JC, Jaffee LF, Ridgeway EB, Reynolds GT (1978). A
free calcium wave traverses the activating egg of the
medaka Oryzias latipes. J Cell Biol 76:448.

Hagiwara S, Byerly L (1981). Calcium channel. Ann Rev
Neurosci 4:69.

Hagiwara S, Chichibu S, Naka K (1964). The effects of
various ions on resting and spike potentials of
barnacle muscle fibers. J Gen Physiol 48:163.

Hagiwara S, Naka K (1964). The initiation of spike potential in barnacle muscle fibers under low intracellular
Ca^{2+}. J Gen Physiol 48:141.

Hagiwara S, Nakajima S (1966a). Differences in Na and Ca
spikes as examined by application of tetrodotoxin, procaine, and manganese ions. J Gen Physiol 49:793.

Hagiwara S, Nakajima S (1966b). Effects of the intracellular Ca ion concentration upon the excitability of the
muscle fiber membrane of a barnacle. J Gen Physiol 49:807.

Hagiwara S, Takahashi K (1967). Surface density of calcium
ions and calcium spikes in the barnacle muscle fiber membrane. J Gen Physiol 50:583.

Herrera AA (1979). Electrophysiology of bioluminescent
excitable epithelial cells in a polynoid polychaete worm.
J Comp Physiol 129:67.

Hitchcock SE (1977). Regulation of motility in nonmuscle
cells. J Cell Biol 74:1.

Kakiuchi S, Yamazaki R, Nakajima H (1970). Properties of a
heat stable phosphodiesterase-activating factor isolated
from brain extract. Proc Japan Acad 46:587.

Kater SB (1977). Calcium electroresponsiveness and its
relationship to secretion in molluscan exocrine gland
cells. Neurosci Symp 2:195.

Katz B, Miledi R (1969). Tetrodotoxin-resistant electrical
activity in presynaptic terminals. J Physiol (Lond) 203:
459.

Kidokoro Y (1975). Spontaneous calcium action potentials in
a clonal pituitary cell line and their relationship to
prolactin secretion. Nature 258:741.

Kohlhardt M, Bauer P, Krause H, Fleckenstein A (1972).
Differentiation of the transmembrane Na and Ca channels in
mammalian cardiac fibers by the use of specific inhibitors.
Pflügers Arch 335:309.

Llinas R, Blinks JR, Nicholson C (1972). Calcium transient
in presynaptic terminal of squid giant synapse: Detection
with aequorin. Science 176:1127.

Matthews EK, Sakamoto Y (1975). Electrical characteristics
of pancreatic islet cells. J Physiol (Lond) 246:421.

Meech RW (1974). The sensitivity of *Helix aspersa* neurones
to injected calcium ions. J Physiol (Lond) 237:259.

Miyazaki S, Takahashi K, Tsuda K (1972). Calcium and sodium
contributions to regenerative responses in the embryonic
excitable cell membrane. Science 176:1441.

Naitoh Y, Eckert R, Friedman K (1972). A regenerative calcium response in *Paramecium*. J Exp Biol 56:667.

Niedergerke R, Orkand RK (1966). The dual effect of calcium on the action potential of the frog's heart. J
Physiol (Lond) 184:291.

Nonomura Y, Hotta Y, Ohashi H (1966). Tetrodotoxin and
manganese ions: effects on electrical activity and tension
in taenia coli of guinea pig. Science 152:97.

Oomura Y, Ozaki S, Maeno T (1961). Electrical activity of

a giant nerve cell under abnormal conditions. Nature 191: 1265.

Reuter H (1967). The dependence of slow inward current in Purkinje fibres on the extracellular calcium-concentration. J Physiol (Lond) 192:479.

Rose B, Loewenstein WR (1975). Permeability of cell junction depends on local cytoplasmic calcium activity. Nature 254:250.

Schanne FAX, Kane AB, Young EE, Farber JL (1979). Calcium dependence of toxic cell death: a final common pathway. Science 206:700.

Tillotson D (1979). Inactivation of Ca conductance dependent on entry of Ca ions in molluscan neurons. Proc Natl Acad Sci USA 76:1497.

Zipser B, Bennett MVL (1973). Tetrodotoxin resistant electrically excitable responses of receptor cells. Brain Res 62:253.

Top: Scientific Session in progress
Bottom (around table): S. Nakajima; J. Heiny; F. Ashcroft;
J. Lansman; D. Junge; R. Gruener; A. Kammer; M.B. Rheuben

The Physiology of Excitable Cells, pages 13-24
© 1983 Alan R. Liss, Inc., 150 Fifth Avenue, New York, NY 10011

Ca CHANNEL IN THE GH$_3$ CELL

Harunori Ohmori and Susumu Hagiwara

Department of Physiology
University of California
Los Angeles, California 90024

Cells of the rat clonal pituitary line GH$_3$ have been
shown to generate action potentials that have both Na and Ca
components (Kidokoro 1975; Biales, Dichter, Tischler 1977).
However, mostly because of the limitation of the small
cell diameter, details of the nature of these ionic channels
have not been studied. On the other hand, the small cell
size is an advantage for the application of patch clamp
techniques to the whole cell in voltage clamp experiments,
since the small surface area of the membrane and the small
ionic currents make it easier to avoid the problems of
adequate space clamp and series resistance (Hagiwara, Ohmori
1982; Fenwick, Marty, Neher 1982a, b). We have used the
whole cell patch clamp technique to analyze the macroscopic
properties of the Ca current and the surface patch recording
technique to study microscopic properties of single Ca
channel currents.

The whole cell voltage-clamp was attained by gentle
continuous suction of about 100 cm H$_2$O to rupture the
patch of the membrane after achievement of the gigaseal
to the membrane. The break of the membrane patch was
indicated by the sudden appearance of a large capacitative
current in response to a small test voltage pulse. When
the membrane is held at the bath potential a large fluctuation in current is generally observed, which is eliminated
when the membrane is hyperpolarized to the level of the
resting potential. We therefore held the membrane at -70mV.
A series of step voltage changes was applied, and the macroscopic currents generated were recorded through the patch
electrode.

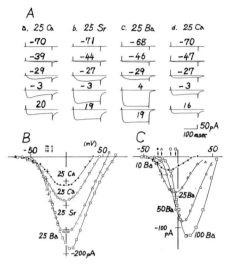

Fig. 1. Whole cell recording from GH$_3$ Ca channel. A, Membrane was held at -75mV. All the traces are averages of 3-4 traces. B, Peak value of the inward current was plotted from A, against voltage. C, External [Ba] was replaced with TEA. Arrows on the abscissa indicate membrane potentials which correspond to the half maximal value of the inward current. 10°C.

When the concentration of the external Ca ion was increased from the normal value of 2.5 mM to 25 mM in Na-free TEA (tetraethyl ammonium) medium, an inward current with slow activation kinetics was observed. The activation kinetics become faster as the membrane is depolarized. Similar activation kinetics were observed when Ca was replaced with either Sr or Ba. From Fig. 1A, B, it is obvious that the permeability sequence is Ba > Sr > Ca.

The inward current carried by Ca and Sr showed time dependent decay during 200 msec command pulses. We do not think that this decay is due to the inactivation of the Ca channel. The following three reasons strongly suggest that this decay is due to the activation of a Ca activated K conductance: (1) In the Ba solution there is no decay in the current (Fig. 1A-C). (2) When isotonic CsCl solution was used in a patch electrode of relatively large tip diameter (1-2 μm) to facilitate the exchange of the CsCl with the intracellular medium, the decay was eliminated even in Ca-containing external solution (data not shown). (3) Small quantities (100-200 μm) of either quinine or quinidine in the extracellular medium effectively eliminated the decay in Ca external medium (data not shown). Quinine or quinidine are known to block Ca-activated K conductance in several preparations (Armand-Hardy, Ellory, Ferreira, Fleminger, Lew 1975; Atwater, Dawson, Ribalet, Rojas 1979; Fishman,

Spector 1981). The kinetics observed in the Ba solution seem to be typical of the GH_3 Ca channel. The following discussion is based on the results obtained in Ba-containing solutions.

When the external Ba concentration was increased by replacement for TEA in Na-free extracellular medium, the amplitude of the Ba current was increased (Fig. 1C). Simultaneously, the I-V relation shifted in the positive direction. This shift of the I-V relation could be due to the screening effect of Ba ions on the negative surface charge of the membrane. In fact, we cannot eliminate the possibility of a binding effect of the divalent cation to the negatively charged ligand on the membrane surface. However, since we could not detect significant difference in the "threshold" of the Ca channel, or in the voltage dependence of the activation kinetics among Ca, Sr and Ba ions, we tentatively neglected the possibility of binding effects of these divalent cations to the negative surface charge of the membrane. From a curve fitting procedure assuming a homogeneously distributed negative surface charge, we can estimate the density as 1 electronic charge per 50 $\mathring{A}^2$. For an accurate estimation of the surface charge density we need to know the maximum value of the surface potential by decreasing the concentration of the divalent cation close to zero (Gilbert 1971; Ohmori, Yoshii 1977). However, it is obviously impossible to get this kind of information from the Ca channel where the permeant cation and the surface charge screening cation are the same.

Ionic currents are assumed to consist of a large number of individual currents through single ionic channels distributed throughout the membrane. This idea has been examined in the Na channel, K channel and in the ACh receptor channel, first by noise analysis and more recently by single channel recording.

From previous noise analyses performed on the Ca current, we predicted that the single channel current should be extremely small for the Ca channel (Akaike, Fishman, Lee, Moore, Brown 1978; Krishtal, Pidoplichko, Shakhovalov 1981). Strategically it would be better to get some idea of the size of the unitary current and the density of the channels on the GH3 membrane before trying single channel recording. We therefore performed noise analysis to estimate the size of the unitary current, the density of the channels,

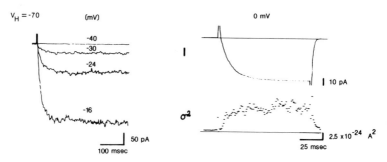

Fig. 2. A, demonstration of noise in single traces of Ba
current. Membrane was held at -70 mV. B, average of 80
current traces and the variance measured at each time point.

and temporal properties of the gating.

　　　Individual traces are shown in Fig. 2A at several different potentials. Immediately after the test potential step
reached the so called "threshold" of this channel, significant
fluctuations of the current appeared. The mean value and the
variance was calculated at each time point from 80 current
traces (Fig. 2B). At the holding potential, the variance is
very small. Immediately after the start of the voltage pulse
the variance increases, and the variance decreases again after
the termination of the pulse. By applying the technique of
ensemble noise analysis, the size of the unitary current
underlying the increase of the variance can be estimated
(Sigworth 1977; Ohmori 1981).

　　　It would seem reasonable to assume that the Ca channel
has two conductance states: open and closed. From this basic
assumption the mean current and the variance would be expressed as the following on the basis of the binomial distribution, where p is the channel open probability, i is the
single channel current and N is the number of channels.

$$I = N\,p\,i \qquad\qquad (1)$$
$$\sigma^2 = N\,i^2 p(1-p) \qquad\qquad (2)$$

By taking the ratio between the average current (I) and the
variance (σ^2) at each time point, the following equation can
be derived.

$$\sigma^2/I = i\,(1-p) = i - I/N \qquad (3)$$

Since the single channel current (i) and the number of channels (N) are considered to be time-independent parameters,
we can expect a linear relation between the ratio (σ^2/I) and

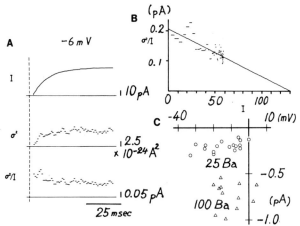

Fig. 3. A, average current, variance, and their ratio. The broken line indicates the start of the voltage pulse. B, the ratio is plotted against average current at each time point. C, unitary currents were plotted on the voltage axis for 25 mM and 100 mM Ba solutions.

the mean current (I). From the plot of this ratio against the average current at the corresponding time, we can estimate the single channel current, i, as the y-intercept, and the number of channels, N, from the reciprocal of the slope. In Fig. 3A, the average current (plotted with reversed polarity), the variance and the ratio between the above two variables are plotted. Fig. 3B shows an almost linear relation between the ratio and the average current when plotted at the corresponding time. The single channel current was -0.2 pA and the number of channels was 650. The regression coefficient was 0.6. The estimation of the single channel Ba current was about -0.2 pA in the 25 mM Ba solution and about -0.6 pA in the 100 mM Ba solution both at 0 mV membrane potential.

From the procedure of curve fitting to the macroscopic current, we have restricted the activation kinetics either to an m^2 or to an m^3 process. By comparison with the activation time constant and the time constant of the tail current, we think it appropriate to conclude that the activation follows an m^2 process of the Hodgkin and Huxley-type (Hagiwara, Ohmori 1982).

The microscopic kinetic properties of the Ca channel can be understood through measurements of the steady state noise spectrum. The case shown in Fig. 4A agrees fairly well with m^2 gating kinetics. However, a closer inspection of the spectrum reveals a systematic deviation of the power at the higher frequency region. This deviation is more obvious in

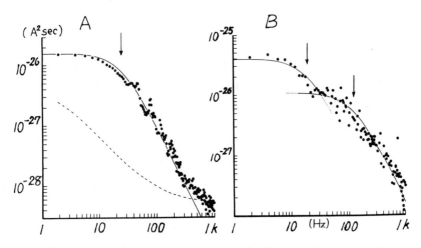

Fig. 4. Power density spectrum of the steady state Ba current noise. Arrows indicate cutoff frequencies. Broken line indicates the spectrum at the holding potential. Both spectra were made in 100 mM Ba, at -8 mV (A) and at 0 mV (B).

the case of Fig. 4B, in which we can detect a second Lorentzian spectrum. The existence of the higher frequency component in the spectrum is inconsistent with m^2-type gating kinetics. This implies the existence of much faster kinetics than an m^2 gating process: An m^2 process predicts a double Lorentzian spectrum, but the difference in cutoff frequencies is a factor of two, which is practically impossible to detect in a power density spectrum. Therefore the question is: What is the real nature of the Ca channel's gating? We have tried to answer this question through recordings of single channel Ba currents.

Since the Ca channel is known to be washed out by intracellular perfusion (Byerly, Hagiwara 1982), we have restricted our single channel recording experiment to the cell-attached patch. The patch electrode contained 100 mM $BaCl_2$ buffered by HEPES-TEA-OH to pH 7.4. By using Na-free, Ca-free, Mn, and TEA with 2 mM quinidine as the bathing solution, the putative activities of Na, Ca and K channels on the membrane facing the bathing medium could be eliminated. The resting potential was approximately -10 mV in this bathing medium. From noise analysis, the density of Ca channels on the GH3 cell membrane was found to be very small; therefore we used electrodes with relatively large tip diameters (2-5 μm) to

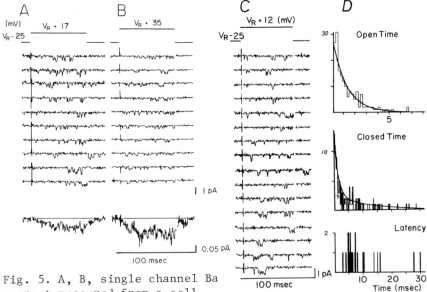

Fig. 5. A, B, single channel Ba current measured from a cell attached patch. Voltage pulse was applied from a resting potential (V_R) of -25 mV to the levels indicated. The bottom traces show the average of 45 and 41 traces, respectively. C, single channel currents and D, open time, closed time and latency histograms calculated from the currents in C.

increase the possibility of having channels within the patch. The large tip diameter inevitable decreased the seal resistance (to 20-50 Gohm), and made the conductance of the patch membrane significant in comparison with the conductance of the rest of the membrane. This leads to a potentiometric error in the transpatch-membrane potential when a step voltage change is applied. However, the error in the estimation of the transpatch-membrane potential would be less than 10% (Hagiwara, Ohmori 1983).

When a step depolarizing voltage change is applied to the patch electrode, we can see pulse-like events, which are shown in Fig. 5A, B, C after subtraction of the capacitative transient and leakage current. At the bottom of columns A and B the average currents are shown. Several features are obvious from these traces: (1) The time course of the activation kinetics of the average current becomes faster in response to a more positive voltage step. (2) There is

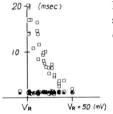

Fig. 6. Voltage dependence of the time constant for the open time (◯) and two time
constants for the closed time histogram (▲,
☐), measured from seven different patches.

no inactivation of the current, at least in this 70 msec
period. (3) The amplitude of the average current becomes
larger with greater depolarization. (4) However, the
amplitude of each pulse-like event becomes smaller as the
membrane is depolarized. These are features expected from
a knowledge of the whole cell clamp and the noise analysis.
Therefore, the pulse-like events observed here can be identified as Ba currents through single Ca channels.

We have measured the open time, closed time and latency
(the time of the first opening after the start of the voltage
pulse) in Fig. 5D. The open time histogram shows a single
exponential decay with a time constant of 1.2 msec. The
closed time histogram clearly indicates more than a single
exponential component. We have fitted the histogram with
two exponential functions with time constants of 1.3 msec
and 11.4 msec. The latency histogram seems to show a peak
at a certain delay after the start of the voltage pulse,
although we cannot show a clear rising phase in the histogram because of the limited number of traces available.

Three statistical parameters were measured at several
membrane potentials. The time constant (mean open time) of
the open time histogram, and the two time constants of the
closed time histogram. These were plotted against patch
potential in Fig. 6. We could detect no potential dependence
in the mean open time and the shorter time constant of the
closed time histogram. The longer time constant of the
closed time histogram showed clear potential dependence,
close to that expected from the m^2-gating kinetics. The
latecny histogram showed a potential dependence similar to
that of the longer time constant of the closed time histogram.
This feature of the open time histogram is totally inconsistent with the gating kinetics of the Hodgkin-Huxley model.
What we can imagine from these experimental results is the

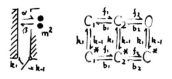

Fig. 7. Proposal of the two independent gating kinetics for Ca channel.

existence of two gating kinetics which are not related to each other. One is slow and may be analogous to the m^2 kinetics of the Hodgkin-Huxley model; the other is a very fast flickering kinetics. The potential dependence of this fast flickering kinetics remains uncertain. Although we could not detect any clear potential dependence in it, a fast decaying tail current is frequently observed in other preparations (Byerly, Hagiwara 1982; Fenwick et al. 1982b), which may indicate some form of potential dependence of the fast flickering kinetics in a range of membrane potentials more negative than we tested.

A possible model of the gating kinetics is shown in Fig. 7. We think it easy to explain our results with two independent gates; a potential dependent slow gate and a relatively potential independent fast flickering gate. The horizontal kinetic pathway is the slow potential dependent gating process and the vertical pathway is the fast flickering kinetics. If what is known about fast tail currents applies in the GH$_3$ cell, most channels would be in state C_1^* at the resting potential, and would flicker between O and C^* in the depolarized state. The fast tail current would be generated by the step from O to C^*, and the slow tail current would be by the closing process from O to C_2. From this gating scheme we can understand the single exponential decay in the open time histogram, and the multiple exponential decay in the closed time histogram. Because of the limited number of events, it is quite possible that we could not have detected more than two exponential processes in the closed time histogram. For the same reason, we might have missed some form of potential dependence in both the open time histogram and the fast decaying portion of the closed time histogram.

We prefer the above gating model to the linear sequential model. The sequential model has an advantage over the above two independent gating kinetics, in that at least one can predict the pattern of the histograms, quantitatively.

Our results might be explained by one of the following sequential models.

$$C_1 \underset{b}{\overset{f}{\rightleftharpoons}} C_2^* \underset{k_{-1}}{\overset{k_1}{\rightleftharpoons}} 0 \qquad (4)$$

$$C_1 \underset{b_1}{\overset{f_1}{\rightleftharpoons}} C_2 \underset{b_2^*}{\overset{f_2^*}{\rightleftharpoons}} C_3^* \underset{k_{-1}}{\overset{k_1}{\rightleftharpoons}} 0 \qquad (5)$$

$$C_1 \underset{b_1}{\overset{f_1}{\rightleftharpoons}} C_2 \underset{b_2}{\overset{f_2}{\rightleftharpoons}} 0 \underset{k_{-1}}{\overset{k_1}{\rightleftharpoons}} C^* \qquad (6)$$

In either model the step between 0 and C^* is presumed to be very fast and will be seen as fast flickering kinetics in single channel recordings. In scheme (4), we can expect a strictly potential independent process for the open time histogram by assuming k_{-1} to be potential independent. However, the fast decaying process of the closed time histogram would become potential dependent to the same extent as the slow decaying process of the closed time histogram. By increasing the number of closed states and adding a very fast transition kinetics (f_2^*, b_2^*) between the last two closed states (C_2 & C_3^*) before going to the open state (5), it may be possible to predict fast potential-independent decay in the closed time histogram. In gating scheme (6), one can make the fast decaying component of the closed time histogram potential independent and the open time histogram potential independent effectively by setting the values k_1 and k_{-1} substantially larger than the other rate constants. However, when the membrane potential is stepped back to the resting potential a significant number of the Ca channels must be in the closed state C^*; the number is proportional to the ratio $k_1/(k_1+b_2)$. The number of the Ca channels in state C^* at the resting state would be much larger than Ca channels in the other closed states, if we assume the existence of a fast tail current is true of the GH_3 Ca channel; this assumption implies that $k_1 \gg k_{-1}$ at negative membrane potentials. Since there is no possibility for the channel in C^* to come back to the other closed states without passing through the open state, in response to the second voltage pulse, a significant portion of the Ca channel must open with very fast kinetics (with a time constant of approximately 500 μsec macroscopically), and in the latency histogram we should notice a fast decaying process with a

time constant similar to the fast component in the closed
time histogram. Because of the presence of the capacitative
transient, the 2 msec period after a step voltage change may
not be reliable in our present records. However, we should
have been able to detect at least the tail of a fast
decaying process in the latency histogram. The absence of
such a tail in the latency histogram in our single channel
measurements and the absence of a very fast rising process
in the macroscopic kinetics implies the existence of a
pathway between the closed state C^* and the other closed
states. We think the model shown in Fig. 7 is most reasonable and realistic based on our present knowledge of the
Ca channel gating. Similar fast flickering kinetics is
observed in the Ca channel of other preparations (Fenwick
et al. 1982b; Lux, Nagy 1981; Reuter, Stevens, Tsien, Yellen
1982) and also in the K channel (Conti, Neher 1980) and in
the Ca-activated K channel (Marty 1981).

REFERENCES

Armando-Hardy M, Ellory JC, Ferreira HG, Fleminger S, Lew VL
    (1975). Inhibition of the calcium-induced increase in the
    potassium permeability of human red blood cells by quinine.
    J Physiol 250:32.
Atwater I, Dawson CM, Ribalet B, Rojas E (1979). Potassium
    permeability activated by intracellular calcium concent-
    ration in the pancreatic β-cell. J Physiol 288:575.
Biales B, Dichter MA, Tishler A (1977). Sodium and calcium
    action potential in pituitary cells. Nature Lond. 267:172.
Byerly L, Hagiwara S (1982). Calcium currents in internally
    perfused nerve cell bodies of Limnea stagnalis. J Physiol
    322:503.
Conti F, Neher E (1980). Single channel recording of K⁺ cur-
    rents in squid axons. Nature Lond. 285:140.
Fenwick EM, Marty A, Neher E (1982a). A patch clamp study
    of bovine chromaffin cells and of their sensitivity to
    acetylcholine. J Physiol 331:577.
Fenwick EM, Marty A, Neher E (1982b). Sodium and calcium
    channels in bovine chromaffin cells. J Physiol 331:599.
Fishman MC, Spector I (1981). Potassium currents in neuro-
    blastoma cells. Proc Natl Acad Sci USA 78:5245.
Gilbert DL (1971). Fixed surface charges. In Adelman WJ Jr.
    (ed): "Biophysics and Physiology of Excitable Membranes",
    NY: Van Nostrand Reihhold.

Hagiwara S, Ohmori H (1982). Studies of calcium channels in rat clonal pituitary cells with patch electrode voltage clamp. J Physiol 331:231.

Hagiwara S, Ohmori H (1983). Studies of single calcium channel currents in rat clonal pituitary cells. J Physiol 336:

Kidokoro Y (1975). Spontaneous calcium action potentials in a clonal pituitary cell line and their relationship to prolactin secretion. Nature Lond. 258:741.

Lux HD, Nagy K (1981). Single channel Ca^{2+} currents in Helix pomatia neurones. Pflügers Arch 391:252.

Marty A (1981). Ca-dependent K channels with large unitary conductance in chromaffin cell membranes. Nature Lond. 291:497.

Ohmori H, Yoshii M (1977). Surface potential reflected in both gating and permeation mechanisms of sodium and calcium channels of the tunicate egg cell membrane. J Physiol 267: 429.

Ohmori H (1981). Unitary current through sodium channel and anomalous rectifier channel estimated from transient current noise in the tunicate egg. J Physiol 311:289.

Reuter H, Stevens CF, Tsien RW, Yellen G (1982). Properties of single calcium channels in cultured cardiac cells. Nature Lond. 297:501.

Sigworth FJ (1977). Sodium channels in nerve apparently have two conductance states. Nature Lond. 270:265.

The Physiology of Excitable Cells, pages 25-38
© 1983 Alan R. Liss, Inc., 150 Fifth Avenue, New York, NY 10011

CALCIUM-MEDIATED INACTIVATION OF CALCIUM CURRENT IN NEURONS
OF *APLYSIA CALIFORNICA*

R. Eckert, D. Ewald and J. Chad

Department of Biology and Ahmanson Laboratory
University of California, Los Angeles
Los Angeles, California 90024

This article, presented in honor of Professor Susumu
Hagiwara on the occasion of his 60th birthday, can begin most
appropriately with reference to certain experiments reported
17 years ago (Hagiwara, Nakajima 1966). These experiments,
carried out with Ca-EGTA buffers in perfused barnacle muscle,
showed that elevation of intracellular free Ca^{2+} causes suppression of calcium spike generation in response to depolarization, thus providing the first indication that intracellular Ca^{2+} may modulate membrane excitability. In more recent experiments on perfused snail neurons and tunicate
eggs it was seen that elevation of intracellular free
Ca^{2+} can block the calcium conductance (Kostyuk, Krishtal
1977; Takahashi, Yoshii 1978). Observations first made on
Paramecium (Eckert 1977; Brehm, Dunlap, Eckert 1978; Brehm,
Eckert 1978a,b) and *Aplysia* neurons (Eckert, Tillotson 1979;
Tillotson 1979; Tillotson, Eckert 1979) suggested that inactivation of the calcium conductance, following its voltage-
dependent activation, depends upon the entry and accumulation
of calcium ions during the flow of the current.

BACKGROUND

The Ca current elicited by a fixed-amplitude test pulse
(V_2) becomes maximally attenuated following prepulse voltages
of mid-range values that produce maximal Ca^{2+} entry (Brehm,
Eckert 1978a,b; Eckert, Tillotson 1979; Tillotson 1979). The
degree of attenuation of the Ca current (i.e., residual inactivation remaining at the time the test pulse is applied)
was shown in *Aplysia* neurons over the range tested to be a

linear function of the calcium carried into the cell by the current elicited during the prepulse. This relationship held regardless of whether Ca entry during the prepulse was altered by changing Ca_o, pulse length, or pulse amplitude (Tillotson, Eckert 1979; Eckert, Tillotson 1981). Under appropriate conditions (i.e., low test pulse potential and/or short pulse-pair interval) the inward current during the test pulse is fully suppressed by a large Ca entry during the prepulse (Eckert, Tillotson, Brehm 1981). Injection of the calcium buffer EGTA or extracellular replacement of Ca with Ba both interfere with inactivation of the Ca conductance (Brehm, Eckert 1978, 1979; Eckert, Tillotson 1979; Tillotson 1979; Brehm, Eckert, Tillotson 1980; Eckert, Tillotson 1981; Plant, Standen 1981), causing a slower rate of relaxation of the inward current, enhanced strength of the steady state inward current, and an increased rate of removal of inactivation.

Several considerations argue against the possibility that relaxation and attenuation of the Ca current results primarily from a reduced driving force due to intracellular accumulation of free Ca^{2+} during current flow. Photometric measurements following Ca and Ba currents in *Aplysia* neurons (Gorman, Thomas 1978) and dorid neurons (Ahmed, Conner 1979) injected with arsenezo III show that the free Ba^{2+} concentration drops more slowly than free Ca^{2+}. Moreover the Ba current equals or exceeds the Ca current. Thus, the entry and accumulation of barium ions should in fact produce a greater reduction in driving force. Nonetheless, the current carried by Ba undergoes markedly less relaxation than that carried by Ca (Eckert, Tillotson 1981). Depletion of Ca from a restricted extracellular space cannot account for the current-dependent relaxation of the Ca current in *Aplysia* neurons, since injection of EGTA causes the Ca current to increase and its relaxation to decrease. Moreover, elevation of Ca_o increases the rate of relaxation (Eckert, Tillotson 1981).

In the more recent work reviewed below, the membrane potential of axotomized *Aplysia* neurons was controlled by conventional two-electrode voltage clamp, while K and Na currents were minimized by TEA and TTX. Holding potentials were -40 mV throughout, and test potentials were generally held to +10 mV or lower to minimize residual K currents and nonlinear outward currents that become prominent at higher potentials (Hagiwara, Byerly 1981). The bath contained 0.045mM TTX, 200mM TEA, 268mM NaCl, 20mM KCl, 45mM $MgCl_2$, 20mM $CaCl_2$ and 15mM Tris-HCl, pH 7.8; temperature, 13°-14^oC.

Tail currents were measured at -40 mV, which was close to E_K, due to elevated K_o. Computer fits of the tails showed that they could be fitted by the sum of two time constants of decay, the shorter being about 0.38 ms, and the longer one about 2.0 ms. Both the fast and slow phases inactivated with increasing test pulse duration, and were eliminated when Co replaced Ca in the physiological saline.

TAIL CURRENT MEASUREMENTS AVOID POSSIBLE K-CURRENT ARTIFACTS

It is unlikely that hidden Ca-dependent K currents account for the Ca-dependent reduction in net inward current, for prior Ca entry in molluscan neurons leads to a decrease in the Ca-dependent K current during a test depolarization (Eckert, Lux 1977; Eckert, Ewald 1982). Furthermore, Ca inactivation becomes weaker as test pulse voltage becomes more positive (Eckert, Ewald 1982b), whereas activation of Ca-dependent K current becomes stronger (Hermann, Gorman 1979). Nonetheless, to avoid possible artifact due to contamination by K currents, experiments on Ca inactivation were repeated using measurements of Ca deactivation tails at membrane potentials close to E_K (Eckert, Ewald 1981 1983a,b). The insets in Figure 1 show deactivation tails recorded at the termination of 10ms test pulses to the various potentials indicated, with the conditioning pulse (ending 1s prior to V_2) alternately "on" or "off." With increasing V_2, the tail currents grew in size to a plateau beyond about +60 mV. Presentation of the conditioning pulse caused reduction in tail current amplitude, shown plotted across the full range of V_2. This inactivation was virtually eliminated by intracellular EGTA, whereas outward current during strong depolarization was unaffected. The reduction in tail current amplitude following presentation of V_1 is independent of K currents, so these measurements provide convincing evidence for inactivation of the Ca conductance.

TEMPORAL CHARACTERISTICS OF INACTIVATION

Under moderate to strong voltage-clamp depolarization, the calcium current during a prolonged voltage step relaxes with a time course containing two exponential phases having time constants τ_1 and τ_2, respectively (Adams, Gage 1979; Eckert, Ewald 1982a; Eckert, Ewald, Chad 1982). Under small depolarization causing weak currents, inactivation is greatly reduced or absent (Eckert, Lux 1975, 1976). The existence of rapid and slow phases of inactivation of Ca current in

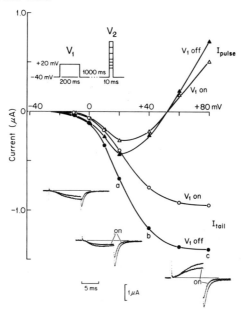

Fig. 1. Inactivation of inward current by prior depolarization. Plotted against the voltage of 10ms test pulse is the
amplitude of the current measured just prior to the end of
the pulse (I_{pulse}) and during the deactivation tail 500 μs
after end of the pulse (I_{tail}). Linear leakage and symmetrical capacitive currents were eliminated by summing current
signals from symmetrical hyper- and depolarizing pulses by
digital means. For each value of V_2, the test pulse was given
both with V_1 "on" and with V_1 "off" (Eckert, Ewald 1983b).

molluscan nerve cells has elicited speculation (Magura 1977)
that these phases reflect functionally distinct calcium conductances having differing inactivation. Thus, it might be
postulated that one process is Ca-mediated, while the other
process is voltage-dependent (Hodgkin, Huxley 1952). As we
shall see below, a single Ca-mediated process can adequately
account for both the rapid and slow phases of inactivation.

The possibility of a dual origin of inactivation of the
calcium current in *Aplysia* neurons (Brown, Morimoto, Tsuda,
Wilson 1981) was investigated in experiments in which we
measured the rates of development of both the early rapid and
the late slow phases of inactivation as functions of voltage
and of current independently. Signals were digitized and
stored so that currents elicited by equivalent depolarizing

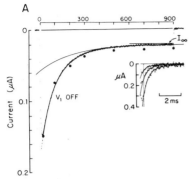
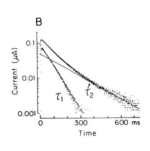

Fig. 2. Inactivation of calcium current during a depolarization. A, Voltage clamp pulse to 0 mV (V_h = -40mV) for 900 ms elicited current that relaxed toward steady-state asymptote, I_∞. The solid thin line is the computer-generated fit for $\tau_2 + I_\infty$, the dashed line is the fit for $\tau_1 + \tau_2 + I_\infty$. τ_1 and τ_2, had time constants of 80 ms and 250 ms, respectively. Inset shows sample of tail currents recorded following pulses of 20, 100, and 500ms duration. They were measured 400 µs (dashed vertical line) from computer-generated fit (Eckert, Ewald 1983a). Normalized tail current amplitudes (x 0.5) are shown as solid circles with current trajectory. B, Determination of τ_1 and τ_2 of the current shown in A. Log I_{Ca} was plotted against time, and τ_2 was determined from slope of the plot between 350 and 900 ms. This was peeled off to determine τ_1. These values were checked by fitting the calculated curve to the current in A (Chad, Eckert, Ewald 1983b).

and hyperpolarizing pulses could be added before processing to cancel symmetrical leakage and capacitive currents after correction for any time-dependent inward rectification. Current trajectories recorded during long pulses were analyzed with the aid of a computer by successive 'peeling off' operations. Inactivation proceeded toward a plateau level (I_∞) with a time course described by the sum of two exponentials, τ_1 and τ_2 (Fig. 2A). These were readily separated in semilogarithmic plots (Fig. 2B).

The filled circles plotted in Figure 2A are proportional (shown at 0.5X relative gain) to the amplitude of tail currents measured in the same neuron following pulses ending at times indicated by the filled circles. These tail currents, samples of which are shown in the insets, were measured as described above and elsewhere (Eckert, Ewald 1981, 1983a,b). Any potassium current activated during the depolarization would have been virtually absent in the tail current mea-

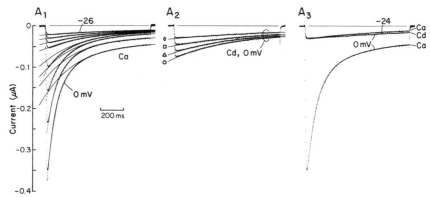

Fig. 3. Current dependence of inactivation. Bath contained
5 mM 4-AP, 200 mM TEA and 45 μM TTX. A, Currents elicited
in control ASW by 900ms pulses to -26, -24, -22, -20, -15,
-10, -5, and 0 mV. Computer fits were obtained as in Fig. 2.
B, Currents elicited by depolarization to 0 mV during progressive block by 0.5 mM Cd^{2+}. The progressively blocked
currents are indicated by the symbols o, △, □, and ◇.
C, Superposition of three traces selected from A and B.
The two smaller traces, one at -24 mV in control ASW, the
other at 0 mV after partial cadmium block, were selected
for similar peak currents. The much larger, rapidly decaying current was recorded at 0 mV before addition of cadmium.

surements made at -40 mV, which was close to E_K (Eckert,
Ewald 1983a,b). Conversely, any inward nonspecific current
(Byerly, Hagiwara 1982) that might have been present during
tail currents measured at -40 mV should have been absent or
greatly reduced at 0 mV, which is close to the reversal potential reported for that current. The essential correspondence between tail current amplitudes and the decay of the
inward current virtually eliminates the possibility that the
relaxation of inward current under the experimental conditions is an artifact of an outward current.

INACTIVATION IS MOST CLOSELY RELATED TO Ca^{2+} ENTRY

Currents measured during 900ms depolarizations ranging
in potential from -26 mV to 0 mV were digitized and stored
(Fig. 3A); 0.5 mM Cd^{2+} was then added to the bath, and the
membrane was again depolarized to 0 mV for 900ms. This was
done several times at 20s intervals while I_{Ca} became progressively smaller due to the growing block by cadmium (Fig. 3B).
The object was to obtain small currents at 0 mV that subse-

quently could be matched with currents of similar initial strength elicited at significantly lower membrane potentials before the Cd block. A close match occurred between peak current of the non-blocked current recorded at -24 mV (second trace from top in Fig. 3A) and a current recorded at 0 mV following 91% block of the peak I_{Ca} (Fig. 3C; the full current recorded at 0 mV prior to application of cadmium is also shown for comparison). In the full Ca current recorded at 0 mV, τ_1 equalled 65 ms and τ_2 equalled 295 ms. After 91% block of the peak Ca current with cadmium, the τ_1 component had disappeared, and τ_2 had increased to over 900 ms. The current recorded at -24 mV prior to the cadmium block exhibited essentially the same peak amplitude, had no τ_1 component, and had a τ_2 of 650 ms. Thus, the value of τ_2 at 0 mV increased progressively as the Ca current became smaller, finally exceeding the time constant recorded at -26 mV before application of cadmium. Since the loss of the fast phase and the increase in time constant of the slow phase both occurred at a fixed voltage during progressive block of I_{Ca}, both phases of inactivation appear to be related primarily to the size of Ca current rather than the membrane voltage.

A COMPUTER MODEL OF CALCIUM-MEDIATED INACTIVATION

A computer model of the Ca current sheds some light on the time course of current relaxation, and shows that the two phases of current relaxation can originate in a feedback interaction between Ca^{2+} entry, calcium buffering kinetics, and the Ca-dependent inactivation of the Ca conductance. The model closely simulates the biphasic time course of inactivation described in the previous section, even though the τ_1 and τ_2 phases are not introduced as separate processes (Eckert, Ewald, Chad 1982; Chad, Eckert, Ewald, 1983a,b).

The basis of the model is a linear system of four channel states as shown in Figure 4A. The voltage-dependent activation is described as an m^2 process (Hodgkin, Huxley 1952) which leads to opening of the calcium channel and influx of Ca ions. The m^2 formulation has been shown to be a good approximation for the activation of Ca-current in perfused neurons where Ca-dependent inactivation has been reduced with internal EGTA so as to have less effect on the time course of the current during the activation phase (Kostyuk, Krishtal, Pidoplichko 1981; Byerly, Hagiwara 1982). Intracellular buffering is modeled as a single compartment which contains a uniform initial concentration of buffer.

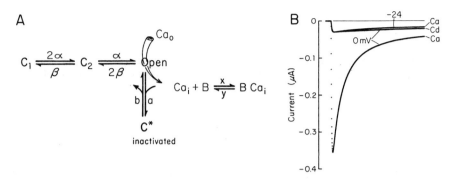

Fig. 4. A, Model of calcium channel transitions. C_1 represents the closed channel at hold potential. Horizontal
transitions are voltage dependent in accordance with m^2
kinetics. Vertical transitions are Ca-dependent. Calcium
is shown entering through the open channel. Intracellular
free Ca can bind to the open channel, producing the inactivated state, C^*. The probability of free calcium being
removed by buffer, B, and hence being unavailable for inactivation of the channel is P_B. B, Modeled Ca currents showing the current dependence of the time course of inactivation
in a simulation of the experiment shown in Figure 3C. The
large current was simulated to match the time course of the
large current recorded at 0 mV in Figure 3C. The parameters
used were τ_m (11 ms), m_∞ (1.0) and K (2.4×10^5 M^{-1}). A single
change simulating the effect of blocking open channels with
Cd^{2+} (0.5 mM) produced the small modeled current labeled Cd
(K_{Cd} = 1×10^{-5} M). The other simulated small current,
labeled Ca, was obtained with parameters identical to the
large current except for the values of m_∞ (0.27) and K
(7×10^5 M^{-1}).

The values used for the buffer concentration (~4×10^{-4} M) and
affinity (~2×10^{-6} M) are within the physiological range.
The resting concentration of free calcium is assumed to be
close to 0.2 μM but can be affected by Ca-loading (i.e.,
Ca entry during a prepulse) or by EGTA injection which
raise or lower, respectively, intracellular free Ca in Helix
neurons (Alvarez-Leefmans, Rink, Tsien 1981). Free intracellular Ca is assumed to be able to inactivate the open
channel, converting the voltage-activated state to the
non-conducting, inactivated state (C^*). We have chosen to
consider the rates of deactivation of the inactivated state
and Ca-mediated inactivation of the deactivated states as

negligible and have omitted these states in the modeling of the currents. Removal of inactivation is assumed to occur via the open state upon removal of Ca. This mechanism of inactivation is analogous to the one suggested for the action of certain non-depolarizing local anaesthetics on the acetylcholine-induced conductance at the neuromuscular junction (Adams 1975; Lambert 1983).

In the model, the fraction of entering Ca^{2+} that remains free to inactivate the channel depends primarily on cytoplasmic buffering of Ca^{2+}. The probability of calcium ions being bound to buffer is represented by the value P_B, which depends on the buffering capacity of the system and on the amount of calcium which has entered. This relationship is approximated with a simple single-buffer compartment consisting of a $1\mu m$ deep layer immediately under the membrane containing a fixed concentration of buffer, $[B]_0$, with binding affinity K_A. At time t the concentration of unbound calcium is $[Ca_i]t$. It can be shown that

$$P_B = \frac{[B]_0}{[B]_0 + [Ca_i]_t + K_A}.$$ (1)

This simple approach does not consider non-equilibrium conditions, but serves as an approximation for a complex system. At long times, loss of Ca^{2+} from the region near the membrane will have a significant effect. Making the assumption that Ca^{2+} diffuses uniformly into an infinite sink, we represent diffusion with a single rate constant, D. Thus, rate of loss of free Ca from the inner surface of the membrane $(d[Ca_i]/dt)$ is given by

$$\frac{d[Ca_i]}{dt} = D[Ca_i].$$ (2)

We can represent Ca_i inactivating the voltage-activated, opened channel as

$$O + Ca_i \underset{b}{\overset{a}{\rightleftharpoons}} C^*$$

open   inactivated,

assuming that Ca_i is at equilibrium inactivating calcium channels, and the total number of voltage activated channels is $[O]_0$, i.e. the number of open channels if there were no inactivation. The number of voltage-activated non-inactivated channels at equilibrium is $[O]_t$. The ratio of rate constants a/b can be represented as a single constant, K. Thus the probability of the channels being open and not inactivated is:

$$Prob = [O]_t/[O]_0 = 1/(1+K \cdot [Ca_i]).$$ (3)

To model I_{Ca}, m^2 activation was multiplied by the probability of channels not being inactivated (Eqn. 3). Free Ca_i can be expressed as the variable S, which is,

$$S = \int_0^t (1 - P_B) \cdot I_{Ca} \cdot \frac{1}{2Fv} - D \cdot S \, dt \qquad (4)$$

where F is Faraday's constant and v is the effective volume into which the Ca diffuses. Thus the calcium current at any time is simply:

$$I_{Ca} = \bar{g}_{Ca} \, (V_m - E_{Ca}) \, [m_\infty - (m_\infty - m_0) \, e^{-t/\tau_m}]^2 \cdot 1/(1+KS). \qquad (5)$$

These equations were solved by iteration with S as a dependent variable to give values for the calcium current through time.

The time courses generated by this model closely match those of calcium currents recorded under conditions described above. As in the recorded currents, the time course of inactivation seen in the modeled currents can be approximated by the sum of two exponentials $(\tau_1 + \tau_2)$ that approach a third non-inactivating component (I_∞) (Figs. 4B, 5). It is significant that two phases of relaxation are generated by the model even though it contains only one process of inactivation.

The modeled currents, like the recorded currents, undergo changes in time course of relaxation that are most closely related to the rate of entry of Ca^{2+}. With reduction of peak current strength, there is preferential loss of the rapid, early phase of inactivation. This can be seen clearly in Figure 4B, which is a simulation of the experiment shown in Figure 3C. The current strength at a given voltage (0 mV) was reduced by a simulated cadmium block of open channels. Cadmium was assumed to compete with calcium for a binding site at the open channel, thus blocking the channel. The degree of blockade was calculated according to Hagiwara and Takahashi (1967). The effect of cadmium (0.5 mM) could be closely simulated by using a K_{Cd} of 1×10^{-5} M. A value of 5.4 mM was used for K_{Ca} (Akaike, Lee, Brown 1978). In the modeled currents, as in the recorded currents, the early phase of inactivation was lost as the block progressed, and the time course of relaxation approached that obtained without the Cd block at a much lower voltage, -24 mV (i.e., lowered value of m_∞). Changing membrane voltage and hence degree of activation (i.e., the value of m_∞) in the modeled current closely simulated the changes in amplitude and relaxation that occurred in currents recorded at different voltages (Fig. 5A).

The modeling of currents in a double-pulse inactivation experiment, shown in Figure 5B along with recorded currents from such an experiment, is a particularly good test of the model, for the effect of the prepulse was assumed to result entirely from Ca entry and accumulation during V_1. This was modeled by a simple rise in ambient free Ca_i present

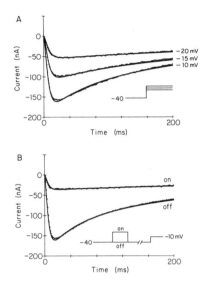

Fig. 5. Modeled fits to calcium currents recorded from two *Aplysia* (L2, L6) neurons. A, Responses to three step potentials and modeled fits (smooth traces). The voltage-dependent variables τ_m and m_∞ predominantly determine the activation phase (-10 mV, τ_m, 11 ms, m_∞, 0.8; -15 mV, τ_m, 12 ms, m_∞, 0.62; -20 mV, τ_m, 13 ms, m_∞, 0.44). The inactivation phase was determined by the efficacy parameter, K (-10 mV, 5.5×10^5 M^{-1}; -15 mV, 4.7×10^5 M^{-1}; -20 mV, 4.4×10^5 M^{-1}). B, Ca currents and modeled fits in response to potential step to -10 mV, in the absence ("off") and presence ("on") of a prepulse to +10 mV for 200 ms ending 1 s before V_2. The effect of V_1 was simulated in the model by an increase in the single value corresponding to the initial internal Ca concentration ("off," 0.3×10^{-6} M; "on," 6.8×10^{-6} M). All other parameters were the same in both traces (τ_m, 9 ms; m_∞, 0.87; K, 7.5×10^5 M^{-1}).

at the onset of V_2, going from 0.3 µM without the prepulse to 6.8 µM with the prepulse.

In summary, several lines of evidence confirm earlier reports that Ca^{2+} entry and accumulation during flow of Ca current leads to an inactivation of the Ca conductance. Recent experiments demonstrate that the temporal characteristics of inactivation are closely related to the rate of Ca^{2+} entry during current flow, and only secondarily influenced by membrane voltage. The kinetics of the Ca current are closely simulated with a simple model that uses only Ca-mediated inactivation in addition to m^2 voltage-dependent activation, while omitting any form of voltage-dependent inactivation. The two phases of inactivation and the late steady-state current in *Aplysia* neurons described above can arise, as demonstrated by this model, from feedback interactions of Ca^{2+} entry and accumulation, Ca-mediated inactivation and the removal of free Ca from the region near the inner face of the membrane. Thus, the complex behavior of the calcium current is consistent with a single, functionally uniform population of channels exhibiting a single process of inactivation. Standen and Stanfield (1982) recently published similar conclusions based on a closely related model of Ca-mediated inactivation.

Supported by USPHS NS 8364, NSF BNS 80-12346.

REFERENCES

Adams DJ, Gage PW (1979). Characteristics of sodium and calcium conductance changes produced by membrane depolarization in an *Aplysia* neurone. J Physiol 289:143.

Adams PR (1975). A model for the procaine end-plate current. J Physiol 246:61P.

Ahmed Z, Conner JA (1979). Measurement of calcium influx under voltage clamp in molluscan neurones using the metallochromic dye arsenazo III. J Physiol 286:61.

Akaike N, Lee KS, Brown AM (1978). The calcium current of *Helix* neuron. J Gen Physiol 71:509.

Alvarez-Leefmans FJ, Rink TJ, Tsien RY (1981). Free calcium ions in neurones of *Helix aspersa* measured with ion-selective micro-electrodes. J Physiol 315:531.

Brehm P, Dunlap K, Eckert R (1978). Calcium-dependent repolarization in *Paramecium*. J Physiol 274:639.

Brehm P, Eckert R (1978a). Ca-dependent inactivation of calcium conductance in *Paramecium*. Soc Neurosci Abstr 4:234.

Brehm P, Eckert R (1978b). Calcium entry leads to inactivation of calcium channel in *Paramecium*. Science 202:1203.

Brehm P, Eckert R (1979). Elevation of internal free Ca^{2+} is required for inactivation of Ca conductance in *Paramecium*. Soc Neurosci Abstr 5:290.

Brehm P, Eckert R, Tillotson D (1980). Calcium-mediated inactivation of calcium current in *Paramecium*. J Physiol 306:193.

Brown AM, Morimoto K, Tsuda Y, Wilson DL (1981). Calcium current-dependent and voltage-dependent inactivation of calcium channels in *Helix aspersa*. J Physiol 320:193.

Byerly L, Hagiwara S (1982). Calcium currents in internally perfused nerve cell bodies of *Limnea stagnalis*. J Physiol 322:503.

Chad J, Eckert R, Ewald D (1983a). Kinetics of calcium current inactivation simulated with an heuristic model. Biophys J 163:398.

Chad J, Eckert R, Ewald D (1983b). Current dependence of early and late phases of inactivation during calcium current flow in voltage-clamped neurons of *Aplysia californica*. (Submitted.)

Eckert R (1977). Genes, channels and membrane currents in *Paramecium*. Nature 268:104.

Eckert R, Ewald D (1981). Ca-mediated Ca channel inactivation determined from tail current measurements. Biophys J 33:145a.

Eckert R, Ewald D (1982). Residual calcium ions depress activation of calcium-dependent current. Science 216:730.

Eckert R, Ewald D (1982a). Fast and slow components of calcium inactivation in *Aplysia* neurons both exhibit current dependence. Soc Neurosci Abstr 8:943.

Eckert R, Ewald D (1982b). Ca-dependent inactivation of Ca conductance in *Aplysia* neurons exhibits voltage dependence. Biophys J 37:182a.

Eckert R, Ewald D (1983a). Calcium tail currents in intact voltage-clamped neurons of *Aplysia californica*. J Physiol (in press).

Eckert R, Ewald D (1983b). Inactivation of calcium conductance characterized from tail current measurements in neurons of *Aplysia californica*. J Physiol (in press).

Eckert R, Ewald D, Chad J (1982). A single Ca-mediated process can account for both rapid and slow phases of inactivation exhibited by a single calcium conductance. Biol Bull 163:398.

Eckert R, Lux HD (1975). A non-inactivating inward current recorded during small depolarizing voltage steps in snail pacemaker neurons. Brain Res 83:486.

Eckert R, Lux HD (1976). A voltage-sensitive persistent calcium conductance in neuronal somata of *Helix*. J Physiol 254:129.

Eckert R, Lux HD (1977). Calcium-dependent depression of a late outward current in snail neurons. Science 197:472.

Eckert R, Tillotson D (1979). Intracellular EGTA interferes with inactivation of calcium current in *Aplysia* neurons. Soc Neurosci Abstr 5:291.

Eckert R, Tillotson D (1981). Calcium-mediated inactivation of the calcium conductance in caesium-loaded giant neurones of *Aplysia californica*. J Physiol 314:265.

Eckert R, Tillotson D, Brehm P (1981). Calcium mediated control of Ca and K currents. Fed Proc 40:2226.

Gorman ALF, Thomas MV (1978). Changes in the intracellular concentration of free calcium ions in a pacemaker neurone, measured with the metallochromic indicator dye arsenazo III. J Physiol 275:357.

Hagiwara S, Byerly L (1981). Calcium channel. Ann Rev Neurosci 4:69.

Hagiwara S, Nakajima S (1966). Effects of the intracellular Ca ion concentration upon the excitability of the muscle fiber membrane of a barnacle. J Gen Physiol 49:807.

Hagiwara S, Takahashi K (1967). Surface density of calcium ions and calcium spikes in the barnacle muscle fiber membrane. J Gen Physiol 50:583.

Hermann A, Gorman ALF (1979). External and internal effects of tetraethylammonium on voltage-dependent and Ca-dependent K^+ current components in molluscan pacemaker neurons. Neurosci Lett 12:87.

Hodgkin AL, Huxley AF (1952). The dual effect of membrane potential on sodium conductance of the giant axon of Loligo. J Physiol 116:497.

Kostyuk PG, Krishtal OA (1977). Effects of calcium and calcium-chelating agents on the inward and outward current in the membrane of mollusc neurones. J Physiol 270:569.

Kostyuk PG, Krishtal OA, Pidoplichko VI (1981). Calcium inward current and related charge movements in the membrane of snail neurones. J Physiol 310:403.

Lambert JJ, Durant NN, Henderson EG (1983). Drug induced modification of ionic conductance at the neuromuscular junction. Ann Rev Pharmacol Toxicol 23:505.

Magura IS (1977). Long-lasting inward current in snail neurons in barium solutions in voltage-clamp conditions. J Memb Biol 35:239.

Plant TD, Standen NB (1981). Calcium current inactivation in identified neurones of Helix aspersa. J Physiol 321:273.

Standen NB, Stanfield PR (1982). A binding-site model for calcium channel inactivation that depends on calcium entry. Proc R Soc Lond B 217:101.

Takahashi K, Yoshii M (1978). Effects of internal free calcium upon the sodium and calcium channels in the tunicate egg analyzed by the internal perfusion technique. J Physiol 279:519.

Tillotson D (1979). Inactivation of Ca conductance dependent on entry of Ca ions in molluscan neurons. Proc Nat Acad Sci USA 77:1497.

Tillotson D, Eckert R (1979). Ca inactivation in Aplysia neurons is quantitatively related to prior Ca entry. Soc Neurosci Abstr 5:296.

The Physiology of Excitable Cells, pages 39-49
© 1983 Alan R. Liss, Inc., 150 Fifth Avenue, New York, NY 10011

CALCIUM INJECTION AND CALCIUM CHANNEL INACTIVATION

T. D. Plant, N. B. Standen and T. A. Ward

Department of Physiology
University of Leicester
University Road, Leicester LE1 7RH, U.K.

Calcium channels have been shown to occur in a wide variety of cells since calcium action potentials were first demonstrated in crustacean muscle by Fatt and Ginsborg (1958). (For reviews see Hagiwara 1973; Reuter 1973; Hagiwara, Byerly 1981.) The study of these Ca channels is of interest and importance since Ca is involved in a number of different cellular processes including excitation-secretion coupling, action potential generation in some tissues, excitation-contraction coupling and the control of membrane potassium permeability (see reviews by Erulkar, Fine 1980; Kostyuk 1980; Hagiwara, Byerly 1981).

The sensitivity of the Ca current to intracellular Ca concentration, $[Ca]_i$, was demonstrated by Hagiwara and Naka (1964) who found that injection of Ca-binding agents into barnacle muscle fibres enabled these cells to produce an all-or-none calcium action potential. In the same preparation, Hagiwara and Nakajima (1966) injected EGTA/Ca buffers and showed that action potentials could be generated when $[Ca]_i$ was less than 8×10^{-8}M, whereas oscillatory responses were seen with $[Ca]_i$ between 8×10^{-8}M and 5×10^{-7}M. Studies using internal dialysis methods to produce long-term changes in $[Ca]_i$ have confirmed that increased $[Ca]_i$ reduces Ca current in *Limnaea* and *Helix* neurones (Kostyuk, Krishtal 1977) and in tunicate egg cells (Takahashi, Yoshii 1978).

Recently it has been suggested by Brehm and Eckert (1978) and Tillotson (1979) that this sensitivity of the Ca channel to $[Ca]_i$ may provide a mechanism for inactivation

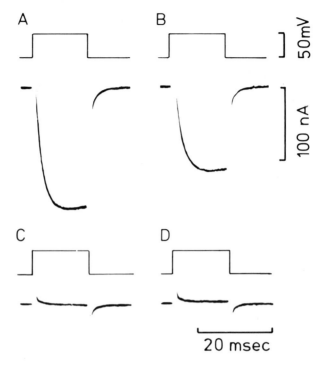

Fig. 1. Records of inward current (lower traces) in response to depolarising voltage clamp steps (indicated above). The saline contained (mM): $SrCl_2$, 25; $MgCl_2$, 5; KCl, 10; TEACl, 75; 4-aminopyridine, 2; Tris-HCl, 5. A conventional 2-electrode clamp was used and Ca was injected by ionophoresis while the cell was clamped at the holding potential of -50mV. A, control. B, 10s after a 100nA, 10s injection of Ca. C, control in 15mM $CoCl_2$ saline (substituted for $SrCl_2$). D, 10s after a 100nA, 10s Ca injection in the same solution as C. T 16°C.

of the channel. Thus Ca entry leads to a rise in Ca concentration near the inner surface of the cell membrane and so causes Ca channel inactivation. This contrasts with the inactivation mechanism of the sodium channel which appears to depend directly on membrane potential. The main evidence for Ca-dependent inactivation of Ca channels comes from two-pulse experiments which show that little inactivation occurs at very depolarized membrane potentials close to E_{Ca}, where there is little Ca entry, and also from the observation that Ca inactivation is slowed and reduced

by intracellular EGTA. Evidence of such Ca-dependent
inactivation has now been found in a variety of tissues
including *Paramecium* (Brehm, Eckert 1978; Brehm, Eckert,
Tillotson 1980), molluscan neurones (Tillotson 1979;
Eckert, Tillotson 1981; Plant, Standen 1981), insect
muscle (Ashcroft, Stanfield 1980, 1982), and vertebrate
cardiac muscle (Marban, Tsien 1981; Fischmeister, Mentrard,
Vassort 1981).

The work we describe in this paper is concerned with
the effects of changes in $[Ca]_i$ on the inactivation of Ca
channels in giant neurones of *Helix aspersa*. We have used
ionophoretic injection of Ca to produce relatively rapid
increases in $[Ca]_i$ and also injected EGTA and EGTA/Ca
buffers to investigate the range of Ca concentrations
over which changes in Ca inactivation occur (Standen 1981;
Plant, Standen, Ward 1983).

A typical response to an intracellular injection of
Ca is shown in Fig. 1. In most experiments we used Sr
solution and short test depolarizations since these conditions enabled us to repeat test pulses at 10s intervals
without any progressive decline in inward current amplitude. Fig. 1A shows the inward current recorded during
a voltage-clamp step to OmV. The inward current was substantially reduced 10s after an injection of Ca (Fig. 1B)
and subsequently recovered over the next 60s. The time
course of the effect of Ca injection was shown in Fig. 2,
which also shows that the degree of reduction in inward
current was graded with the size of the injection. Recovery of the inward current after injection could be
approximated by a single exponential in most cells; the
mean τ was 18.0 ± 1.2s (n = 17). The reduction in inward current was not an artefact of the injection technique itself, since injections of K or Mg ions had little
or no effect. The leakage current measured using a hyperpolarizing voltage-clamp step was not affected by Ca
injection. The effect of Ca injection could be enhanced
by application of 2μM carbonyl cyanide m-chlorophenyl-
hydrazone, a metabolic uncoupler which blocks mitochondrial Ca uptake. The effect could also be abolished by
prior injection of EGTA.

A problem in attempting to demonstrate Ca-dependent
inactivation of Ca channels is that such a mechanism is
often hard to separate experimentally from the effects of

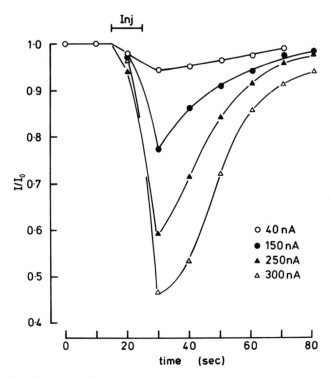

Fig. 2. Effect of amplitude of injected current. 10s Ca
injections at the current indicated. Ordinate: inward
current expressed relative to its pre-injection value
(I/I_0). Abscissa: time .25 Sr solution. Holding potential
-40mV. Test potential 0mV. T 18°C.

activation of a Ca-dependent K current, $I_{K(Ca)}$, (Hagiwara,
Byerly 1981; Plant, Standen 1981). We have therefore been
concerned to eliminate the possibility that the reduction
in inward current which we see on injection of Ca in fact
represents an activation of $I_{K(Ca)}$. Fig. 1, C, D shows
that no such K current is revealed when the inward current
is blocked by 15mM $CoCl_2$. As a further test, we have
repeated Ca injections under conditions where the size
of the inward current is changed, either by reducing $[Sr]_o$
to 2.5mM or by partial block using $5 \times 10^{-5}M$ $CdCl_2$. If
Ca injection reduces inward current by causing Ca-channel
inactivation we would expect a given injection to

inactivate the same proportion of channels and so reduce
the inward current by the same proportion in 25 Sr and 2.5
Sr solutions. On the other hand, if Ca injection activates
$I_{K(Ca)}$ then this should cause a much greater apparent
reduction in inward current when the inward current is
small because of reduced $[Sr]_o$ or partial block. Experimentally we observed that a given Ca injection reduced the
inward current by a fixed proportion under these different
conditions, providing further evidence against any role of
$I_{K(Ca)}$ activation in the response.

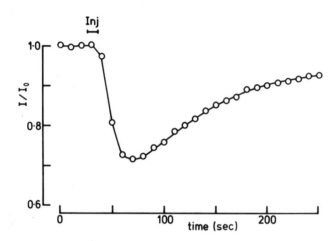

Fig. 3.   Effect of Ba injection (200nA, 10s).   Ordinate:
normalized inward current (I/I_o).   Abscissa: time.   25 Sr
solution.   Holding potential −50mV.   Test potential   0mV.
T 20°C.

We also think that it is very unlikely that Ca
injection reduces the inward current by causing a
reduction in electrochemical driving force sufficient to
reduce the current flow through open Ca channels.   The
injections which we normally used would raise the total
intracellular [Ca] by 2–4 x 10^{-4}M.   In *Aplysia* neurones
less than 1% of the total injected Ca remains free (Gorman,
Thomas 1980), so that it is likely that very little of
the injected Ca remained ionized in our experiments.   The
peak ionized $[Ca]_i$ measured using Ca-sensitive electrodes
was between 10^{-6} and 10^{-5}M.   Either a constant-field or an

energy barrier model for Ca permeation predicts that increasing $[Ca]_i$ to $10^{-5}M$ would reduce the inward current by less than 0.1% (Plant, Standen, Ward 1983). It seems therefore, that Ca injection reduces the inward current by causing inactivation of Ca channels and it seems likely that this action is mediated by Ca binding to a site at, or close to, the inner membrane surface and which forms part of the Ca channel.

We have found that injections of Sr and Ba ions also caused inactivation. The response to Sr was very similar to that for Ca, whereas Ba injection caused an inactivation which both developed and recovered more slowly (Fig. 3). Sr and Ba are less strongly buffered in cytoplasm than is Ca (Tillotson, Gorman 1982), so that these ions may reach higher concentrations after injection and could act at the same site as does Ca. It is also possible that these ions might displace Ca from some internal pool and thus cause a rise in $[Ca]_i$. This second alternative seems quite probable for Ba in view of the slow response and the fact that such a Ca-displacing effect has been shown by Meech and Thomas (1980).

Ca-sensitive microelectrodes of the type described by Tsien and Rink (1980) were used to measure the ionized $[Ca]_i$ in 25 Ca, Na-free saline. The mean pCa was 6.73 ± 0.13 (n = 6) corresponding to an ionized [Ca] of $1.9 \times 10^{-7}M$. This value is close to that reported by Alvarez-Leefmans, Rink and Tsien (1981) in *Helix* neurones bathed in a normal Na-containing saline and, in agreement with these workers, we found no rise in $[Ca]_i$ on changing from Na-containing saline to Na-free saline. Alvarez-Leefmans et al (1981) have also shown that EGTA injection causes a sustained fall in $[Ca]_i$. We therefore studied the effects of EGTA on the inward Ca current. Fig. 4 shows that ionophoretic injection of EGTA caused an increase in Ca current, the effect being maximal after about 2 minutes. The mean increase in I_{Ca} at +20mV was $31.1 \pm 2.4\%$ (n = 6). Taking the transport number of EGTA as 0.5 to give an upper estimate of the amount of EGTA injected the calculated final $[EGTA]_i$ would be between 4 and 12mM.

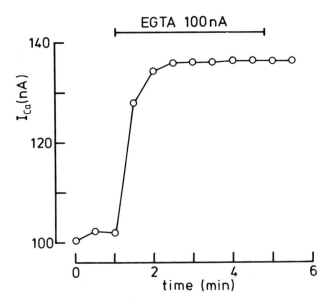

Fig. 4. Effect of EGTA injection. Ordinate: Ca current measured with a test pulse to +20mV. Abscissa: Time. 25 Ca solution. Holding potential –40mV. T 10°C.

    In an attempt to investigate the range of $[Ca]_i$ over which changes in Ca channel inactivation occur we have also injected EGTA/Ca buffers using a pressure injection method. We accepted only those results where the I_{Ca} reached a new steady level after injection. The results for a number of such experiments are shown in Fig. 5. Buffers with [Ca] greater than $8.9 \times 10^{-7}M$ decreased I_{Ca}, whereas buffers with [Ca] of $1.8 \times 10^{-7}M$ or less increased I_{Ca}. In addition both EGTA alone and EGTA/Ca buffers slowed the rate of inactivation of Ca currents as has been reported previously (Eckert, Tillotson 1981; Brown, Morimoto, Tsuda, Wilson 1981) but, even with maximal injections, did not remove inactivation entirely.

    Our results provide evidence that Ca injection reduces I_{Ca} by causing Ca channel inactivation. They also support the suggestion that Ca currents are inactivated as a consequence of an increase in $[Ca]_i$ caused by Ca entry (Brehm, Eckert 1978; Tillotson 1979). A simple model based on the assumption that Ca accumulates in a submembrane compartment

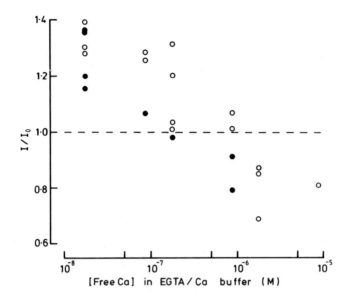

Fig. 5. Effects of injection of EGTA/Ca buffers of different free [Ca]. Each point is from a different cell.
Ordinate: Ca current after injection relative to its pre-
injection value (I/I_0). Abscissa: [Ca] in buffer. The
injected solution contained 62.5mM EGTA/Ca, 125mM KCl and
either 10mM HEPES (O) or 500mM HEPES (●) at pH 7.3.
25 Ca solution. Holding potential −40mV. T, 9−11°C.

and binds to a membrane site to cause Ca-channel inactivation can account for a number of experimental observations
including non-exponential decline of I_{Ca} and the continuing
inward current seen during prolonged depolarizations (Plant,
Standen, Ward 1983; Standen, Stanfield 1983). It is clear,
however, that Ca-dependent inactivation is not a universal
property of Ca channels since Ca channel inactivation in
egg cells depends directly on membrane potential (Hagiwara,
Ozawa, Sand 1975; Fox 1981).

The effects of injection of EGTA and EGTA/Ca buffers
suggest that there is a steady-state level of Ca channel
inactivation present at the resting $[Ca]_i$ of around 2 x
10^{-7}M in these cells, so that inactivation is sensitive
to changes in $[Ca]_i$ in the physiological range. This
regulation of inactivation by $[Ca]_i$ may represent one of

several ways in which Ca channels are affected by cellular metabolism. In cardiac muscle there is evidence that catecholamines increase the number of functional Ca channels by increasing the intracellular level of cyclic AMP (Tsien 1973; Reuter 1974) and it has been suggested that a similar cyclic AMP dependence may underlie the progressive washout of Ca current seen in perfused molluscan neurones (Byerly, Hagiwara 1982).

ACKNOWLEDGEMENTS

This work was supported by the M.R.C. and the Wellcome Trust.

REFERENCES

Alvarez-Leefman FJ, Rink TJ, Tsien, RY (1981). Free calcium ions in neurones of *Helix aspersa* measured with ion-sensitive microelectrodes. J Physiol Lond 315:531.

Ashcroft FM, Standfield PR (1980). Inactivation of calcium currents in skeletal muscle fibres of an insect depends on calcium entry. J Physiol Lond 308:36.

Ashcroft FM, Standfield PR (1982). Calcium inactivation in skeletal muscle fibres of the stick insect, *Carausius morosus*. J Physiol Lond 330:349.

Brehm P, Eckert R (1978). Calcium entry leads to inactivation of calcium channel in *Paramecium*. Science 202:1203.

Brehm P, Eckert R, Tillotson D (1980). Calcium-mediated inactivation of calcium current in *Paramecium*. J Physiol Lond 306:193.

Brown AM, Morimot K, Tsuda Y, Wilson DL (1981) Calcium current-dependent and voltage-dependent inactivation of calcium channels in *Helix aspersa*. J Physiol Lond 320:193.

Byerly L, Hagiwara S (1982). Calcium currents in internally perfused nerve cell bodies of *Limnea stagnalis*. J Physiol Lond 322:503.

Eckert R, Tillotson D (1981). Calcium-mediated inactivation of the calcium conductance in caesium-loaded giant neurones of *Aplysia californica*. J Physiol Lond 314:265.

Erulkar SD, Fine A (1980). Calcium in the nervous system. Rev Neuroscience 4:179.

Fatt P, Ginsborg BL (1958). The ionic requirements for the production of action potentials in crustacean muscle fibres. J Physiol Lond 142:516.

Fischmeister R, Mentrard D, Vassort G (1981). Slow inward current inactivation in frog heart atrium. J Physiol Lond 320:27.

Fox AP (1981). Voltage dependent inactivation of a calcium channel. Proc Natl Acad Sci USA 78:953.

Gorman ALF, Thomas MV (1980). Potassium conductance and internal calcium accumulation in a molluscan neurone. J Physiol Lond 308:287.

Hagiwara S (1973). Ca spike. Adv Biophys 4:71.

Hagiwara S, Byerly L (1981). Calcium channel. Rev Neurosci 4:69.

Hagiwara S, Naka K (1964). The effect of various ions on the resting and spike potentials of barnacle muscle fibres. J Gen Physiol 48:163.

Hagiwara S, Nakajima S (1966). Effects of the intracellular Ca ion concentration upon the excitability of the muscle fibre membrane of a barnacle. J Gen Physiol 49:807.

Hagiwara S, Ozawa S, Sand O (1975). Voltage clamp analysis of two inward current mechanisms in the egg cell membrane of a starfish. J Gen Physiol 6:617.

Kostyuk PG (1980). Calcium ionic channels in electrically excitable membrane. Neurosci 5:945.

Kostyuk PG, Krishtal OA (1977). Effects of calcium and calcium-chelating agents on the inward and outward current in the membrane of mollusc neurones. J Physiol Lond 270:569.

Marban E, Tsien RW (1981). Is the slow inward calcium current of heart muscle inactivated by calcium? Biophys J 33:143.

Meech RW, Thomas RC (1980). Effect of measured calcium chloride injections on the membrane potential and internal pH of snail neurones. J Physiol Lond 298:111.

Plant TD, Standen NB (1981). Calcium current inactivation in identified neurones of Helix aspersa. J Physiol Lond 321:273.

Plant TD, Standen.NB, Ward TA (1983). The effects of injection of calcium ions and calcium chelators on calcium channel inactivation in Helix neurones. J Physiol Lond (in press).

Reuter H (1973). Divalent cations as charge carriers in excitable membranes. Prog Biophys Mol Biol 26:1.

Reuter H (1974). Localization of beta adrenergic receptors, and effects of noradrenaline and cyclic nucleotides on action potentials, ionic currents and tension in mammalian cardiac muscle. J Physiol Lond 242:429.

Standen NB (1981). Ca channel inactivation by intracellular

Ca injection into *Helix* neurones. Nature Lond 293:158.
Standen NB, Stanfield PR (1983). A binding-site model for calcium channel inactivation that depends on calcium entry. Proc R Soc Lond B (in press).
Takahashi K, Yoshii M (1978). Effects of internal free calcium upon the sodium and calcium channels in the tunicate egg analysed by the internal perfusion technique. J Physiol Lond 279:519.
Tillotson D (1979). Inactivation of Ca conductance dependent on entry of Ca ions in molluscan neurones. Proc Ntl Acad Sci USA 76:1497.
Tillotson D, Gorman ALF (1982). Spatial profile of divalent ion buffering in nerve cell bodies measured with arsenazo III. Biophy J 37:58.
Tsien RW (1973). Adrenaline-like effects of intracellular iontophoresis of cyclic AMP in cardiac Purkinje fibres. Nature New Biol 245:120.
Tsien RY, Rink TJ (1980). Neutral carrier ion-sensitive microelectrodes for measurement of intracellular free calcium. Biochim Biophys Acta 559:623.

The Physiology of Excitable Cells, pages 51–63
© 1983 Alan R. Liss, Inc., 150 Fifth Avenue, New York, NY 10011

Ca CHANNELS IN THE MEMBRANE OF THE HYPOTRICH CILIATE
STYLONYCHIA

Joachim W. Deitmer

Abteilung für Biologie
Ruhr-Universität
D-4630 Bochum, F.R.G.

INTRODUCTION

Since the first intracellular potential recording in
the ciliate *Paramecium* by Kamada (1934), numerous studies
on the electrical excitability of ciliates as related to
ciliary motor activity and cellular behavior have been
carried out (for recent reviews see Machemer, de Peyer
1977; Eckert, Brehm 1979; Kung, Saimi 1982). Electrical
excitability in *Paramecium* has been found to consist of a
graded, Ca-dependent action potential (Naitoh, Eckert 1968;
Naitoh, Eckert, Friedman 1972). This Ca action potential
initiates the reversal of ciliary beating, resulting in
backward movement of the cell (see Eckert, Naitoh, Machemer
1976). Behavioral mutants, unable to perform ciliary reversal and thus backward swimming, were shown to lack
Ca-dependent excitability (Kung, Eckert 1972). Recent
voltage-clamp studies indicate that the ciliate membrane
has a voltage-dependent Ca channel (Brehm, Eckert 1978;
Satow, Kung 1979; de Peyer, Deitmer 1980; Deitmer, Machemer
1982), similar in properties to those described in many
multicellular organisms (for review see Hagiwara 1975;
Hagiwara, Byerly 1981).

Another type of Ca-dependent membrane process has been
described in ciliates, i.e. the Ca receptor potential seen
in response to mechanical stimulation of the cell anterior
(Naitoh, Eckert 1969; de Peyer, Machemer 1978). This
depolarizing mechanoreceptor potential normally reaches
threshold to trigger a Ca action potential.

In the hypotrich ciliate *Stylonychia* a more complex form of membrane excitability has been described (de Peyer, Machemer 1977). In this evolutionarily more advanced ciliate the action potential contains two components, one graded and one 'all-or-none,' both being dependent on extracellular Ca. In the present account some new voltage-clamp experiments on the voltage-dependent and the mechanosensitive Ca channels of this unicellular organism are described.

MATERIAL AND METHODS

The experiments were carried out on the hypotrich ciliate *Stylonychia mytilus* Syngen I. The cells were cloned and cultured in Pringsheim solution. They were fed every three days with the phytomonad alga *Chlorogonium elongatum*. Diagrams of the *Stylonychia* cell as viewed from its lateral and ventral side are shown in Fig. 1A,B. The cell is approximately 220 μm in length, 40 to 50 μm wide and 10 to 15 μm thick. The cell is highly organized structurally with a pronounced asymmetry. This is particularly conspicuous in respect to its ciliary organelles, consisting of 15 to 75 single cilia, each of them resembling in structure those of *Paramecium*. The membranellar band, a velum-shaped ciliary organelle, runs anteriorly to the ventral side towards the oral groove. The beating of the membranelles presumably provides a water current to support feeding. The marginal and ventral cirri mainly function as locomotor organelles, i.e. for swimming and walking.

The cells were transferred into the experimental solution at least 30 min prior to experimentation. The standard experimental solution contained 1 mM $CaCl_2$, 1 mM KCl and 1 mM Tris-HCl (or 1 mM HEPES) at pH 7.4 $\pm$ 0.1. The temperature was kept constant at 17°C.

Electrical potential recording and current injection were performed using two intracellular microelectrodes. The membrane potential was measured as the difference between an intracellular and an extracellular microelectrode, both filled with 1 M KCl (resistance 20 to 60 MΩ). A third microelectrode, filled with 2 M K-citrate, was inserted into the cell for current injection. Voltage-clamp of the cell membrane was performed by means of a conventional feedback system using a high-gain differential

amplifier (AD171K). The membrane current was monitored by
a current-voltage converter connected to the bath.

Mechanical stimulation of the cell surface was
performed by a glass stylus driven by a ceramic phonocartridge, to which DC pulses of different waveforms were
applied. The stylus was a fine-tipped glass capillary
(1 to 5 μm tip) of 1 mm in diameter. Details of electrical
recording and mechanical stimulation have been described
elsewhere (de Peyer, Machemer 1977, 1978; Deitmer 1981).

RESULTS AND DISCUSSION

The resting membrane potential of *Stylonychia* is
relatively constant at -50 mV. Depolarization of the
membrane by brief outward current pulses evokes an action
potential that characteristically contains two components,
a large, graded peak and a second smaller, all-or-none
peak or 'shoulder' (Fig. 1C,D). A very similar compound
action potential can be elicited by mechanically stimulating the anterior part of the cell (de Peyer, Machemer
1978; de Peyer, Deitmer 1980). In the latter case a
depolarizing mechanoreceptor potential gives rise to the
action potential.

The action potential triggers marginal and ventral
cirri to beat, with the power stroke directed towards the
cell anterior. The senso-electrical coupling and its
behavioral sequence is analogous to that described in
Paramecium (see Eckert, Naitoh, Machemer 1976). In
contrast to *Paramecium*, however, which moves forward at
resting membrane potential due to continuous beating of the
cilia towards the cell posterior, the locomotor organelles,
i.e. marginal and ventral cirri, in *Stylonychia* are quiescent at resting membrane potential. A detailed analysis
of the motor responses in *Stylonychia* has been performed
using high-speed cinematography (250 images/s) to monitor
ciliary beating while holding the membrane under voltage-
clamp control (Machemer, de Peyer 1982; Deitmer, Machemer,
Martinac 1983).

Both components of the action potential are dependent
upon extracellular Ca (de Peyer, Machemer 1977). The
amplitude of the first and the second peak varies with the
extracellular Ca concentration. If extracellular Ca is

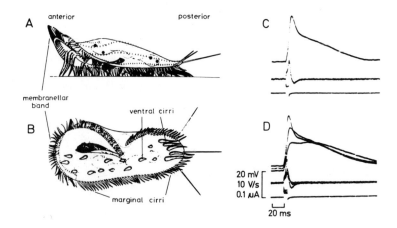

Fig. 1. Outline of the cell as viewed from the lateral
(A) and ventral (B) side, after Machemer. The different
ciliary organelles are indicated. Single (C) and superimposed (D) action potentials (upper traces), their first
derivatives (middle traces) and the current pulses injected (lower traces) are shown. In (D) pulses of three different amplitudes were injected to demonstrate the graded
character of the first peak and the apparent all-or-none
character of the second peak ('shoulder').

replaced by Mg or Mn, no action potentials can be elicited,
either by electrical or by mechanical stimulation (de Peyer,
Deitmer 1980). If, however, Sr or Ba replaces Ca, action
potentials of increased amplitude and duration are generated. In the presence of Co or Mn and Ca, the excitability
is reduced. The action potential remains unaffected by
changes or removal/addition of extracellular Na or Cl.
All these properties of membrane excitability meet the
criteria for Ca spikes (Hagiwara 1975). Voltage-clamp
experiments on the ionic selectivity of the early inward
current (de Peyer, Deitmer 1980) have confirmed this conclusion: the inward current is carried by Ca, Sr or Ba,
but not by Mg, Mn or Co.

CURRENT-VOLTAGE RELATIONSHIPS

The course of a typical current-voltage relationship as obtained in voltage-clamp is shown in Fig. 2. The peaks of the fast early current and the steady-state current have been plotted versus the membrane potential. Small membrane depolarizations of only 2 to 3 mV elicit a transient inward current. The amplitude of this current increases with further depolarization and produces a transient peak in the inward current-voltage relationship at -45 to -40 mV. The maximum of the inward current-voltage relationship is at around -17 mV, the peak net inward current being on average -24 nA. A zero net inward current is measured near +25 mV, presumably due to overlapping inward and outward currents. This interception of the X-axis does therefore not represent a true Ca-equilibrium potential. It is, however, conceivable that the Ca equilibrium potential, calculated to be +115 mV at rest with an assumed intracellular Ca activity of 10^{-7} M, becomes considerably more positive during larger depolarizations, as a transient rise in intraciliary Ca is to be expected following voltage-gated Ca influx into the small volume of the cilia. If a transient accumulation of intraciliary Ca of up to 5×10^{-5} M is assumed (Naitoh, Kaneko 1972), the Ca equilibrium potential may indeed be as low as +40 mV at the time of peak current flow.

A time-dependent steady-state outward current is activated by depolarizations of +20 mV and more. A prominent increase of the outward rectification occurs at -20 to -15 mV, i.e. at potentials of maximum Ca inward current. It has been suggested that a large fraction of this outward current is activated by Ca, as has been described in many nerve and muscle cells (Meech 1978), since in Ca-free, Mg-containing solution, where the fast early inward current is suppressed, little or no outward rectification has been observed (de Peyer, Deitmer 1980).

Hyperpolarization of the membrane produces a time-dependent and a steady-state inward-going rectification, which is inhibited by Ba (Ballanyi, Deitmer, in preparation, see also Hagiwara et al. 1978). This inward rectifier, however, has not yet been studied in any detail in *Stylonychia*.

Fig. 2 also shows the inward current-voltage relationship obtained when extracellular Ca is replaced

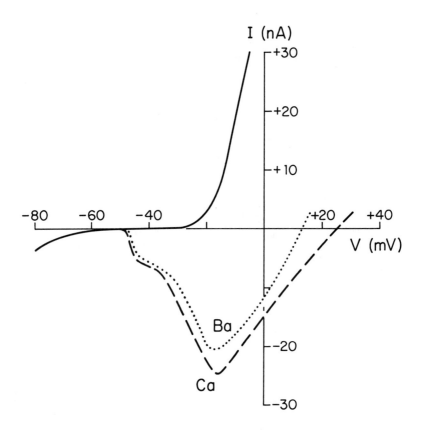

Fig. 2. Current-voltage relations: early peak inward current in the standard Ca solution (broken line), and when Ca was replaced by Ba (dotted line), and steady-state current at the end of a 60 to 80 ms pulse (continuous line) in the Ca solution. Holding (resting) potential at -50 mV.

by Ba. Both transient and maximum peaks are present in the Ba solution, but are reduced by 10 to 30%. The inward current-voltage relations in a solution with Sr as the only divalent cation present are very similar to those in Ca and Ba solution (de Peyer, Deitmer 1980).

'INACTIVATION' OF Ca AND Ba CURRENTS

A characteristic difference in the voltage-dependent inward currents carried by Ca or Ba is their time-dependent decay during a depolarizing pulse (Fig. 3). In Ca the net inward current decays very fast, while in Ba there is little decay of the inward current, even in the course of seconds. If two or more depolarizing steps are applied consecutively, the second and following evoked inward currents are much reduced in Ca, but only little decreased in Ba (Fig. 3). The fast decay of the net inward current in Ca might be indicative for (1) a Ca-induced Ca inactivation (see e.g. Brehm, Eckert 1978; Tillotson 1979), and/or (2) the activation of an outward current, short-circuiting the inward current, and/or (3) a voltage-dependent inactivation. In Ba a fast outward current might be absent or blocked, and a Ca-induced inactivation mechanism might not be present. A pure voltage-dependent inactivation is unlikely, since this would be expected to affect Ba currents similarly as Ca currents. As there is no build-up

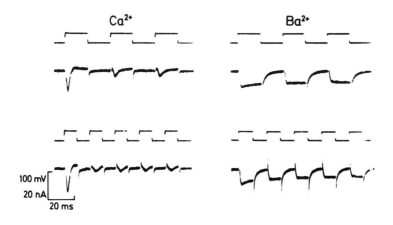

Fig. 3. Net inward currents in Ca and Ba: A series of three 18 ms and five 10 ms depolarizing pulses from -50 to -20 mV in the standard Ca solution, and when Ca was replaced by Ba. Note the differences in the time course of net inward currents in Ca and in Ba. The intervals between the pulses were chosen to produce an apparent 'steady-state inactivation' after the first pulse in Ca (indicated by the similar amplitude of the following net inward currents).

of an outward current (or outward current tail) between the
consecutively evoked pulses, an increased outward current
during the series of depolarizing pulses does not seem to
be responsible for the decreased net inward current. Thus,
Ca-induced inactivation of Ca inward current appears to be
the most likely mechanism responsible for the decrease in
net inward current in this cell. This has been further
substantiated by double-pulse experiments (Deitmer, unpublished), although the other mechanisms mentioned above
cannot yet fully be discarded.

MECHANORECEPTOR CURRENT

Ion selectivity. When the anterior of the cell is
mechanically stimulated, a net inward current is recorded.
This mechanoreceptor inward current is carried by Ca. The
reversal potential of this current changes by about 24 mV/
ten-fold change in the external Ca concentration (de Peyer,
Deitmer 1980). This inward current is also present if Ca
is replaced by Sr, Ba or Mg (Fig. 4). The reversal potential of the current changes by 21 mV/tenfold change in the
external Mg concentration. The mechanoreceptor inward current can thus be carried by Mg ions, in distinct contrast
to the early voltage-dependent inward current. Mn or Co
cannot carry charge through the mechanoreceptor channel.

The anterior mechanoreceptor channel may also be
slightly permeable to K ions. In general, its ion selectivity appears to be less specific than the voltage-
dependent Ca channel, and also less specific than the
posterior mechanoreceptor channel, which is highly selective for K ions (Machemer, de Peyer 1978; Deitmer 1981).

Is there interaction between voltage-dependent and
mechanoreceptor Ca currents? It has been shown in
Paramecium that deciliation eliminates the voltage-
dependent Ca current (Ogura, Takahashi 1976; Dunlap 1977;
Machemer, Ogura 1979), but leaves the mechanoreceptor
responses unaffected (Ogura, Machemer 1980). It has been
inferred from these results that the voltage-dependent Ca
channels reside in the ciliary membrane, while the mechanoreceptor channels are in the somatic cell membrane. In
Stylonychia, mechanical stimulation of the cirri or membranelles does not elicit mechanoreceptor responses, suggesting that the mechanoreceptor channels also in this cell

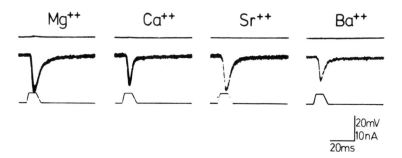

Fig. 4. Mechanoreceptor current recordings (from different cells) in the presence of Mg, Ca, Sr or Ba, as the nominally only divalent cation. Holding potential (upper traces) is −50 mV; the voltage applied to drive the phonocartridge for mechanical stimulation is shown in the lower traces (from de Peyer, Deitmer, unpublished).

probably reside in the somatic membrane. Even so, inflow of Ca into the cell via either the voltage-dependent or the mechanoreceptor channel might cause a build-up of intracellular Ca, which might cause mutual interactions between these channels. In a preliminary set of experiments this was tested by giving depolarizing voltage pulses and mechanostimulation consecutively. As shown in Fig. 5, the two types of Ca currents do not appear to influence each other. The voltage-dependent Ca current does not affect the mechanoreceptor Ca current (Fig. 5A), and the mechano-receptor Ca current does not change the voltage-dependent Ca current to any significant extent (Fig. 5B). The two Ca currents are thus independent of one another, possibly due to a spatial separation of the two channels involved. The intracellular build-up of Ca accompanying the mechano-receptor current must be too small and/or too far away from the voltage-dependent Ca channels to give rise to any Ca-induced inactivation of those channels. The decay of the mechanoreceptor Ca current itself does not appear to be produced by accumulation of intracellular Ca or Ca influx (Deitmer, unpublished).

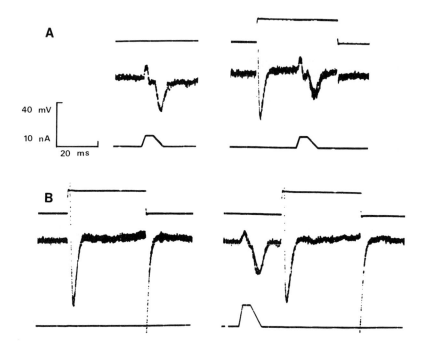

Fig. 5. Voltage-dependent and mechanoreceptor inward
currents. (A) A Ca mechanoreceptor current alone and
during a +20 mV depolarizing step eliciting a voltage-
dependent Ca inward current. (B) A Ca inward current
during a +20 mV depolarizing step alone, and when preceded
by a Ca mechanoreceptor current. No interaction between
the two types of Ca currents is apparent. All traces consist of two superimposed recordings, and consist of membrane voltage (upper traces), membrane current (middle
traces) and driving pulse for mechanical stimulation
(lower traces).

CONCLUSION

The membrane of *Stylonychia* appears to contain at
least three types of Ca channels, i.e. two voltage-
dependent Ca channels and one Ca channel at the cell
anterior activated by mechanical stimulation. Inactivation of the voltage-dependent, but not the mechanoreceptor
channel, seems to be due to intracellular Ca accumulation.
Voltage-dependent and mechanoreceptor currents do not

interfere with each other. The ion-selectivity of voltage-
dependent and anterior mechanoreceptor channels differ in
so far as Mg ion pass the mechanoreceptor channel but not
the voltage-dependent channels. At present, the development and distribution of these various Ca channels during
the cell cycle are being investigated.

ACKNOWLEDGEMENTS

I would like to thank Drs. H. Machemer and J.E. de Peyer
for many helpful discussions during the course of this
work, which was supported by the DFG, SFB 114, TP A5.
This article was written during my stay in the Department
of Biology, University of California, Los Angeles, and I
thank Dr. R. Eckert for his hospitality, and the Max-Kade
Foundation, New York, for a fellowship.

REFERENCES

Brehm P, Eckert R (1978). Calcium entry leads to inactivation of calcium channel in *Paramecium*. Science 202:1203.
Deitmer JW (1981). Voltage and time characteristics of
the potassium mechanoreceptor current in the ciliate
Stylonychia. J Comp Physiol 141:173.
Deitmer JW, Machemer H (1982). Osmotic tolerance of Ca-
dependent excitability in the marine ciliate *Paramecium
calkinsi*. J Exp Biol 97:311.
Deitmer JW, Machemer H, Martinac B (1983). Simultaneous
recording of responses of membranelles and cirri in
Stylonychia under membrane voltage-clamp. J Submicr
Cytol 15:285.
Dunlap K (1977). Localization of calcium channels in
Paramecium caudatum. J Physiol 271:119.
Eckert R, Brehm P (1979). Ionic mechanisms of excitation
in *Paramecium*. Ann Rev Biophys Bioeng 8:353.
Eckert R, Naitoh Y, Machemer H (1976). Calcium in the
bioelectric and motor functions of *Paramecium*. Symp Soc
Exp Biol 30:233.
Hagiwara S (1975). Ca-dependent action potential. In:
Membranes-A Series of Advances, Vol. 3, ed. G. Eisenman,
3:359, New York: Dekker.
Hagiwara S, Byerly L (1981). Calcium channel. Ann Rev
Neurosci 4:69.

Hagiwara S, Miyazaki S, Moody W, Patlak J (1978). Blocking effects of barium and hydrogen ions on the potassium current during anomalous rectification in the starfish egg. J Physiol 279:167.

Kamada T (1934). Some observations on potential differences across the ectoplasm membrane of *Paramecium*. J Exp Biol 11:94.

Kung C, Eckert R (1972). Genetic modification of electric properties in an excitable membrane. Proc Natn Acad Sci USA 69:93.

Kung C, Saimi Y (1982). The physiological basis of taxes in *Paramecium*. Ann Rev Physiol 44:519.

Machemer H, Ogura A (1979). Ionic conductances of membranes in ciliated and deciliated *Paramecium*. J Physiol 296:49.

Machemer H, Peyer de JE (1977). Swimming sensory cells: Electrical membrane parameters, receptor properties and motor control in ciliated protozoa. Verh Dtsch Zool Ges, p 86; Stuttgart: Fischer-Verlag.

Machemer H, Peyer de JE (1982). Analysis of ciliary beating frequency under voltage clamp control of the membrane. Cell Motility Suppl 1:205.

Meech RW (1978). Calcium-dependent potassium activation in nervous tissues. Ann Rev Biophys Bioeng 7:1.

Naitoh Y, Eckert R (1968). Electrical properties of *Paramecium caudatum*: all-or-none electrogenesis. Z vergl Physiol 61:453.

Naitoh Y, Eckert R (1969). Ionic mechanism controlling behavioral responses in *Paramecium* to mechanical stimulation. Science 164:963.

Naitoh Y, Eckert R, Friedman K (1972). A regenerative calcium response in *Paramecium*. J Exp Biol 56:667.

Naitoh Y, Kaneko H (1972). Reactivated triton-extracted models of *Paramecium*: Modification of ciliary movement by calcium ions. Science 176:523.

Ogura A, Machemer H (1980). Distribution of mechanoreceptor channels in the *Paramecium* surface membrane. J Comp Physiol 135:233.

Ogura A, Takahashi K (1976). Artificial deciliation causes loss of calcium-dependent responses in *Paramecium*. Nature 264:170.

Peyer de JE, Deitmer JW (1980). Divalent cations as charge carriers during two functionally different membrane currents in the ciliate *Stylonychia*. J Exp Biol 88:73.

Peyer de JE, Machemer H (1977). Membrane excitability in *Stylonychia*: Properties of the two-peak regenerative Ca-response. J Comp Physiol 121:15.

Peyer de JE, Machemer H (1978). Hyperpolarizing and depolarizing mechanoreceptor potentials in *Stylonychia*. J Comp Physiol 127:255.

Satow Y, Kung C (1979). Voltage sensitive Ca-channels and the transient inward current in *Paramecium tetraurelia*. J Exp Biol 78:149.

Tillotson D (1979). Inactivation of Ca conductance dependent on entry of Ca ions in molluscan neurons. Proc Natn Acad Sci USA 76:1497.

Top: C. Baud; W. Moody, Jr.; J. Deitmer
Bottom: S. Krasne; M. Barish; H. Ohmori; S. Ciani

The Physiology of Excitable Cells, pages 65–72
© 1983 Alan R. Liss, Inc., 150 Fifth Avenue, New York, NY 10011

HYDROGEN ION CURRENTS IN DEPOLARISED MOLLUSCAN NEURONES

R.W. Meech and R.C. Thomas

Department of Physiology
University of Bristol
University Walk, Bristol, BS8 1TD, U.K.

INTRODUCTION

This paper provides evidence for a hydrogen ion channel in excitable cells, and describes how the membrane conductance for hydrogen ions increases during depolarisation (see Thomas and Meech, 1982). The physiological significance of this pathway is not known but a number of possibilities appear to be worth exploring:

1. We find that at potentials above about OmV cells become more alkaline than the external medium. It is possible that this contributes to regulation of the internal pH (pH_i) in cells which generate large depolarising action potentials.

2. At the resting potential pH_i regulation involves a mechanism which exchanges external Na^+ and HCO_3^- ions for internal Cl^- and H^+ ions (Thomas 1977). Under its operation changes in pH_i take place relatively slowly. In contrast there could be rapid changes of pH_i in the vicinity of a H^+ channel which could influence a number of different membrane properties such as the membrane potassium conductance (see Brown and Meech, 1975, 1979; Meech 1979; Wanke, Carbone and Testa, 1979; Moody 1980; Moody and Hagiwara, 1982), sodium current inactivation (Brodwick and Eaton, 1978), light activated channels (Brown and Meech, 1975; 1979), electrical coupling between cells (Turin and Warner, 1977), as well as other cellular functions such as the assembly of microtubules (Regula, Pfeiffer and Berlin, 1981) and microfilaments (Begg and Rubhun, 1979) or the regulation of glycolysis (Ui, 1966).

3.   Physiologically generated changes in pH_i have been
reported in different tissues, (Brown, Meech and Thomas,
1976; Brown and Meech, 1979; McDonald and Jobsis, 1976;
Shen and Steinhardt, 1978; Webb and Nuccitelli, 1981). It
remains to be seen whether any of them are generated by
hydrogen ion movements through specific hydrogen ion channels.

METHODS

     Experiments were done on exposed cells in isolated
suboesophageal ganglia from the snail, *Helix aspersa*. The
neuronal cell body was penetrated with four electrodes. One
was to measure pH (Thomas 1978), a second was a reference
liquid ion-exchanger electrode to measure the membrane
potential - E_m (Thomas & Cohen, 1981), a third supplied
feed-back current to voltage-clamp the cell (Thomas 1969)
and a fourth was a micropipette for iontophoretic HCl
injection. The preparation was superfused with saline
buffered to pH 7.5 either with HEPES alone or with HEPES
and 4 mM $NaHCO_3$. The bicarbonate saline was equilibrated
with 0.5% CO_2 in air.

RESULTS

     Figure 1 shows that at depolarised potentials pH_i
recovers from acid injections by a mechanism clearly
different to that operating normally. In this experiment
the membrane potential was held first at -50 mV during two
control injections of HCl, and was then shifted to +15 mV
for three more injections. After the first injection pH_i
recovered within 10 mins. To inhibit this recovery the
bathing medium was changed to a bicarbonate-free saline.
Recovery from the second injection was slow and before it
was complete the membrane was depolarised to +15 mV. This
caused a rapid and unexpected increase in pH_i to a steady
level at pH 7.6. (The normal pH_i is between 7.3 and 7.5).
After a third HCl injection the pH_i again returned to this
steady level and the recovery was faster than the control
in normal saline. In other experiments normal pH_i regulation
was abolished by using Na-free saline or the inhibitor SITS
and the cell was depolarised either under voltage-clamp or
with isotonic KCl (Thomas 1979). In each case there was
a rapid recovery of pH_i following an acid load.

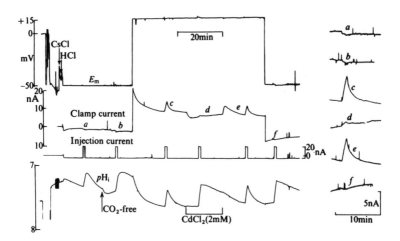

Fig. 1. The effect of HCl injections on the intracellular pH (pH_i) and voltage-clamp current recorded from an identified snail neurone. The four traces are pen recordings of membrane potential (E_m), clamp current, injection current and pH_i. At the right of the figure sections of the current record (a to f) are shown in more detail. The voltage-clamp current electrode was filled with CsCl so that Cs^+ ions were injected when the cell was depolarised. This had the effect of reducing the outward current and gave improved voltage control. The arrows above the top trace indicate the point at which the current microelectrodes were inserted.

In depolarised cells pH_i recovery can be blocked by the presence of heavy metals in the bathing medium. Fig. 1 shows that the recovery of pH_i following an acid load at +15 mV was slowed by saline containing 2 mM $CdCl_2$. When the Cd^{++} ions were removed there was a significant increase in the rate of H^+ loss. Other heavy metals such as 1 mM $CuCl_2$ or 1 MM $LaCl_3$ had a similar effect.

We had tested the way in which the membrane potential
level influenced the pH_i recovery from an acid load. The H^+
loss was unaffected by hyperpolarising the membrane to -70
mV for 5 mins. but it increased when the membrane was depolarised to -30 mV. With each further depolarisation there
was a rapid pH_i increase to a new steady state. The relationship between membrane potential and pH_i is shown in
Fig. 2. The solid line which is drawn through the points
shows the slope expected if Ph_i changes by 1 pH unit for a
58 mV change in membrane potential. The interrupted line
shows the expected result if the membrane potential was at
the hydrogen ion equilibrium potential. If this were the
case pH_i would equal the external pH (i.e., pH 7.5) when the

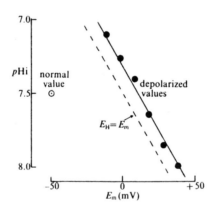

Fig. 2. Graph showing the relationship between membrane
potential (E_m) and steady state intracellular pH (pH_i).
The normal pH regulating system had been blocked by SITS
(4-acetamide-4'-isothiocyanatostilbene-2'-disulphonic
acid). The solid line has a slope of 58 mV per pH unit
and has been positioned to give the best fit to the results.
The interrupted line shows the relationship expected if H^+
ions were passively distributed across the cell membrane.

membrane potential was held at 0 mV. In fact, as Fig. 2
shows pH_i is close to 7.3 at 0 mV. In four experiments the
average difference between the internal and external pH at
0 mV was 0.33 ± 0.09 (S.D.) pH unit. In experiments in
which SITS was not used the difference was 0.07 ± 0.04
(S.D.) pH unit (n = 4). It is possible that at least part
of this difference is because the pH of the unstirred layers
near the cell membrane is different to that found in bulk
solution.

At depolarised potentials the membrane appears to
become so highly permeable to H^+ ions that their concentration inside the cell is determined simply by the external
pH and the potential across the cell membrane. If hydrogen
ions are the only ion species moving it should be possible
to detect an outward current associated with their passage
across the cell membrane. Such a current can be seen in the
clamp current trace in Fig. 1. Injections a), b) and f)
which were carried out at -50 mV produced no obvious change
in the current trace but c) produced a large outward current
which declined in an exponential fashion once the injection
was terminated. The presence of 2 mM Cd^{++}, which blocked
rapid pH_i recovery, also blocked the appearance of the outward current (see injection d). The current reappeared
once the Cd^{++} had been removed (injection e). The time
constant for the decline in the injection dependent current
is about 1.4 min. which is half the time constant for the
associated pH_i recovery. One possible reason for this
difference is that our measurements of pH_i changes are
limited by the response time of the pH electrode.

If all the injected H^+ ions leave the cell by crossing
the cell membrane and there are no other ion movements then
the injection-induced charge movement should equal the
charge used to inject the H^+ ions in the first place. However the charge movement generated by injection c) was only
about 66% of the charge used to inject HCl (assuming a
transport number for H^+ of 0.94, Thomas 1976) and so it is
possible that the H^+ movement which we observe is coupled
to the movement of other ions. Recent experiments (Meech
and Thomas, unpublished) indicate that the charge movement
generated by identical injections depends on both the membrane potential and on the steady value of pH_i. It is
increased at positive membrane potentials and at low pH.

In perfused molluscan neurones there is an outward "residual current" which appears upon depolarisation even when K^+ ions have been removed from both sides of the membrane. (Byerly and Hagiwara, 1982; Kostyuk and Krishtal, 1977). This current is probably carried by hydrogen ions because its associated "tail" seen after repolarising the membrane, reverses at a potential which depends on the internal and external pH of the cell (Byerly, Meech and Moody, unpublished). Like the H^+ current, the residual current is reduced by Cd^{++} and other calcium channel blocking agents. In *Limnea* it is half activated in about 5 msec during pulses to +20 mV at room temperature and so it could have a significant effect on pH values close to the membrane in actively firing cells. It seems likely that a similar current exists in striated muscle fibres, (Keynes, Rojas, Taylor and Vergara, 1973 - see figure 6; Palade and Almers, 1978).

This work was supported by the MRC and by the Wellcome Trust. Preliminary experiments by R.W.M. were carried out in the Physiology Department, University of Utah, supported by N.I.H. Grant EY 02290 and by R.C.T. in the Physiology Department of Yale University partly supported by USPHS grant AM 17322. We thank Michael Rickard and Toni Gillet for technical help. Figs. 1 and 2 from Thomas and Meech, (1982).

REFERENCES

Begg DA, Rebhun LI (1979). pH regulates the polymerization of actin in the sea urchin egg cortex. J Cell Biol 83:241.
Brodwick MS, Eaton DC (1978). Sodium channel inactivation in squid axon is removed by high internal pH or tyrosine-specific reagents. Science 200:1494.
Brown HM, Meech RW (1975). Effects of pH and CO_2 on large barnacle photoreceptors. Biophys J 15:276a.
Brown HM, Meech RW, Thomas RC (1976). pH changes induced by light in large *Balanus* photoreceptors. Biophys J 16:33a
Brown HM, Meech RW (1979). Light induced changes of internal pH in a barnacle photoreceptor and the effect of internal pH on the receptor potential. J Physiol Lond 297:73.
Byerly L, Hagiwara S (1982). Calcium currents in internally perfused nerve cell bodies of *Limnea stagnalis*. J Physiol Lond 322:503.
Keynes RD, Rojas E, Taylor RE, Vergara J (1973). Calcium

and potassium systems of a giant barnacle muscle fibre under membrane potential control. J Physiol Lond 229:409.

Kostyuk PG, Krishtal OA (1977). Separation of sodium and calcium currents in the somatic membrane of mollusc neurones. J Physiol Lond 270:545.

McDonald VW, Jobsis FF (1976). Spectrophotometric studies on the pH of frog skeletal muscle. pH changes during and after contractile activity. J Gen Physiol 68:179.

Meech RW (1979). Membrane potential oscillations in molluscan "burster" neurones. J Exp Biol 81:93.

Moody W Jr (1980). Appearance of calcium action potentials in crayfish slow muscle fibres under conditions of low intracellular pH. J Physiol Lond 302:335.

Moody WJ, Hagiwara S (1982). Block of inward rectification by intracellular H^+ in immature oocytes of the starfish *Mediaster aequalis*. J Gen Physiol 79:115.

Palade GJ, Almers W (1978). Slow sodium and calcium currents across the membrane of frog skeletal muscle fibers. Biophys J 21:168a.

Regula CS, Pfeiffer JR, Berlin RD (1981). Microtubule assembly and disassembly at alkaline pH. J Cell Biol 89:45.

Shen SS, Steinhardt RA (1978). Direct measurement of intracellular pH during metabolic derepression of the sea urchin egg. Nature 272:253.

Thomas RC (1969). Membrane current and intracellular sodium changes in a snail neurone during extrusion of injected sodium. J Physiol Lond 201:495.

Thomas RC (1976). The effect of carbon dioxide on the intracellular pH and buffering power of snail neurones. J Physiol Lond 255:715.

Thomas RC (1977). The role of bicarbonate chloride and sodium ions in the regulation of intracellular pH in snail neurones. J Physiol Lond 273:317.

Thomas RC (1978). "Ion-sensitive Intracellular Micro-electrodes." London: Academic Press.

Thomas RC (1979). Recovery of pH in snail neurones exposed to high external potassium. J Physiol Lond 296:77P.

Thomas RC, Cohen CJ (1981). A liquid ion-exchanger alternative to KCl for filling intracellular reference microelectrodes. Pflugers Arch Ces Physiol 390:96.

Thomas RC, Meech RW (1982). Hydrogen ion currents and intracellular pH in depolarized voltage-clamped snail neurones. Nature Lond 299:826.

Turin L, Warner AE (1977). Carbon dioxide reversibly abolishes ionic communication between cells of early

amphibian embryo. Nature, Lond 270;56.
Ui M (1966). A role of phosphofructokinase in pH-dependent
regulation of glycolysis. Biochim Biophys Acta 124:310.
Wanke E, Carbone E, Testa PL (1979). K^+ conductance modified by a titratable group accessible to protons from the
intracellular side of the squid axon membrane. Biophys
J 26:319.
Webb DJ, Nuccitelli R (1981). Direct measurement of intracellular pH changes in *Xenopus* eggs at fertilization
and cleavage. J Cell Biol 91:562.

The Physiology of Excitable Cells, pages 73–82
© 1983 Alan R. Liss, Inc., 150 Fifth Avenue, New York, NY 10011

A KINETIC MODEL FOR INWARD RECTIFICATION IN THE EGGS OF THE
POLYCHEATE NEANTHES ARENACEODENTATA

R. Gunning and S. Ciani

Jerry Lewis Neuromuscular Research Center
UCLA School Of Medicine
Los Angeles, CA 90024

This paper presents evidence that the steady-state and
kinetic properties of the inward rectifier in the eggs of
the marine polychaete Neanthes arenaceodentata are
consistent with a three-state model for the pore, assuming
that only one state is conductive and ohmic.

Inward rectification is a K-selective component of the
ionic permeability of certain cell membranes which is more
conductive when the current is driven toward the interior
of the cell. Its discovery in skeletal muscle (Katz, 1949)
was followed by several studies on the same tissue, some of
which anticipated the currently popular idea that both
voltage and external potassium effect the membrane
permeability (Hodgkin & Horowicz, 1959; Adrian & Freygang,
1962a,b). A confirmation of these concepts was given by
Hagiwara & Takahashi (1974) in one of the most complete
characterizations of the steady-state properties of inward
rectification as a function of voltage and external K. From
current clamp studies on the eggs of a starfish, the
following equation was derived empirically for the steady-
state conductance

$$G_{ss} = \frac{B \; K^{1/2}}{1 + \exp\left[(\Delta V - V_h)/v\right]} \qquad ;(1)$$

where [K] is the concentration of external K, ΔV is the
displacement of the membrane voltage from its resting
value, and B, Vh and v are constants. A dependence on ΔV
was also found for the activation kinetics of the

conductance: hyperpolarization gave rise to exponential increases in the current, with time constants decreasing with more negative values of ΔV (Hagiwara, Miyazaki & Rosenthal,1976).

Although the hypothesis that inward rectification is mediated by ionic channels has been widespread for some time, convincing evidence was provided by noise analysis studies (Ohmori, 1978), as well as by more recent single channel recordings (Ohmori, Yoshida & Hagiwara, 1981; Fukushima, 1981, 1982). Thus, in what follows, the terms "pore", and "channel", will be used to denote the unitary conductive element of inward rectification.

RESULTS OF VOLTAGE CLAMP EXPERIMENTS ON THE EGGS OF NEANTHES ARENACEODENTATA

Experiments have been carried out with the main objective of characterizing and modeling the steady-state and the kinetic properties of inward rectification. Using a two-microelectrode voltage clamp, application of hyperpolarizing and depolarizing voltage steps resulted in currents approaching asymptotic levels with time courses following single exponentials. Three parameters were studied as functions of voltage and external K: 1)the apparent instantaneous conductance, G_1; 2) the steady-state conductance, G_{ss}; and 3) the time constant of relaxation, τ. Since the results are presented in detail elsewhere (Gunning, 1983), only a summary will be given here, emphasizing those aspects that most decisively effected our choice of the model.

Both the instantaneous, G_1, and the steady-state, G_{ss}, conductances were found to increase sigmoidally with increasing hyperpolarization and could be described by the following empirical equations:

$$G_1 = \frac{G_1^{max}}{1+\exp\ [(\Delta V - V')/v]} \ ; \ (2) \ , \ G_{ss} = \frac{G_{ss}^{max}}{1+\exp\ [(\Delta V - V_h)/v]} \ ; \ (3)$$

with

$$G_1 = B'[K]^{\alpha} \ ; \ (4) \ , \ G_{ss} = B[K]^{\alpha} \ ; \ (5)$$

Here, ΔV and [K] have the same meanings as in eq(1), and

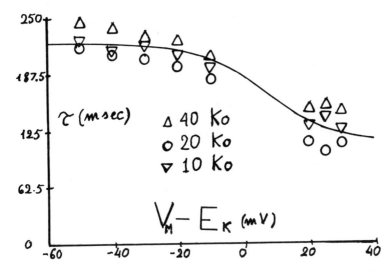

Fig.1. Time constants of activation plotted as functions of ΔV. The solid line is drawn to eq.(6) with $\tau^{max} = .223$ sec., A=.51, v=11mV, V_h=+5mV.

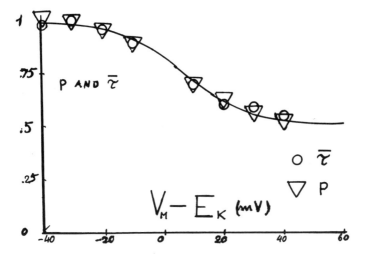

Fig.2. $p=\bar{G}_{ss}/\bar{G}_1$ and $\bar{\tau} = \tau/\tau^{max}$ as a function of ΔV. The solid line is drawn to eq.(6) divided by τ^{max}. The values of the parameters are the same as in Fig. 1. [K]=10mM. A similar identity between p and τ/τ^{max} was also found for [K]=20mM and 40mM.

the remaining parameters are constants. Fitting the data
with the above equations gave the following approximate
values for the constant parameters: $.55 < \alpha < .75$, $v=11mV$,
$V_h=+5mV$, $V'=V_h+5mV$. B and B'vary somewhat from cell to cell
(e.g. with the size of the egg) and B' is typically 10-25%
smaller than B. The order of magnitude of the conductance
in artificial sea water is 1 μS (0.2 mS/cm^2). A sigmoidal
increase with increasing hyperpolarization and a similar
dependence on ΔV, with the same steepness factor, v, was
found also for the time constant of relaxation, τ . The
data, illustrated in Fig.1, are fit quite accurately by the
empirical equation

$$\tau = \tau^{max} \left(A + \frac{1-A}{1+\exp[(\Delta V - V_h)/v]} \right) \quad ; \ (6)$$

where A is a positive constant smaller than unity, and
τ^{max} is the limiting value of τ for $\Delta V \to +\infty$. The most
striking result is illustrated in Fig.2, where the plot of $\tau/$
τ^{max} is found to be superimposable on that of the ratio of
the steady-state to the instantaneous conductance, each
normalized to its maximum value.

$$(\bar{G}_{ss}/\bar{G}_1) = (G_{ss}/G_{ss}^{max}) \cdot (G_1^{max}/G_1) = \tau/\tau^{max} \quad ; (7)$$

In the next section we will show that the results can
all be accounted for by a simple kinetic model, according
to which the pore can exist in three states, only one being
conductive and voltage independent.

KINETIC MODEL

In the attempt to formulate a kinetic model, we have
tried to minimize untestable hypotheses, such as details
about the energy barriers in the permeation pathway and
about the mechanisms of "gating". Thus, while it is
impossible to avoid a certain degree of arbitrariness,
which, in our case, consists of assuming that the
proportionality to $[K]^{\alpha}$ in eqs.(4) and (5) reflects
properties of the single channel, and that the exponentials
in eqs.(2) and (3) describe equilibria between the states
of the pore, no specific assumptions have been made about

the particular physical mechanisms which generate those terms.

Since only one exponential is seen experimentally, a two-state model of the type

$$C \underset{y}{\overset{z}{\rightleftharpoons}} 0 \; , \qquad\qquad ; \; (8)$$

where 0 is the only conductive state, is the most natural starting point for interpreting the data. Although it is formally possible to fit the experimental results with such a scheme, one is forced to a number of implausible conclusions. For example, two quantities of seemingly different nature, γ, the single channel conductance, and , τ, the relaxation time constant, would have to be almost identical functions of ΔV. This coincidence seems fortuitous and difficult to justify.

A surprisingly simple and consistent description can, instead, be obtained by the following three-state model

$$I \underset{x}{\overset{w}{\rightleftharpoons}} C \underset{y}{\overset{z}{\rightleftharpoons}} 0 \qquad\qquad ; \; (9)$$

where I, C and 0 stand for "inactivated", "closed" and "open", respectively. The following simplifying assumptions are made:

(i) 0 is the only conductive state, the single channel conductance, γ, being voltage-independent and proportional to $[K]^{\alpha}$, $.55 < \alpha < .75$.

(ii) x and w are constants, independent of both voltage and $[K]$.

(iii) $y+z \gg x+w$ in the entire range of the experimental values of ΔV.

Assumption (i) implies that when external K is fixed, conductance changes with voltage and time reflect changes in the number of open channels, N_0:

$$G(t, \Delta V) = \gamma N_0(t, \Delta V) \qquad\qquad ; \; (10)$$

Assumption (iii) is meant to imply that the early current after a voltage step reflects a new equilibrium, characterized by the new values of the rate constants z and y, between the pores originally present in C and 0, while exchange with the state I has not yet started.

Integrating the conventional kinetic equations for scheme (9), using the three approximations above and recalling eq.(10), the conductance is given by

$$G(t) = (G_1 - G_{ss}) e^{-t/\tau} + G_{ss} \qquad \cdot (11)$$

where the apparent "instantaneous" G_1 conductance, is

$$G_1 = G_1^{max}(1+z/y)^{-1}; \quad (12), \text{ with } \quad G_1^{max} = \frac{\gamma N^T(1+z_0/y_0)}{1+(z_0/y_0)(1+w/x)} \quad ; \quad (13)$$

N^T being the total number of pores and the subscript o in eq.(13) indicating that the rate constants are evaluated at the holding potential. The steady-state conductance, G_{ss}, is

$$G_{ss} = \frac{G_{ss}^{max}}{1+(z/y)(1+w/x)} \quad ; \quad (14), \text{ with } \quad G_{ss} = \gamma N^T \quad ; \quad (15)$$

and the relaxation time constant, τ, is

$$\tau = \tau^{max} \frac{1+z/y}{1+(z/y)(1+w/x)} \quad ; \quad (16), \text{ with } \tau^{max} = x^{-1} \quad ; \quad (17)$$

From eqs.(12), (14) and (16), the experimentally observed equality between the ratio of the normalized conductances and the normalized time constant, eq.(7), can be deduced immediately. Moreover, introducing the definitions

$$z/y = \exp\left[(\Delta V - V')/v\right] \qquad ; \quad (18)$$

$$1+w/x = A^{-1} = \exp\left[(V'-V_h)/v\right] \qquad ; \quad (19)$$

only elementary rearrangements of eqs.[(12) to (19)] are required to recast eqs.(12), (14) and (16) in the same form as the empirical expressions (2), (3) and (6). Note also that eqs.(13) and (15) represent much more than merely formal definitions. For example, the right-hand side of eq.(13), divided by γ , gives the number of channels that are either in the C or the O state at the holding potential. Since any exchange with state I is, by assumption, a slow process, it is easy to recognize that such a number represents the maximum number of channels that can be detected "instantaneously" for sufficiently large hyperpolarizing pulses. Also definition (15) is consistent with the fact that G_{ss} corresponds to the case in which all the pores are in the open state.

Thus, the three-state model (9) accounts for the data in <u>Neanthes</u> with excellent consistency and with surprisingly few assumptions.

COMPARISON WITH OTHER SYSTEMS

The instantaneous and steady-state conductances in <u>Neanthes</u> are basically similar to those found in starfish eggs (Hagiwara & Takahashi, 1974; Hagiwara, Miyazaki & Rosenthal, 1976) and in muscle (Hestrin, 1981; Leech & Stanfield, 1981). However, a qualitative discrepancy exists for the relaxation time constant, which, as the membrane is hyperpolarized, increases in the <u>Neanthes</u> preparation [see Fig.1], but decreases in the others. This discrepancy can be resolved by simply adding to scheme (9) a direct path between I and O. This modified model is illustrated below

$$; (20)$$

where, due to microscopic reversibility, the two additional rate constants, u and s, are related to the others by

$$u/s = yx/zw. \qquad ; (21)$$

Using the same assumptions (i) and (ii), and modifying assumption (iii) by requiring that $y+z \gg x+w+u+s$, one reobtains eqs.[(12) to (15)], while eq.(16) for τ becomes

$$\tau = (x+u)^{-1} \cdot \frac{1+z/y}{1+(z/y)(1+w/x)} \qquad ; (22)$$

Eq.(21) shows that, if x and w are constant, as required by assumption (ii), the ratio, u/s, is proportional to y/z and, consequently, to $\exp[-\Delta V/v]$. Although eq.(21) is insufficient to establish whether the dependence on $\Delta V/v$ is relegated to u or to s, or whether it is shared between the two, the finding that the time constant is a function of ΔV is compatible with eq.(22) only if u is either constant or proportional to $\exp[\Delta V/v]$. The first alternative would lead us back to the case of Neanthes. However, in the second case, using eqs.(18) and (19), τ becomes

$$\tau = \frac{1}{x+u_0\exp(-\Delta V/v)} \cdot \frac{1+\exp[(\Delta V-V')/v]}{1+\exp[(\Delta V-V_h)/v]} \quad ; \quad (23)$$

where u_0 is a constant. Eq.(23) predicts a decrease of τ, when ΔV increases in the negative direction, in qualitative agreement with the data in starfish eggs and skeletal muscle. This result shows that both types of behaviors for the time constant can be accounted for by the triangular model, the transition between the two being simply obtainable by shifts in the relative values of the rate constants x and u .

CONCLUSIONS

The simplest and the most consistent interpretation of our data on the inward rectifier current in the eggs of Neanthes is provided by the three state kinetic model (9) with the assumption that only state 0 is conductive and that the instantaneous conductance reflects a fast, so far unresolved, transition between the states C and 0. Not only is this scheme consistent with the experimental data on the instantaneous and steady-state conductances, but it also predicts the behavior of the activation time constant, and the unexpected proportionality between this time constant and the ratio of the normalized conductances.

Although we have not carried out a quantitative description of the inward rectifier data in other cells, such as starfish eggs and muscle, we suggest that the difference in the behavior of the time constants between these systems and Neanthes may be accounted for by a simple extension of the model whereby a direct path is added between the states I and O. As shown by eq.(23), this extended model, with only one additional adjustable parameter, possesses sufficient flexibility to predict different behaviors of the time constants. A qualitative agreement with the data on muscle and starfish is found when the rate constants for the direct transition between the states I and O are comparable to, or greater than, those for the transition between I and C. On the other hand, a qualitative and quantitative agreement with the data of Neanthes is obtained when the transition between I and O is so slow that the closed model (20) can be approximated by the linear scheme (9).

The inference, brought forward by this analysis, that the simple triangular model might be capable of providing a unified description of the inward rectifier data in different preparations, notwithstanding the remarkable discrepancy in the dependence of the activation time constant on ΔV, is one of the most satisfying results of this investigation.

ACKNOWLEDGEMENTS:

This work has been supported by the grant USPHS GM 27042-04 and by a grant from the Muscular Dystrophy Association through the Jerry Lewis Neuromuscular Research Center.

REFERENCES

Adrian RH, Freygang WH (1962a). The potassium and chloride conductance of frog muscle membrane. J Physiol 163:61.
Adrian RH, Freygang WH (1962b). Potassium conductance of frog muscle membrane under controlled voltage. J Physiol 163:104.
Fukushima Y (1981). Single channel potassium currents of the anomalous rectifier. Nature Lond 294:368.
Fukushima Y (1982). Blocking kinetics of the anomalous potassium rectifier of tunicate egg studied by single channel recording. J Physiol 331:311.
Gunning R (1983). Kinetics of inward rectifier gating in the eggs of the marine polycheate, Neanthes arenaceodentata. J Physiol (in press).

Hagiwara S, Takahashi K (1974). The anomalous rectification and cation selectivity of the membrane of a starfish egg cell. J Memb Biol 18:61.

Hagiwara S, Miyazaki S, Rosenthal NP (1976). Potassium currents and the effect of cesium on this current during anomalous rectifion of the egg cell membrane of a starfish. J gen Physiol 67:621.

Hestrin S (1981). The interaction of potassium with the activation of anomalous rectification in frog muscle membranes. J Physiol 317:497.

Hodgkin AL, Horowicz P (1959). The influence of potassium and chloride ions on the membrane potential of single muscle fibers. J Physiol 148:127.

Katz B (1949). Les constantes electriques de la membrane du muscle. Arch Sci Physiol 3:285.

Leech CA, Stanfield PR (1981). Inward rectification in frog skeletal muscle fibers and its dependence on membrane potential and external potassium. J Physiol 319:295.

Ohmori H (1978). Inactivation kinetics and steady-state current noise in the anomalous rectifier of tunicate egg cell membranes. J Physiol 281:77.

Ohmori H, Yoshida S, Hagiwara S (1981). Single K channel currents of anomalous rectification in cultured rat myotubes. Proc Natn Acad Sci U.S.A. 78:4960.

The Physiology of Excitable Cells, pages 83-96

A MODEL FOR ION-COMPOSITION-DEPENDENT PERMEABILITIES OF K[+] AND T1[+] DURING ANOMALOUS RECTIFICATION

Sally Krasne[1], Sergio Ciani[1], Susumu Hagiwara[1], and Shun-ichi Miyazaki[2]

Department of Physiology
[1]UCLA Medical School, Los Angeles, Calif. 90024
[2]Jichi Medical School, Tochigi-ken, Japan

In a previous paper (Ciani et al., 1978), a theoretical model was proposed to account for the steep (third power) dependence of the conductance of inward rectifying potassium channels in starfish egg cell membranes on the displacement, ΔV, of the transmembrane voltage from the resting potential (Hagiwara and Takahashi, 1974). This model assumed that this steep dependence came from two independent phenomena required for channel opening, the binding of three external potassium ions to a "gating site" in the channel-forming molecule and the movement of three charges through the membrane's electric field. This theoretical model is refered to as the "Electrochemical-Gating" hypothesis and is in contrast to the "Blocking Particle" hypothesis originally formulated by Armstrong (1975) and subsequently elaborated by Hille and Schwartz (1978). In this latter hypothesis, the inward rectification results from the presence, inside the cell, of large cations that can move part way into the pores but are unable to cross them completely; thus ion conduction ceases because of internal blockage of the ion-permeation pathway by an ion from the intracellular solution.

A number of physical models could correspond to the "Electrochemical Gating" hypothesis. For example, three negative charges in the channel-forming molecule could migrate from the inner to the outer membrane surface under the influence of an electric field, and then three K[+] ions could bind to these charged groups, causing channel opening; or three K[+] ions could bind to neutral sites on the channel-forming molecule at the outer membrane surface, and the electric field could cause migration of these ion-site complexes to the inner membrane surface, thus causing channel opening; or three K^+ ions could migrate through the channel-forming molecule to a site near the inner membrane surface, and their binding to the site could cause channel opening. For the sake of convenience in terminology, the derivation presented by Ciani et al. (1978) was formulated in terms of the first physical model, and we shall follow this procedure here as well. Thus, the effect of the electric field is treated as influencing the orientation of a channel-forming molecule and the external K^+ ions are treated as binding to the external surface of "oriented" channel-forming molecules. It should be made clear, however, that for the steady-state current and voltage phenomena derived here, the conclusions are the same for any of the physical models described above corresponding to the "Electrochemical-Gating" hypothesis.

One prediction of the "Electrochemical-Gating" hypothesis made by Ciani et al. (1978) was that, since the dependence of conductance on ΔV arises from a dependence upon external potassium and transmembrane voltage, changes in internal potassium would not result in a ΔV dependence but rather would show the same dependence upon total trans-membrane voltage. This prediction has since been borne out (Hagiwara and Yoshii, 1979). In addition, Ciani (1982) has since deduced the kinetic behaviors expected for the inward rectification currents for the "Electrochemical Gating" and the "Blocking Particle" hypotheses and has shown that the former, but not the latter, hypothesis was consistent with the experimental data.

In the present chapter, we aim to extend the "Electro-chemical-Gating" model to account for the observations (Hagiwara and Takahashi, 1974; Hagiwara et al., 1977) that the K-Tl selectivity during anomalous rectification varies as a function of the molar fraction of Tl [defined as $c_{Tl}/(c_K + c_{Tl})$] in the bathing medium; this change of selectivity can be observed as a minimum in either the conductance vs. Tl molar fraction or zero-current potential vs. Tl molar fraction relationship (Hagiwara et al., 1977) as illustrated later in this chapter by the data points in Figs. 1A and B, respectively.

ADDITIONAL ASSUMPTIONS OF THE MODEL

In formulating a model to describe the conductance and zero-current potential behaviors in K-T1 mixtures, we have made some additional assumptions to those of the previous paper (Ciani et al., 1978). These assumptions are as follows:

1) Both K^+ and Tl^+ are capable of binding to the "gating site" (as necessary for channel opening), and the predominant stoichiometry of this ion binding in K-T1 mixtures is the same as that observed (Hagiwara et al., 1977; Ciani et al., 1978) in pure K and pure T1 bathing solutions, namely three ions per "gating site". More specifically, the steady-state, membrane conductance observed (Hagiwara et al., 1976; 1977) in both K and T1 solutions obeys the empirical equation deduced by Hagiwara and Takahashi (1974)

$$G_i = \frac{B_i \, (c_i)^{1/2}}{1 + \exp\left(\frac{\Delta V - \Delta V_h^i}{v}\right)} \tag{1}$$

in which c_i is the external concentration of ion $\underline{i}$ (either K^+ or Tl^+), B_i and V_h are constants for a given ion $\underline{i}$, and v is a constant which was found to equal approximately 8.5mV over at least the concentration range between 10mM and 100mM K^+ (Hagiwara and Takahashi, 1974; Ciani et al., 1978) or Tl^+ (Hagiwara et al., 1977). According to the model in the previous paper (Ciani et al., 1978), v is related to the number of ions, n, which must bind to the "gating site" in channel opening by the relationship

$$v = \frac{RT}{nF} \tag{2}$$

where RT/F is 25mV. Therefore, n equals 3 for both K and T1 bathing media, and, we assume here, for K-T1 mixtures.

2) The permeability properties of a channel depend upon which ions are bound to the "gating site". Thus, we can visualize four different states of the channel, each with potentially different permeability properties, depending upon whether the "gating site" contains 3 K's, or 2 K's and 1 Tl, or 1 K and 2 Tl's, or 3 Tl's.

3) Tl^+ can have voltage-dependent blocking effects in channels for which 3 K's are bound to the "gating site" as was observed experimentally (Hagiwara et al., 1977) for effects of small amounts of Tl added to K-containing bathing media. Since no analogous blocking effects were found for small amounts of K in Tl-containing bathing media, and since our theoretical fits to the data do not require consideration of any additional blocking, we have assumed that only this one type of blocking occurs.

4) The voltage-dependence of ion permeation through <u>single</u> channels is given by a function of voltage which is independent of which ions are bound to the "gating site", appearing as a multiplicative factor in the expression for the fluxes and thus cancelling out in the permeability ratios. This assumption is supported by the data in Fig. 1A of Hagiwara et al., (1977) in which the relationships for external ion concentration vs. zero-current potential are parallel for solutions containing either K^+ or Tl^+ or a 1:1 K^+-Tl^+ mixture.

THEORETICAL RESULTS

The purpose of this section is to derive the expressions for conductances and zero-current potentials expected from a "four-state" model. Note that superscripts to symbols generally denote the combination of ions bound to the "gating site"; the first number in the superscript denoting the number of K^+'s and the second number denoting the number of Tl^+'s bound to the "gating site"; for example, $[A_n^{r,(n-r)*}]$ is the density, per unit surface, of channels having n ions bound to the "gating site" of which r are K^+'s and (n-r) are Tl^+'s.*

----

*Note that in the case of ion concentrations, c_i, or Tl^+ molar fractions, y, a superscript denotes the power to which the argument should be raised.

Conductance–Voltage Behaviors in K–Tl Mixtures

In general, the expression for the total net flux of K^+ and Tl^+ through a membrane containing any combination of channel states can be written

$$
J_{K-Tl}^{net} = \frac{-I}{zF} = \left[c_K \sum_{r=0}^{n} P_K^{r,(n-r)} + c_{Tl} \sum_{r=0}^{n} P_{Tl}^{r,(n-r)} \right] e^{-\frac{\phi}{2}} -
\tag{3}
$$

$$
\left[c_K' \sum_{r=0}^{n} P_K^{r,(n-r)} + c_{Tl}' \sum_{r=0}^{n} P_{Tl}^{r,(n-r)} \right] e^{\frac{\phi}{2}}
$$

$$
\frac{-I}{zF} = \sqrt{\lambda \lambda'} \; 2 \sinh \left[\frac{\phi - \phi_o}{2} \right]
\tag{4}
$$

where λ and λ' are the coefficients of $e^{-\phi/2}$ and $e^{\phi/2}$ in eq. (3), and

$$
\phi_o = \ln \frac{\lambda}{\lambda'}
\tag{5}
$$

ϕ and ϕ_o are the membrane potential and the membrane potential at zero current, respectively, in units of $RT/z_i F$; namely

$$
\phi = \frac{z_i FV}{RT} \quad \text{and} \quad \phi_o = \frac{z_i FV_o}{RT}
\tag{6}
$$

V and V_o denote potential differences (inside minus outside) and positive current is defined as a net outward flux of positive charge. $P_i^{r,(n-r)}$ denotes the permeability coefficient of ion i in a channel with r K's and n–r Tl's bound to the "gating site"; c_i is the external concentration and c_i' the internal concentration of ion i.

The permeabiity coefficients for a particular type of channel in eq. (3) are equal to the product of the ion

permeability coefficient in a single channel and the
density of empty channels.

$$P_i^{r,(n-r)} = f_i^{r,(n-r)}(\emptyset) \cdot [A_n^{r,(n-r)*}(e)], \qquad (7)$$

where $f_i^{r,(n-r)}(\emptyset)$ is the permeability coefficient for ion $\underline{i}$
in a single channel binding r K's and n-r Tl's and
$[A_n^{r,(n-r)*}(e)]$ is the density of these channels which are
empty (i.e., not "blocked"). According to assumption (4)
above (as well as the implications of the data of Fig. 1A
of Hagiwara et. al., 1977), the voltage dependence of
single-channel permeation is the same for K^+ and Tl^+ in any
given state of the channel. Therefore, we can conclude
that the voltage-dependences of the single-channel permeabilities for any two permeant ions in any type of channel
are simply related by a proportinality constant. Thus

$$f_i^{r,(n-r)}(\emptyset) = p_i^{r,(n-r)} \cdot f'(\emptyset) \qquad (8)$$

where the single-channel permeability coefficient, $p_i^{r,(n-r)}$, is independent of transmembrane potential and $f'(\emptyset)$ is
the same for all permeant ions in all channels.

According to assumption (3) above (as well as the data
of Hagiwara et al., 1977), the only case in which a
significant degree of "blocking" may occur is for Tl^+
blocking of K-stabilized channels. Since the present model
assumes (see Ciani et al., 1978) that a single channel can
contain only one ion at a time, the total density of empty
channels with 3 K's bound to the gating site is thus given
by

$$[A_n^{n,0*}(e)] = \frac{[A_n^{n,0*}]}{1 + c_{Tl}K_{Tl}^{n,0}e^{-u\emptyset}} \qquad (9)$$

where μ is the fraction of the transmembrane potential drop
which occurs between the outer membrane-solution interface
and the Tl^+ "blocking site" within the channel and $K_{Tl}^{n,0}$ is
the binding constant for Tl at the "blocking site" at zero
transmembrane voltage.

Combining eqs. 7, 8 and 11, we can write the overall permeability coefficients for K^+ and for Tl^+, respectively, as follows:

$$\sum_{r=0}^{n} P_K^{r,(n-r)} \tag{10}$$

$$= f'(\phi) \left\{ P_K^{(n,0)} \frac{\left[A_n^{(n,0)*}\right]}{1 + K_{Tl}^{(n,0)} c_{Tl} e^{-\mu\phi}} + \sum_{r=0}^{n-1} P_K^{r,(n-r)} \left[A_n^{r,(n-r)*}\right] \right\}$$

and

$$\sum_{r=0}^{n} P_{Tl}^{r,(n-r)} \tag{11}$$

$$= f'(\phi) \left\{ P_{Tl}^{(n,0)} \frac{\left[A_n^{(n,0)*}\right]}{1 + K_{Tl}^{(n,0)} c_{Tl} e^{-\mu\phi}} + \sum_{r=0}^{n-1} P_{Tl}^{r,(n-r)} \left[A_n^{r,(n-r)*}\right] \right\}$$

The total densities of each state of the channel can be related to the total density of channels, $[A]^{Tot}$, assuming the same equilibrium as denoted by eqs. (5-7) of the previous paper (Ciani et.al., 1978) and in addition including assumption 1 of the present paper, in which case

$$[A_o^*] = e^{zn\phi}[A_o] \tag{12}$$

$$[A_s] = K_s[A_o]c_i^{'s} \quad (s = 1, 2, ...,n) \tag{13}$$

$$\left[A_n^{r,(n-r)}\right] = K_{r,(n-r)}^* [A_o^*] c_K^r c_{Tl}^{(n-r)} \quad (r = 0, 1, ..., n) \tag{14}$$

where $[A_o]$ is the density of channel molecules with no ions bound to the gating site and which have not been "reoriented" by the electric field, $[A_o^*]$ is the density of channel molecules with no ions bound to the gating site but which have been "reoriented" by the electric field, $[A_n^{r,(n-r)*}]$ is the density of the open channel which has r K ions and n-r Tl ions bound to the gating site, and

$\gamma e^{zn\phi}$, K_s, and $K_K^{r,(n-r)*}$ are the equilibrium constants for the following reactions:

$$A_o \xleftarrow{\quad \gamma e^{zn\phi} \quad} A_o^*$$ (15)

$$A_o + s \cdot I' \xleftarrow{\quad K_s \quad} A_s$$ (16)

$$A_o^* + r \cdot c_K + (n-r)c_{T1} \xrightleftharpoons{\quad K_{r,(n-r)}^* \quad} [A_n^{r,(n-r)*}]$$ (17)

where z is the valency of the charged, orientable groups within the channel-forming molecule (z=-1 in the present case), n is the number of such groups per molecule as well as being the maximum number of ions which can bind to the "gating site" of the channel (n=3 in the present model), and I' is an internal ion capable of binding to the unoriented aggregate (I' is assumed to represent internal K^+ in the present model). The conservation equation (which only includes predominant species; recall assumption 1) is

$$[A]^{Tot} = [A_o] + \sum_{s=1}^{n} [A_s] + \sum_{r=0}^{n} [A_n^{r,(n-r)*}]$$ (18)

Combining equations 12 thru 14 with equation 18 and recalling that z=-1 yields the density of oriented, uncomplexed channels, A_o^*

$$[A_o^*] = \frac{[A]^{Tot}}{\dfrac{1 + \sum\limits_{s=1}^{n} K_s c_i^{'s}}{\gamma} e^{n\phi} + \sum\limits_{r=0}^{n} [K_{r,(n-r)}^* \, c_K^r \, c_{T1}^{(n-r)}]}$$ (19)

(where $0^0=1$). Substituting equation (19) into (14) yields

$$[A_n^{r,(n-r)*}] = \frac{[A]^{Tot} \dfrac{K_{r,(n-r)}^*}{K_{(n,0)}^*} e^{-n\phi} \, c_K^r \, c_{T1}^{(n-r)}}{\chi + e^{-n\phi} \sum\limits_{s=0}^{n} \left[\dfrac{K_{s,(n-s)}^*}{K_{(n,0)}^*} \, c_K^s \, c_{T1}^{(n-s)} \right]}$$ (20)

where χ is a constant, provided that the internal ion composition is unaltered and is given by

$$\chi = \frac{1 + \sum\limits_{s=1}^{n} K_s c_i'^s}{\gamma \, K^*_{(n,0)}} \qquad (21)$$

By substituting equation (20) into (10) and (11), inserting these in turn into equation (3) (and rearranging in the form of (4)) and dividing both sides by $V-V_o$, one can obtain the expression for the conductance of the membrane in the presence of any mixture of K^+ and Tl^+ in the bathing medium (omitted here because of space limitations) as an explicit function of the permeant ion concentrations, the equilibrium binding constants for ions to the "gating site" and for Tl^+ blocking of channels with 3 K's at the "gating site", and the single-channel permeability coefficients for each ion. More useful, however, is the expression for this conductance normalized to that obtained in pure K bathing media (i.e. $c_{Tl}=0$) which gives the relationship

$$\frac{G_{K-Tl}}{G_K} = \frac{\chi + c_o^n e^{-n\phi}}{\chi + c_o^n e^{-n\phi}[(1-y)^n + \sum\limits_{r=0}^{n-1} [\,\mathfrak{L}_{r,(n-r)}(1-y)^r y^{(n-r)}]}\, .$$

$$\cdot \left[\frac{(1-y)^n}{1 + K_{Tl}^{(n,0)} yc_o e^{-\mu\phi}} + \sum\limits_{r=0}^{n-1} \eta_{r,(n-r)}(1-y)^r y^{(n-r)}\right] . \qquad (22)$$

$$\cdot \left\{(1-y) + y \frac{\dfrac{\alpha(1-y)^n}{1 + K_{Tl}^{(n,0)} yc_o e^{-\mu\phi}} + \sum\limits_{r=0}^{n-1} \beta_{r,(n-r)}(1-y)^r y^{(n-r)}}{\dfrac{(1-y)^n}{1 + K_{Tl}^{(n,0)} yc_o e^{-\mu\phi}} + \sum\limits_{r=0}^{n-1} \eta_{r,(n-r)}(1-y)^r y^{(n-r)}}\right\}^{1/2}$$

where

$$\alpha = \frac{P_{Tl}^{n,o}}{P_K^{n,o}} \tag{23}$$

$$\vartheta_{r,(n-r)} = \frac{K_{r,(n-r)}^*}{K_{n,o}^*} \tag{24}$$

$$\beta_{r,(n-r)} = \frac{P_{Tl}^{r,(n-r)}}{P_K^{n,o}} \cdot \vartheta_{r,(n-r)} \tag{25}$$

$$\eta_{r,(n-r)} = \frac{P_K^{r,(n-r)}}{P_K^{n,o}} \cdot \vartheta_{r,(n-r)} \tag{26}$$

Zero-Current Membrane Potential Behavior in K-Tl Mixtures

The zero-current potential behavior can be deduced by substituting equations (10), (11), and (20) into (3). However, for the purposes of this chapter, it is more relevant to examine the deviation of the zero-current potential, $\Delta\phi_o$, as the molar fraction of K and Tl ($y=Tl/K+Tl$) is varied at a constant, total external concentration of permeant ions, c_o. This quantity is given by subtracting the zero-current potential expected for pure K bathing media from that in K-Tl mixtures and is equal to

$$\Delta\phi_o = y \frac{(\alpha-1)(1-y)^n + \left[\sum_{r=0}^{n-1}(\beta_{r,(n-r)} - \eta_{r,(n-r)})(1-y)^r y^{(n-r)}\right] \cdot Q}{(1-y)^n + \left[\sum_{r=0}^{n-1}\eta_{r,(n-r)}(1-y)^r y^{(n-r)}\right] \cdot Q} \tag{27}$$

where $\quad \Delta\phi_o = \dfrac{\Delta V_o F}{RT} \quad$ and $\quad Q = 1+K_{T1}^{(n,0)}yc_o e^{-\mu\phi_o^K} \qquad (28)$

and $\Delta\phi_o$ and $\mu\Delta\phi_o$ are small compared to 1 (as appears to be true for the experimental values of potential in Figure 1B and the value deduced for μ in Table 1) so that the equation can be linearized.

"FOUR-STATE" MODEL

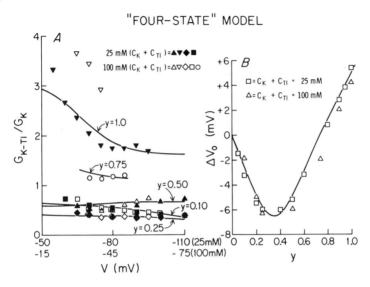

Figure 1. Conductance and zero-current potential behaviors in K-Tl mixtures. A. The ordinate indicates the ratio between the conductance observed at a given molar fraction of Tl^+, y, and a given transmembrane voltage, V (abscissa), and that observed at the same transmembrane voltage and permeant ion concentration when K^+ is the only permeant ion in the external solution (i.e., y= 0). The upper values of voltage on the abscissa refer to the filled symbols and the lower values refer to the open ones. The molar fractions of Tl^+ in the bath are as follows y=0.1, □; y=0.25,◇ ; y=0.5,△; y=0.75,○; y=1.0, ▽). The solid curves represent the theoretical behaviors predicted by equation (22) and the parameters listed in Table 1. B). The ordinate gives the difference between the zero-current potential at the value of y indicated by the abscissa and that in pure K bathing media (y=0) at the same total concentration. The curve indicates the theoretical behavior predicted by equation (28) for the values of parameters listed in Table 1.

TABLE 1.

"Four-site" model Parameters	Best-Fit Values for Parameters	"Four-site" model Constants	Best-Fit Values for Constants
α	0	$\dfrac{P_{T1}^{(3,0)}}{P_K^{(3,0)}}$	0
$\beta_{(2,1)}$	0	$\dfrac{P_{T1}^{(2,1)}}{P_K^{(3,0)}}$	0
$\beta_{(1,2)}$	0	$\dfrac{P_{T1}^{(1,2)}}{P_K^{(3,0)}}$	0
$\beta_{(0,3)}$	3.4	$\dfrac{P_{T1}^{(0,3)}}{P_K^{(3,0)}}$	1.76
$\eta_{(2,1)}$	0.4		
$\eta_{(1,2)}$	1.3	$\dfrac{P_K^{(2,1)}}{P_K^{(3,0)}}$	0.8
$\eta_{(0,3)}$	2.84	$\dfrac{P_K^{(1,2)}}{P_K^{(3,0)}}$	0.43
$\ell_{(2,1)}$	0.5		
$\mathcal{J}(1,2)$	3.0	$\dfrac{P_K^{(0,3)}}{P_K^{(3,0)}}$	1.47
$\mathcal{J}(0,3)$	1.93		
χ	0.085 M^3	$\dfrac{K_{(2,1)}^{*}}{K_{(3,0)}^{*}}$	0.5
$K_{T1}^{(3,0)}$	9.8 M^{-1}	$\dfrac{K_{(1,2)}^{*}}{K_{(3,0)}^{*}}$	3
μ	1		
n	3	$\dfrac{K_{(0,3)}^{*}}{K_{(3,0)}^{*}}$	1.93

DETERMINATION OF PARAMETERS

By fitting limiting cases of equations (22) and (28) a number of parameters can be determined unambiguously. Thus, α can be determined from equation (28) in the limit of very small Tl concentrations and $\beta_{0,3}/\eta_{0,3}$ can be determined from this equation for the case of pure Tl bathing media. μ, $K_{Tl}^{3,0}$, and χ can be determined from the behavior of equation (22) in the limit of very small Tl concentrations and from the voltage dependence of the ratio of this equation for two different values of c_0 for very small Tl concentrations (assuming that the zero-current potential essentially obeys the Nernst equation, which has been observed by Hagiwara et al., 1977, and requires that Q, defined in equation (28), be approximately constant, as appears to be the case). $\beta_{0,3}$, $\eta_{0,3}$, and $\vartheta_{0,3}$ can be assessed independently from equation (22) for pure Tl bathing media once the ratio $\beta_{0,3}/\eta_{0,3}$ has been determined. Having determined these parameters, the remaining six parameters, $\beta_{1,2}$, $\eta_{1,2}$, $\vartheta_{1,2}$, $\beta_{2,1}$, $\eta_{2,1}$, and $\vartheta_{2,1}$ have been determined by curve fitting equations (22) and (28) to the data of Figures 1A and B. The values of all of the parameters and constants thus determined are listed in Table 1.

DISCUSSION

A number of conclusions can be drawn from the values of parameters in Table 1 necessary to fit the data of Figures 1A and B. First, the minimum in the normalized conductance and ΔV_0 vs. y behaviors can be most easily understood, qualitatively, by recognizing that only one state of the channel, namely that in which 3 Tl's are bound to the "gating site", is permeable to Tl^+. Thus, as small amounts of Tl^+ are added to the bathing solution and Tl^+'s are bound to the "gating site", the membrane permeability tends to follow the K^+ concentration in the bath (which is decreasing) until a sufficient amount of Tl^+ is added to form a significant fraction of channels with 3 Tl's bound to the "gating site", at which point the membrane permeability begins to increase again. Second, the "gating site" binds 3 Tl's preferentially to 3 K's, as seen by the fact that $\vartheta_{0,3} > 1$; this in turn is deduced from the voltage dependence of the normalized conductance in pure Tl^+ bathing media which decreases with increasingly negative potentials. Finally, there is some form of "cooperativity" in the binding of ions to the "gating site". Thus, in the absence of "cooperativity", if there were no selectivity of this site between K^+ and Tl^+, the value of $\vartheta_{r,(n-r)}$ would be expected to be 3, 3, 1 for r = 1, 2, 3, respectively; whereas, if the individual, charged groups in the "gating site" were all equivalent and Tl^+ were prefered over K^+, as is indicated in Table 1, then each value of $\vartheta_{r,(n-r)}$ would be greater than that in the absence of selectivity. That the values deduced for $\vartheta_{(2,1)}$ and $\vartheta_{(1,2)}$ are less than that expected for the selectivity shown by $\vartheta_{(0,3)}$ suggests that the "gating site" binds 3 K's or 3 Tl's in preference to mixtures of the two ions.

REFERENCES

Armstrong C (1975). K pores of nerve and muscle membranes. In Eisenman G (ed):"Membranes - A Series of Advances", New York: Marcel Dekker, Inc. 3:325-358.

Ciani S (1982). Theoretical study of the relaxation currents in voltage-clamp conditions as deduced from two models for 'inward' rectification: The 'blocking particle' and the 'electrochemical gating' hypotheses. J Theoret Neurobiol 1:228-248.

Ciani S, Krasne S, Miyazaki S, Hagiwara S (1977). A model for anomalous rectification: Electrochemical-potential-dependent gating of membrane channels. J Memb Biol 44:103-134.

Hagiwara S, Miyazaki S, Krasne S, Ciani S (1977). Anomalous permeabilities of the egg cell membrane of a starfish in K^+-Tl^+ mixtures. J Gen Physiol 70:269-281.

Hagiwara S, Miyazaki S, Rosenthal N P (1976). Potassium current and the effect of cesium on this current during anomalous rectification of the egg cell membrane of a starfish. J Gen Physiol 67:621-638.

Hagiwara S, Takahashi K (1974). The anomalous rectification and cation selectivity of the membrane of a starfish egg cell. J Memb Biol 18:61-80.

Hagiwara S, Yoshii, M (1979). Effects of internal potassium and sodium on anomalous rectification of the starfish egg as examined by internal perfusion. J Physiol 292:215-251.

Hille B, Schwartz W (1978). Potassium channels as multi-ion single file pores. J Gen Physiol 72:409-442.

The Physiology of Excitable Cells, pages 97–108
© 1983 Alan R. Liss, Inc., 150 Fifth Avenue, New York, NY 10011

THE INFLUENCE OF PERMEANT IONS ON THE KINETICS OF INWARD
RECTIFICATION IN SKELETAL MUSCLE

Frances M. Ashcroft & Peter R. Stanfield

University Laboratory of Physiology, Oxford, and
Department of Physiology, Leicester, U.K.

INTRODUCTION

The opening and closing, or gating, of ion channels in
excitable membranes was originally thought to be independent
of the concentration and species of the permeant ion. Recent
evidence, however, has established that permeant ions
influence the gating of both transmitter-activated and
voltage-activated channels. For example, effects of both
concentration (Dubois & Bergman, 1977) and species (Swenson
& Armstrong, 1981) of permeant ion have been described for
the K-permeability of nerve membranes. A similar dependence
of gating on the permeant ion species is also found for Ca
channels (Brehm & Eckert, 1978; Ashcroft & Stanfield, 1981),
for inward rectification (Hagiwara & Yoshii, 1979; Hagiwara,
Miyazaki, Krasne & Ciani, 1977; Hestrin, 1981; Leech &
Stanfield, 1981; Stanfield, Ashcroft & Plant, 1981) and for
acetylcholine-activated channels (Ascher, Marty & Neild,
1978).

Perhaps the most striking example of the interaction
between permeant ions and gating is provided by the kinetics
of the inwardly-rectifying potassium permeability of skeletal
muscle in potassium and in thallium solutions (Stanfield et
al., 1981). This K-permeability allows K^+ to move inward
across the fibre membrane more easily than outward, and it is
primarily permeable to K^+ and Tl^+. A remarkably similar inward
rectifier is found in the membranes of certain egg cells
(Hagiwara & Takahashi, 1974). In skeletal muscle, replacing
external K^+ by Tl^+ results in the usual K-dependent
activation of the inward currents being replaced by a

Tl-dependent inactivation. This paper describes the effects of the concentration and species of the permeant ion, in both the internal and external solutions, on the kinetics of inward rectification.

METHODS

Experiments were carried out on sartorius muscle fibres of the frog Rana temporaria using a three-electrode voltage-clamp to control membrane potential (Adrian, Chandler & Hodgkin, 1970), and in solutions containing (mM): 8 $CaSO_4$, 2 Tris-maleate (pH 7.2), 113 sucrose and either 40 K_2SO_4 (80mM K^+) or 40 Tl_2SO_4 (80mM Tl^+). In some experiments, (as indicated in the text) membrane currents were recorded from semitendinosus fibres of R. catesbiana using a three vaseline gap voltage-clamp (Hille & Campbell, 1976; Hestrin, 1981). In these experiments the external solution contained (mM): 75 Tl_2SO_4, 8 $CaSO_4$, 5 Tris-HEPES (pH 7.4) plus sucrose to 300 mosmoles. Different Tl-concentrations were obtained by substituting Tl_2SO_4 with $Tris_2SO_4$. Internal solutions had either Tl^+ or K^2 as the major cation (120mM), were adjusted to 300 mosmoles, pH 7.1, and made up of (mM): 30 Tris-EGTA, 5 $MgSO_4$, 5 Na_2ATP, 5 Na_2PCr, and either 50 K_2SO_4, 20 KMOPS (K^+, 120mM) or 60 Tl_2SO_4, 20 Tris-MOPS (Tl^+, 120 mM). All experiments were carried out at low temperature.

RESULTS

Fig.1 summarises the gating behaviour of the inward rectifier of the frog sartorius when K^+ is the permeant ion. Superimposed records of membrane potential (above) and membrane currents (below) for hyperpolarizations increasing in amplitude in 10mV steps are shown in A. Over about the first 50msec, the currents increase in size along an exponential time course with a rate that becomes faster with increasing hyperpolarization. The time constants for this activation process are an exponential function of membrane potential (B) and decrease e-fold every 18mV. This voltage-dependence is similar to that found for activation of the inward rectifier of the starfish egg (Hagiwara, Miyazaki & Rosenthal, 1976).

In both skeletal muscle (Hestrin, 1981; Leech & Stanfield, 1981) and in the starfish egg (Hagiwara & Yoshii, 1979), the

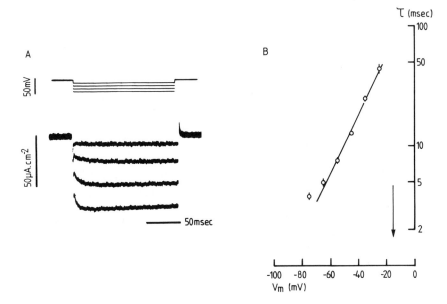

Fig.1. Gating of the inward rectifier in 80mM K-Ringer.
A. Records of membrane potential (above) and membrane current
(below) obtained in response to hyperpolarizations of 10, 20
30 and 40mV. Resting potential (RP), -14mV. Holding potential
(HP), -14mV. Temp. 4 ^{o}C. B. Relationship between membrane
potential and the time constants for activation of K-currents
in 12 fibres. The line is drawn to an exponential which
increases e-fold every 18mV. Temp. 2-5 ^{o}C.

activation time constants are also a function of the external
K-concentration, the relationship between τ and membrane
potential being shifted along the voltage axis in parallel
with the shift in the K-equilibrium potential (V_K). The
steepness of the relationship was unaltered by changing $[K]_o$.
This dependence of τ on both membrane potential and $[K]_o$
implies an interaction between K^+ and the gating mechanism
of the inward rectifier channel.

When Tl^+ is the permeant ion the gating of the inward
rectifier is strikingly different from that found in K-Ringer,
for the Tl-currents inactivate on hyperpolarization (Fig.2).
The decline of the Tl-current appears to consist of two
processes: a very slow component which is probably due to Tl^+
depletion from the lumen of the T-tubules (Gay, 1981), and a
much faster, exponential process. We corrected for depletion
by fitting a straight line to the current between 250-400msec

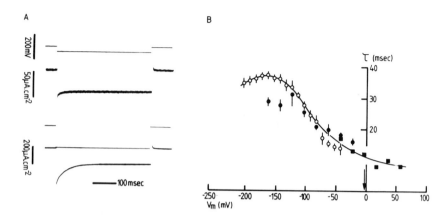

Fig.2 Gating of the inward rectifier in 80mM Tl-Ringer.
A. Records of membrane potential and membrane current for
hyperpolarizations of amplitude 40mV (above) and 140mV (below)
RP, -2mV. HP, -2mV. Temp. 1.2 °C. B. Relationship between
membrane potential and time constants for inactivation of Tl-
currents during a single pulse (O , n=9) from recovery of the
Tl-current at the given membrane potential following a 180mV
hyperpolarization for 150msec (● , n=4) or from the rate of
recovery from inactivation at the given membrane potential
obtained by applying a 150msec test pulse of 180mV a variable
interval after an identical inactivating pulse (■ , n=1). Line
drawn by eye. Reprinted with permission from Nature 289, 509.

and extrapolating to zero time to obtain the steady state
current value: instantaneous currents were then obtained by
fitting the corrected current to a single exponential
(Stanfield et al., 1981).

    Fig.3 compares steady state and instantaneous current-
voltage relations in 80mM K$^+$ (A) and in 80mM Tl$^+$ (B). In K-
Ringer steady state currents were larger than instantaneous
currents at all potentials, whereas in Tl-solutions inactivation
results in instantaneous currents that are larger than those
in the steady state. Further, since inactivation increases
with hyperpolarization, the steady state Tl current-voltage
relation has a region of negative slope. The extent of steady
state inactivation is plotted as a function of membrane
potential in Fig.4A, and increases e-fold for a 48mV
hyperpolarization. This voltage-dependence would be that
expected if the site at which inactivation occurs experienced
0.5 of the membrane voltage field.

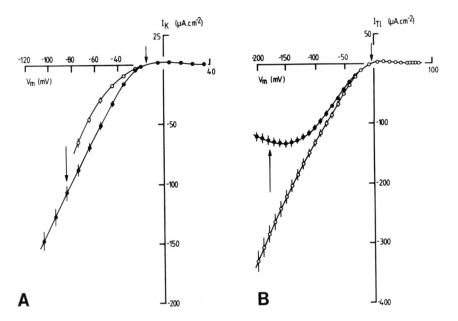

Fig.3 Mean instantaneous (O) and steady state (●) current-voltage relations for 12 fibres in 80mM K-Ringer (A) and for 9 fibres in 80mM Tl-Ringer (B). Arrows indicate the holding potentials. Mean RP, -14.8 + 0.6mV (K-Ringer); -2.7 + 0.4mV (Tl-Ringer). Leak currents subtracted (cf. Adrian & Freygang, 1962; Standen & Stanfield, 1978). Reprinted from Nature 289, 509, with permission.

Time constants of inactivation increase as the membrane potential is made more negative, reaching a maximum value at around -170mV, and then fall with further hyperpolarization (Fig.2B, O). Recovery from inactivation also followed a single exponential time course, with time constants that are of similar amplitude to those obtained for the onset of inactivation at the same potential (Fig.2B, ● ■). Thus inactivation may be described by simple first order kinetics of the following form:

$$\text{open} \underset{k_{-1}}{\overset{k_1}{\rightleftharpoons}} \text{inactivated}$$

where the time constant for the inactivation (τ) is given by $\tau = 1/(k_1 + k_{-1})$ and the extent of inactivation ($\Delta I/I_\infty$) by $\Delta I/I_\infty = k_1/k_{-1}$. The rate constants for inactivation (k_1, k_{-1}) obtained from these equations are plotted against membrane

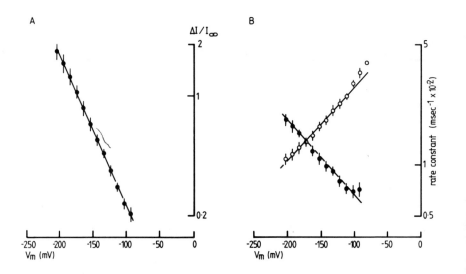

Fig.4A Relationship between membrane potential and the extent
of inactivation ($\Delta I/I_\infty$) for 9 fibres in 80mM Tl-Ringer.
The extent of inactivation is expressed as $\Delta I/I_\infty$ where I_∞
is the steady state current and ΔI the difference between the
steady state and instantaneous current (cf. Ohmori, 1981).
The line fitted through the experimental points is fitted by
regression analysis and increases e-fold every 48mV. The
potential at which $\Delta I/I_\infty$ = 1.0 (where $k_1 = k_{-1}$) was -172mV.
B. Relationship between membrane potential and the rate
constants for inactivation in 80mM Tl-Ringer. ●,k_1. o, k_{-1}
The lines were fitted by regression analysis and vary e-fold
for a 98mV hyperpolarization, k_1 increasing and k_{-1} decreasing.

potential in Fig.4B and vary e-fold for a 98mV
hyperpolarization, k_1 increasing and k_{-1} decreasing with
hyperpolarization.

As we have previously described (Stanfield et al., 1981),
it is unlikely that inactivation represents a voltage-
dependent block of Tl-currents by a cation present in the
external solution, or that it results from depletion of Tl^+
from the T-tubule lumen. Rather inactivation appears to be a
consequence of an interaction between Tl^+ and the inward
rectifier channel.

The effect of external Tl on the time constants for
inactivation of the inward rectifier and on the extent

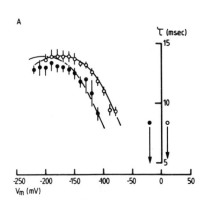
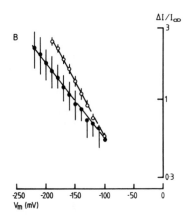

Fig.5 Effects of [Tl]$_0$ on the gating of the inward rectifier. A. Relationship between membrane potential and time constants for inactivation for 3 fibres in 50mM Tl-Ringer (●) and for 3 fibres in 150mM Tl-Ringer (O). The line through the points in 150mM Tl$^+$ is drawn by eye: the same line is shown shifted by 27mV in the negative direction (equivalent to the shift in V_{Tl} on changing from 150mM to 50mM Tl$^+$). Temp. 7-10 ^{o}C. B. Relationship between the extent of inactivation ($\Delta I/I_\infty$) and membrane potential for 3 fibres in 150mM Tl$^+$ (O) and 50mM Tl$^+$ (●). The lines are fitted by regression analysis and increase e-fold every 86mV in 50mM Tl$^+$ and 60mV in 150mM Tl$^+$. The potential at which $\Delta I/I_\infty$ =1.0 was -150mV (50mM Tl$^+$) and -133mV (150mM Tl$^+$).

of steady state inactivation is shown in Fig.5. These experiments were carried out on semitendinosus fibres of R. catesbiana using a vaseline gap voltage clamp method (Hestrin 1981) at around 10 ^{o}C. (Higher temperatures were necessary in these experiments to prevent condensation forming on the vaseline seals and subsequent short-circiuting of the voltage clamp). Under these conditions, inactivation of Tl-currents was qualitatively similar to that previously described, but τ's were increased at all potentials because of the higher temperature.

Decreasing [Tl]$_0$ caused a decrease in the time constants for inactivation, and shifted the relationship between τ and membrane potential along the voltage axis by an amount equivalent to the shift in V_{Tl}. This shift in the relation between the time constants for inactivation and membrane

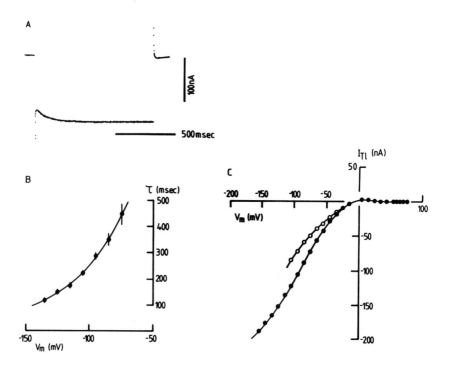

Fig.6. Gating of the inward rectifier in 100mM Tl-Ringer
(external) with an internal solution of 120mM Tl$^+$. A. Record
of membrane current obtained in response to a hyperpolarization
of 100mV. HP, -5mV. Temp. 11°C. B. Relationship between
membrane potential and the time constants for activation of
the Tl-currents in 9 fibres. The line is drawn to an
exponential which increases e-fold every 44mV. Temp. 8-11 $^\circ$C.
C. Instantaneous (O) and steady state (●) current-voltage
relations for the fibre shown in A. Leak currents subtracted.

potential resembles that found for the effect of [K] on the
K-dependent activation of the inward rectifier (Hestrin, 1981;
Leech & Stanfield, 1981). The extent of inactivation was
reduced by decreasing external Tl$^+$ (Fig.5B), and the steepness
of the relationship was also reduced, an e-fold increase in
$\Delta I/I_\infty$ being produced by a 60mV hyperpolarization in 150mM
Tl and by an 86mV hyperpolarization in 50mM Tl-Ringer.

These effects of [Tl]$_o$ on the rate and extent of
inactivation are associated with a significant reduction in
the amplitude of the back rate constant (k_{-1}), at all

potentials, with increasing external Tl^+. The amplitude of k_1 was little changed. In other words, the main result of increasing $[Tl]_o$ is to reduce the rate at which channels move from the inactivated to the open state.

The gating of the inward rectifier in Tl-Ringers is also influenced by the internal ion species. In all the experiments described above, the major internal cation was K^+. Fig.6 illustrates the kinetics of inward rectification when internal K^+ was replaced by Tl^+, for fibres in 100mM Tl-Ringer. Under these conditions, Tl-currents no longer undergo inactivation but instead increase exponentially over the first few hundred msec when the membrane is hyperpolarized. Time constants for this activation process were very much slower than those found with external K^+ and internal K^+ (Fig.1B), and they decreased less steeply with potential (e-fold for a 44mV hyperpolarization: Fig.6B). Steady state and instantaneous current-voltage relations are plotted in Fig.6C. In general, the steady state Tl-currents appeared to be about one third the amplitude of the instantaneous, and half the amplitude of steady state Tl-currents obtained at the same external Tl-concentration when K^+ was the internal ion.

It is probably unlikely that this activation process represents the recovery of channels inactivated by internal Tl^+ because the time constants for activation of Tl-currents are very much slower than are those for recovery from inactivation (compare Figs.2B and 6B).

DISCUSSION

The results described here indicate that the gating of inward rectification in skeletal muscle is dependent on the concentration and species of permeant ion in both external and internal solutions. When K^+ (but not Tl^+) is present in the external and internal solutions, the inward currents undergo a K-dependent activation. Activation is also found when Tl^+, but not K^+, is present on both sides of the membrane but it occurs more slowly and shows a different voltage-dependence to that of K-activation. However, in the presence of external Tl^+ and internal K^+, Tl-currents inactivate on hyperpolarization, and the extent to which they do so is dependent of the external Tl-concentration.

Perhaps the simplest explanation for these results is that activation is present in both K^+ and Tl^+ solutions but that in Tl-Ringer, when K^+ is the internal ion, activation is obscured by an additional inactivation process. Because of the marked differences in the amplitude and voltage-dependence of the time constants for activation it is likely that Tl^+ and K^+ interact with the gating mechanism at different sites.

The mechanism of inactivation of the Tl-currents is unclear. One possibility is that it results from a complex interaction, within the channel, between Tl^+ coming from the external solution and K^+ entering from the internal solution. Such an interaction might be expected to decrease the Tl-currents because when Tl^+ and K^+ are both present in the external solution inward currents are greatly reduced, as if movement of one permeant ion through the channel is blocking the movement of the other (Ashcroft & Stanfield, unpublished). However, similar anomalous mole-fraction conductance effects are also found for inward rectification in the starfish egg but the Tl-currents do not show inactivation (Hagiwara et al. 1977). Furthermore, it is difficult to account for the relatively slow time constants for inactivation with such a hypothesis, or for the fact that inactivation increases (rather than decreases) with external Tl^+.

Another possibility is that Tl^+ stabilises a voltage-dependent conformational change which results in closure of the channel. The fact that increasing external Tl^+ reduces the rate at which channels move from an inactivated to an open state (assuming a first order kinetic model) supports this idea. Further, if the inactivation site were accesible to intracellular Tl^+ it could also account qualitatively for the effects of internal Tl^+, as under these conditions some degree of steady state inactivation would be present thus enabling the activation process to be observed and decreasing the amplitude of the Tl-currents.

In summary, the experiments described here demonstrate that both external and internal ions interact with the gating mechanism of the inward rectifier channel in skeletal muscle. The precise nature of this interaction remains unclear.

ACKNOWLEDGEMENTS

The experiments described in Figs.1-4 were done in collaboration by the authors.   Those in Figs.5,6 were carried
out in Dr. S. Hagiwara's laboratory. F.M.A. thanks Dr.S.
Hagiwara for the privilege of working in his laboratory, and
Dr. R. Eckert for computing facilities.

REFERENCES

Adrian RH,Freygang, WH (1962). Potassium conductance of frog
    muscle membrane under controlled voltage. J Physiol 163:104
Adrian RH, Chandler WK, Hodgkin AL (1970). Voltage clamp
    experiments in striated muscle fibres. J Physiol 208:607.
Ascher P, Marty A, Neild TO (1978). Lifetime and elementary
    conductance of the channels mediating the excitatory
    effects of acetlycholine in Aplysia. J Physiol 278:177.
Ashcroft FM, Stanfield PR (1981). Calcium dependence of the
    inactivation of calcium currents in skeletal muscle fibres
    of an insect. Science 213:224.
Brehm P, Eckert RO (1978). Calcium entry leads to inactivation
    of calcium channel in Paramecium. Science 202:1203.
Dubois JM, Bergman C (1977). The steady state potassium
    conductance of the Ranvier node at various external K-
    concentrations. Pflügers Archiv 370:185.
Gay L (1981). Thallium and potassium permeability mechanism
    of resting frog sartorius muscle. J Physiol 312:39P.
Hagiwara S, Takahashi K (1974). The anomalous rectification
    and cation selectivity of the membrane of the starfish
    egg. J Membr Biol 18:61.
Hagiwara S, Yoshii M (1979) Effects of internal potassium
    and sodium on the anomalous rectification of the starfish
    egg as examined by internal perfusion. J Physiol 292:251.
Hagiwara S, Miyazaki s, Rosenthal,P (1976). Potassium current
    and the effect of cesium on this current during anomalous
    rectification of the egg cell membrane of a starfish.
    J Gen Physiol 67:621.
Hagiwara S, Miyazaki S, Krasne S, Ciani S (1977). Anomalous
    permeabilities of the egg cell membrane of a starfish in
    K^+-Tl^+ mixtures. J Gen Physiol 70:269.
Hestrin S (1981). The interaction of potassium with the
    activation of anomalous rectification in frog muscle
    membrane. J Physiol 317:497.
Hille B, Campbell DT (1976). An improved vaseline gap voltage

clamp for skeletal muscle fibers. J Gen Physiol 67:265.

Leech CA, Stanfield PR (1981). Inward rectification in frog skeletal muscle fibres and its dependence on membrane potential and external potassium. J Physiol 319:295.

Ohmori H (1981). Dual effects of K ions on the inactivation of the anomalous rectifier of the tunicate egg cell membrane. J Membr Biol 53:143.

Standen NB, Stanfield PR (1978). A potential and time-dependent blockade of inward rectification in frog skeletal muscle fibres by barium and strontium ions. J Physiol 280:169.

Stanfield PR, Ashcroft FM, Plant TD (1981). Gating of a muscle potassium channel and its dependence on the permeating ion species. Nature 289:509.

Swenson RP, Armstrong CM (1981). K^+ channels close more slowly in the presence of external K^+ and Rb^+. Nature 291: 427.

The Physiology of Excitable Cells, pages 109-125
© 1983 Alan R. Liss, Inc., 150 Fifth Avenue, New York, NY 10011

EFFECT OF EXTERNAL POTASSIUM CONCENTRATION ON OUTWARD-
CURRENT CHANNEL KINETICS IN HELIX NEURONES

Douglas Junge, Ph.D.

School of Dentistry and Department of Physiology
University of California
Los Angeles, California 90024, USA

INTRODUCTION

My interest in outward-current channels in molluscan
neurons stems rather directly from having been a postdoc in
Dr. Hagiwara's laboratory from 1965-67. For he, of course,
had been studying the ionic properties of such channels
and the blocking action of TEA for the previous decade
(Hagiwara, Watanabe 1955; Hagiwara, Saito 1959). In
addition, Hagiwara, Kusano and Saito, in 1961, first
noticed the early outward, or "A" currents seen in
Onchidium neurons after a hyperpolarizing pre-pulse.

In the classical view, outward-current conductances
have been analyzed in terms of non-interacting gating and
permeation processes. The gating mechanism is considered to
depend on time and voltage, but not on internal or external
ion concentrations. Recently, however, some studies have
appeared in which the gating of outward current channels
was affected by ion concentrations (Dubois, Bergman 1977;
Stühmer, Conti 1979; Begenisich 1979; Dubois 1981). In
each of these cases the time-constant for activation of
outward-current was reduced by increasing $(K)_o$. I have
seen a similar effect in perfused snail neurons; some of
these results have been published previously (Junge 1982).

METHODS

Subesophageal ganglia were isolated from specimens of
Helix aspersa and pinned in a silastic-lined dish under
normal saline solution. The upper cell mass was then

dissected out using a stripping procedure with small forceps, and transferred to the 0.5cc recording chamber. The cell mass was pinned down and treated with 0.2% trypsin in normal saline at room temperature for 15 min, after which it was possible to remove a single unidentified cell body (diam. = 60-100 μm) with a suction pipette arrangement similar to that used by Lee, Akaike and Brown (1978). This is a modification of the internal perfusion technique of Kostyuk, Krishtal, and Pidoplichko (1975), and allows exchange of some internal ions and voltage-clamping with a low access resistance (typically 0.5megΩ). Fifty-two cells were examined with this method, principally during the spring and summer.

The perfusion pipette typically had a diameter of 25 μm. Potentials were recorded either in the pipette holder with a calomel cell or in the cell body with a microelectrode. Electronic leak and series resistance compensation were used routinely, and currents were passed through a 1-kHz low-pass filter before being displayed on the oscilloscope. The normal saline solution had the composition NaCl 75 mM, KCl 5 mM, CaCl$_2$ 10 mM, MgCl$_2$ 15 mM and tris-HCl (pH 7.5) 5 mM. The standard internal solution was K-aspartate 105 mM, EGTA 3 mM and HEPES (pH 7.3) 5 mM. K-free external solutions were prepared by replacing all of the K with Na or tris. In high-K solutions some of the Na was replaced with K. Na-free external solutions had an equimolar amount of tris as cation, and contained 0.5 mM CdCl$_2$ to block inward Ca-currents. All experiments were performed at 15-17°C.

RESULTS

Figure 1 shows the reduction of outward currents seen in a large (diam. $\sim$ 100 μm) cell when external potassium was replaced with sodium. The upper traces are currents and the lower ones potentials measured with a microelectrode in the cell body. At potentials close to +130mV, the final outward current was reduced by about 16%, and recovered to 104% of the initial value upon replacement of external K-ions. In addition, the onset of outward current was considerably slowed by K-replacement. The outward tail currents following the end of the depolarizing pulse were somewhat slowed in K-free solution. However, this may have been due to decreased accumulation of K ions outside the cell (Frankenhaeuser, Hodgkin 1956), and was not considered a reliable indicator of changes in relaxation kinetics. In order to minimize the effects of inactivating outward currents seen in these cells (Neher, Lux 1971), depolarizing commands were always repeated at 1/sec, at which frequency the inactivating current was reduced essentially to zero.

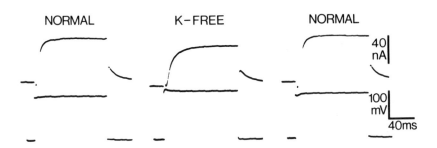

Fig. 1. Effect on outward currents of replacement of external K with Na. Top traces currents; bottom traces potentials measured with microelectrode. Cell 29.

Other possible causes of the reduction of outward current include: 1) an increase in the leak current, which was electrically subtracted from the total membrane currents. However, when inward currents were blocked with Na-free solution containing 0.5 mM cadmium ions and outward currents were blocked by perfusing internally with 105 mM cesium aspartate, the remaining uncompensated leak currents were not sensitive to removal of external potassium. Also 2) the series resistance was compensated electrically and could conceivably have increased with K-replacement, which would decrease the transmembrane potential. However, the series resistance measured under constant-current conditions was not affected by replacement of external K with Na.

It was also necessary to determine which of several outward currents operating in these cells was affected by K-substitution: 1) The calcium-activated outward current (Meech, Standen 1975) was apparently not involved, since the effect was seen with external solutions containing

0.5 mM $CdCl_2$, and the internal solution always contained 3 mM EGTA. 2) An electrogenic sodium pump in these cells is known to be blocked by K-replacement (Thomas 1969). If the pump produced more outward current at depolarized levels, then blockage should reduce the currents, as seen. However, when the experiment was repeated in the presence of 0.1 mM ouabain or strophanthidin, the outward currents were reduced by amounts similar to that in Figure 1 upon removal of external potassium. 3) The possible contribution of early outward currents was reduced by holding the cell potentials at −28 to −40 mV, where these currents are completely inactivated (Neher 1971; Connor, Stevens 1971). Hence, the effect of substitution of sodium for external potassium was apparently confined to the "delayed rectifier" channels in the membrane.

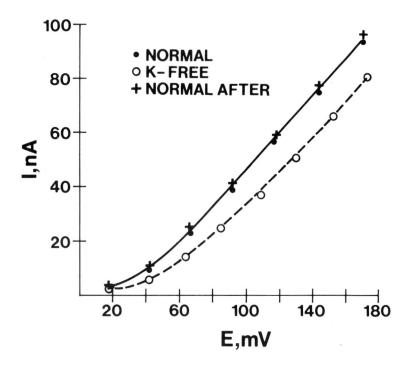

Fig. 2. Steady-state I-V curves in normal, K-free and normal solution. Cell 29.

The effect of replacement of external K on the steady-state I-V relationship is shown in Figure 2. The solid curve was drawn through points obtained in normal saline before and after K-removal, and the broken curve in K-free solution. Outward currents were smaller at all potentials tested in the absence of external K, but the percentage reduction was least at the most positive potentials. (This indicates some sensitivity of the effect to electric field strength in the membrane.) In nine cells subjected to K-replacement, the percentage reduction of outward current at +100 mV was 15 $\pm$ 11% (mean $\pm$ S.D.) and the recovery was 99 $\pm$ 12% upon returning to K-containing solution.

An instantaneous I-V curve obtained when the potential was changed to various levels following an initial step to +145 mV (pipette potential) is shown in Figure 3. The reversal potential was -64 mV in normal saline and -78 mV

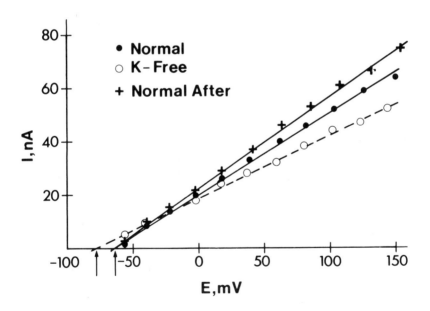

Fig. 3. Instantaneous I-V relation in normal, K-free and normal solution. Reversal potentials indicated by arrows. Cell 21.

in K-free, and the slope of the I-V line decreased with K-removal. The reversal potentials measured in this way were relatively insensitive to the amplitude of the conditioning pulse, unlike reversal potentials in the node of Ranvier (Dubois 1981). Linear instantaneous I-V relationships have also been reported for these cells by Reuter and Stevens (1980).

Conductance-voltage curves could be calculated from data of this type, as shown in Figure 4, using the same cell as in Figure 3. Here, the conductance was taken as $g_K = I / (E - E_{rev})$, where E_{rev} was the zero-current potential in the instantaneous I-V relationship. The conductance decreased about 16% at +100 mV, and recovered upon returning to normal saline. The curves nearly reached maximal values ($\bar{g}_K$) at the most positive potentials, indicating

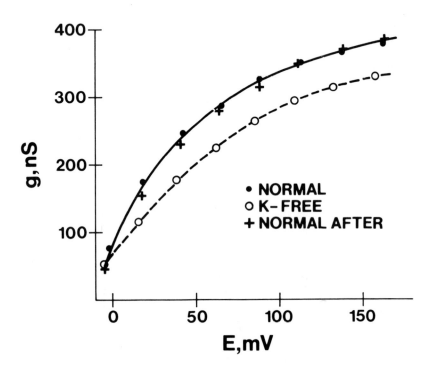

Fig. 4. Conductance-voltage curves in normal, K-free and normal solution. Cell 21.

that almost all available K-channels were activated. The values of $\bar{g}_K$ estimated for this cell by extrapolation were 405 nS in normal saline and 360 nS in K-free. Using noise analysis, Reuter and Stevens (1980) found the number of K-channels per unit area in these cells was 7.2/ μm^2, and the single-channel conductance was 2.4 pS. This implies a peak available conductance of 1.73 mS/cm^2, or about 440 nS for a spherical cell 90 μm in diameter, which agrees roughly with the above measurement of $\bar{g}_K$.

A modified Hodgkin-Huxley (1952) model was used to analyze the effects of potassium replacement on outward currents, and particularly on the slowing of activation of these currents. The predicted time-course of outward currents in response to a depolarizing step is

$$I_K = \bar{g}_K \ (E - E_K) \left[\frac{\alpha_{\infty}}{\alpha_{\infty} + \beta_{\infty}} \ (1 - e^{-(\alpha_{\infty} + \beta_{\infty})t}) \right]^2 \quad (1)$$

where $\bar{g}_K$ = peak conductance, estimated from g-V curves

E_K = potassium equilibrium potential, taken as reversal potential of instantaneous I-V curve

α_{∞} and β_{∞} = opening and closing rate constants

Exponents of 2, 3, and 4 were tried with 2 giving the best least-squares fit to the observed outward currents. Figure 5 shows computer-generated fits to outward currents obtained in Na-free solutions containing 0.5 mM CdCl$_2$ (where inward currents were blocked) with and without external potassium ions (additional tris substituted for K). The onset of current was slower at all potentials in the absence of external K, and the currents reached lower final values. Outward currents recovered to the original level upon restoration of potassium to the solution (not shown). The values of α_{∞} and β_{∞} which gave the best fits at various potentials are shown in Figure 6. After $\bar{g}_K$ was estimated from conductance-voltage curves in K-containing and K-free solutions, it was held constant at either of these values in the model. The values of α_{∞} and β_{∞} determined not only the rate of rise of the currents but also the final values. Thus, the model also accounted for the steady-state I-V relationships in the two

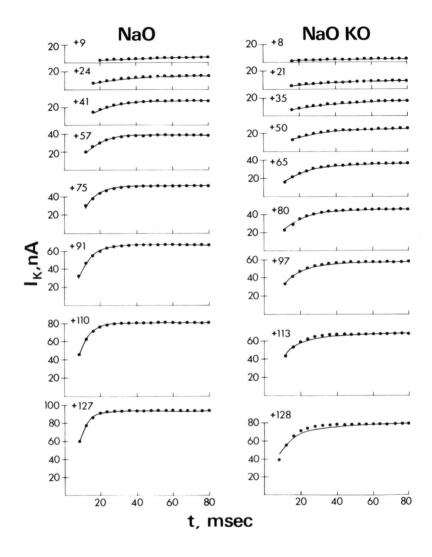

Fig. 5. Fits of three-state kinetic model (dots) to outward currents in K-containing and K-free solution. Na 0: Na-free solution containing 0.5 mM $CdCl_2$ and 5 mM KCl; Na 0 K 0: Na-free solution with 0.5 mM $CdCl_2$, zero K. Voltage commands indicated in each graph. Cell 30.

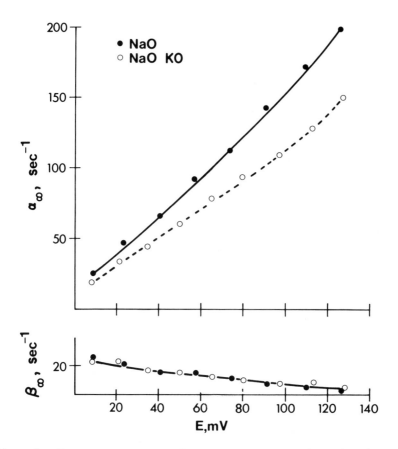

Fig. 6. Rate constants for channel opening (a_{∞}) and closing (β_{∞}) obtained from curve fits of Figure 5.

solutions tested. The result of this analysis was that replacement of external potassium decreased the rate constant for opening of outward-current channels and did not affect the closing rate constant in this range of potentials.

In order to determine the effects of increasing the potassium concentration above the normal (5 mM) level, solutions with concentrations of 20 mM, 40 mM and 80 mM K (in exchange for Na) were applied to a cell. Steady-state

and instantaneous I-V relations were obtained, and the
curves of conductance-vs.-time shown in Figure 7 were
calculated. The membrane was depolarized to +78 mV in each
solution, and the reversal potential, E_{rev}, was -40 mV in
K5, -28 mV in K 20, -14 mV in K 40 and 0 mV in K 80. (This
was approximately a 48 mV change/10-fold change in K-
concentration for the three highest concentrations.)
Although the final level of conductance was relatively
insensitive to external potassium, the rate of rise of conductance increased strongly with increasing K-concentration.
The same result has also been seen in the squid axon by

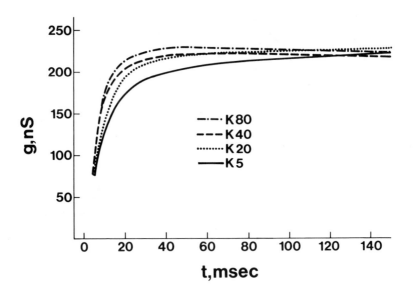

Fig. 7. Effect of increasing external K-concentration on
conductance-time curves. Cell 25.

Stühmer and Conti (1979), upon increasing external
K-concentration from 10 mM to 460 mM. Note that the final
outward current in the above example decreased as the
potassium concentration was raised above the normal
level. In this cell, as in the cell of Fig. 4, reducing the

external K-concentration to zero did cause a reduction of outward current. In several cells, the current tended to reach a maximum near the normal K-concentration, and decreased at higher concentrations. In Figure 7, when external K was elevated, the current decreased by the same fraction as the driving force, $E - E_{rev} = I_K/g_K$, since the final conductance remained the same. This behavior was not always seen in cells tested in this way; in some cases the final conductance increased or decreased at higher potassium concentrations.

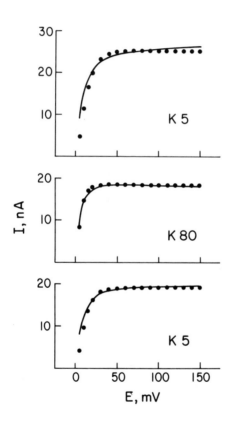

Fig. 8. Fits of kinetic model to outward currents in normal (K 5), high-potassium (K 80) and normal solutions. Potential held at -40 mV, then stepped to +80 mV. Cell 25.

Figure 8 shows fits of Equation 1 (dots) to the rising phases of outward currents obtained in solutions containing 5 mM K and 80 mM K (in exchange for Na). The outward current in 80 mM K reached a lower final value than in 5 mM K but developed more rapidly. The corresponding rate constants obtained from the fitted curves are shown in Table 1. As was the case with potassium replacement experiments (Figure 6), increasing external K-concentration increased the rate constant for channel opening, but in this case it also decreased that for channel closing.

Table 1. Effect of increasing external K-concentration on rate constants and energy barriers in Eyring model of activation.

$(K)_o$	a_∞	β_∞	ΔG_a	ΔG_β
5 mM	99 s^{-1}	11.8 s^{-1}	14.2 kcal/M	15.4 kcal/M
80	210	1.2	13.8	16.8

In order to model this effect of ion concentrations on activation kinetics, an Eyring-type model (Eyring, Henderson, Stover, Eyring 1964; Hille 1975) shown in Figure 9 was used. The n^2 kinetics which fit the data on activation are predicted by assuming that channels may be in any of three states, C_1 and C_2 (closed) or O (open). Channels convert between states with the following rate constants:

$$C_1 \underset{\beta}{\overset{2a}{\rightleftharpoons}} C_2 \underset{2\beta}{\overset{a}{\rightleftharpoons}} O$$

The individual rate constants are related to the free-energy change for channel opening and closing by the expressions

$$a = \frac{\kappa_i kT}{h} e^{-\Delta G_a /RT} \qquad (2)$$

$$\beta = \frac{\kappa_i kT}{h} e^{-\Delta G_\beta /RT} \qquad (3)$$

where κ_i = "transmission coefficient", ≈ 1

k = Boltzmann constant

T = temperature, $^{\circ}$K

h = Planck constant

ΔG_{α} = height of energy barrier for going from C_2 to 0

ΔG_{β} = height of energy barrier for going from C_2 to C_1

R = universal gas constant; RT = .573 kcal/M at 15 $^{\circ}$C

$$\frac{\kappa_i kT}{h} = 6 \times 10^{12} \text{ s}^{-1} \text{ at 15 } ^{\circ}\text{C}$$

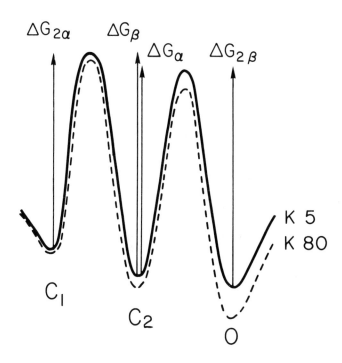

Fig. 9. Energy-barrier model of K-ion action on opening and closing rate constants. Abscissa shows state of channels: C_1 and C_2, closed; 0, open. ΔG_{α} = energy to pass from C_2 to 0, and ΔG_{β} = energy to pass from C_2 to C_1. External K⁺ ions decrease ΔG_{α} and increase ΔG_{β} as shown. Further details in text.

Thus, the shapes of the curves of ΔG in Figure 9 are uniquely determined by the values of α_∞ and β_∞ found in high- and low-K conditions. The effect of increasing $(K)_o$ is to lower the barrier height for channel opening and increase that for channel closing. As a heuristic model, K^+ ions may "relieve" the channel, i.e., partially open it or facilitate the opening process by interaction with channel proteins and external membrane fixed charges. This interaction would also tend to inhibit closure. Although no consistent effects on tail currents were seen in these experiments, it is interesting that elevation of $(K)_o$ has been shown to increase the relaxation time-constant in other preparations (Adelman, Senft 1968; Stühmer, Conti 1979; Århem 1980).

DISCUSSION

The results obtained with decreasing and increasing external potassium concentration indicate that gating of outward-current channels in snail neurones is affected by K^+ ions. Neither Na^+ nor $tris^+$ ions can substitute for potassium in this capacity. The outward current is clearly carried by K^+ ions, since 1) the reversal potential for tail currents varies with the potassium equilibrium potential (this report and Reuter, Stevens 1980), and 2) the outward currents are blocked by internal perfusion with cesium ions (this report and Lee, Akaike, Brown 1978).

Swenson and Armstrong (1981) have suggested that closing of outward-current channels in the squid axon is prevented by entry of external K^+ ions into the channel. Kohler (1977) postulated that external K^+ ions compete with membrane-bound blocking particles for sites in the channel, and thus affect outward-current kinetics. In their study of the node of Ranvier, Dubois and Bergman (1977) suggested that external K^+ ions bind to specific "receptors," which regulate the activation of outward-current channels.

The effect of external K^+ ions on outward-current kinetics may in part account for the "Cole-Moore effect," where conditioning hyperpolarizations delay the onset of these currents (Cole, Moore 1960; Palti, Ganot, Stämpfli 1976; Begenisich 1979). Such conditioning pulses would be expected to decrease the local extracellular

K-concentration, and could thus reduce the rate constant for activation.

It is possible that external K$^+$ ions might increase outward-current conductance by stabilizing channels in the open form, and increasing the mean open time. This hypothesis could be tested in principle by measurement of corner frequencies of the spectral density function (Conti, De Felice, Wanke 1975), in solutions with different K-concentrations. Although two corner frequencies in the spectral density function are predicted for the three-state model discussed above (Hill, Chen 1972), it is unlikely that they could be resolved as they differ only by a factor of two. A more direct answer is provided by recordings from isolated patches of squid axon membrane (Conti, Neher 1980): They observed a mean K-channel open time of 12 ms in high-potassium solution, whereas the value obtained from noise analysis in artificial seawater was close to 5 ms at the same temperature (Conti, De Felice, Wanke, 1975). It seems likely that further studies will also reveal a dependence of the rate of opening and closing of single channels on external K-concentration.

REFERENCES

Adelman WJ, Senft JP (1968). Dynamic asymmetries in the squid axon membrane. J gen Physiol 51:102s.

Århem P (1980). Effects of rubidium, caesium, strontium, barium and lanthanum on ionic currents in myelinated nerve fibers from Xenopus laevis. Acta Physiol Scand 108:7.

Begenisich T (1979). Conditioning hyperpolarization-induced delays in the potassium channels of myelinated nerve. Biophys J 27:257.

Cole KS, Moore JW (1960). Potassium ion current in the squid giant axon: dynamic characteristic. Biophys J 1:1.

Connor JA, Stevens CF (1971). Voltage clamp studies of a transient outward membrane current in gastropod neural somata. J Physiol 213:21.

Conti F, Neher E (1980). Single channel recordings of K$^+$ currents in squid axons. Nature 285:140.

Conti F, De Felice LJ, Wanke E (1975). Potassium and sodium ion current noise in the membrane of the squid giant axon. J Physiol 248:45.

Dubois JM (1981). Simultaneous changes in the equilibrium potential and potassium conductance in voltage clamped Ranvier node in the frog. J Physiol 318:279.

Dubois JM, Bergman C (1977). The steady-state potassium conductance of the Ranvier node at various external K-concentrations. Pflügers Arch 370:185.

Eyring H, Henderson D, Stover BJ, Eyring EM (1964). "Statistical Mechanics and Dynamics." New York: Wiley, p. 453.

Frankenhaeuser B, Hodgkin AL (1956). The after-effects of impulses in the giant nerve fibres of Loligo. J Physiol 131:341.

Hagiwara S, Kusano K, Saito N (1961). Membrane changes of Onchidium nerve cell in potassium-rich media. J Physiol 155:470.

Hagiwara S, Miyazaki S, Rosenthal NP (1976). Potassium current and the effect of cesium on this current during anomalous rectification of the egg cell membrane of a starfish. J gen Physiol 67:621.

Hagiwara S, Saito N (1959). Voltage-current relations in nerve cell membrane of Onchidium verruculatum. J Physiol 148:161.

Hagiwara S, Watanabe A (1955). The effect of tetraethyl-ammonium chloride on the muscle membrane examined with an intracellular microelectrode. J Physiol 129:513.

Hill TL, Chen Y (1972). On the theory of ion transport across the nerve membrane IV: noise from the open-close kinetics of K^+ channels. Biophys J 12:948.

Hille B (1975). Ionic selectivity, saturation, and block in sodium channels: a four-barrier model. J gen Physiol 66:535.

Hille B (1978). Ionic channels in excitable membranes: current problems and biophysical approaches. Biophys J 22:283.

Hodgkin AL, Huxley AF (1952). A quantitative description of membrane current and its application to conduction and excitation in nerve. J Physiol 117:500.

Junge D (1982). External K+ ions increase rate of opening of outward current channels in snail neurons. Pflügers Arch 394:94.

Kohler H-H (1977). A single-file model for potassium transport in squid giant axon. Biophys J 19:125.

Kostyuk PG, Krishtal OA, Pidoplichko VI (1975). Effect of internal fluoride and phosphate on membrane currents during intracellular dialysis of nerve cells. Nature 257:691.

Lee KS, Akaike N, Brown AM (1978). Properties of internally perfused, voltage-clamped, isolated nerve cell bodies. J gen Physiol 71:489.

Meech RW, Standen NB (1975). Potassium activation in Helix aspersa neurones under voltage clamp: a component mediated by calcium influx. J Physiol 249:211.

Neher E (1971). Two fast transient current components during voltage clamp on snail neurons. J gen Physiol 58:36.

Neher E, Lux HD (1971). Properties of somatic membrane patches of snail neurons under voltage clamp. Pflügers Arch 322:35.

Palti Y, Ganot G, Stämpfli R (1976). Effect of conditioning potential on potassium current kinetics in the frog node. Biophys J 16:261.

Reuter H, Stevens CF (1980). Ion conductance and ion selectivity of potassium channels in snail neurones. J memb Biol 57:103.

Stühmer W, Conti F (1979). The effect of high extracellular potassium on the kinetics of potassium conductance of the squid axon membrane. In Adam G, Stark G (eds): "Ann. Meet. Deutsche Gesellschaft Biophysik," New York: Springer, p.84.

Swenson RP, Armstrong CM (1981). K⁺ channels close more slowly in the presence of external K⁺ and Rb⁺. Nature 291:427.

Thomas RC (1969). Membrane current and intracellular sodium changes in a snail neurone during extrusion of injected sodium. J Physiol 201:495.

Top: S. Krasne; M. Letinsky; S. Hestrin; D. Junge;
J. Patlak; W. Moody, Jr.; F. Ashcroft; S. Chichibu
Bottom: R. Meech; S. Hagiwara; R. Gruener

The Physiology of Excitable Cells, pages 127–138
© 1983 Alan R. Liss, Inc., 150 Fifth Avenue, New York, NY 10011

MECHANISM OF ACETYLCHOLINE-INDUCED MEMBRANE
POTENTIAL FLUCTUATIONS IN RAT
ADRENAL CHROMAFFIN CELLS

Y. Kidokoro

The Salk Institute
San Diego, California

INTRODUCTION

The secretion of catecholamines from adrenal
chromaffin cells, like that of many other hormones,
requires Ca ions in the extracellular solution (Douglas,
Rubin 1961; Ishikawa, Kanno 1978). Ca ions enter the
cell and elevate cytoplasmic Ca concentration to a ½M
range to initiate secretion (Knight, Baker 1982). Since
in the resting state the cell membrane is not permeable
to Ca ions, the neurotransmitter, acetylcholine (ACh),
must be stimulating hormone release by altering the
membrane permeability to Ca ions. A number of mechanisms
are known which control membrane Ca permeability. One
mechanism which has been extensively studied in various
systems involves a voltage dependent Ca permeability
increase (Hagiwara, Byerly 1981). A strong suggestion
that voltage dependent membrane permeability increase may
be involved in hormone secretion, comes from a classical
observation that in many secretory systems depolarization
of the cell with excessive K ions leads to hormone
secretion and that external Ca ions are required for this
stimulation (Douglas, Rubin 1961).

ACh receptor channels constitute an alternative
pathway for Ca ions to enter the cell. It is known that
Ca ions enter the frog muscle cell through these
channels, which are mainly permeable to Na and K ions
(Takeuchi, 1963; Adams, Dwyer, Hille 1980; Bregestowsky,
Miledi, Parker 1979). When chromaffin cells were
completely depolarized in isotonic K-sulphate saline and
then exposed to ACh, additional secretion was observed

(Douglas, Rubin 1963). Under these conditions ACh does
not increase Ca entry by further depolarization of the
cell membrane. Instead, additional Ca ions are likely to
to enter the cell through ACh receptor channels. It is
plausible, therefore, that these two mechanisms of Ca
entry, voltage dependent Ca channels and ACh receptor
channels, are operating together during ACh-induced
catecholamine secretion.

In this article I will describe the effect of ACh on
the adrenal chromaffin cell and the underlying mechanism.
I shall briefly review the literature and present some
data recently obtained using single channel current
recording techniques.

ACh depolarizes adrenal chromaffin cells

ACh released at the splanchnic nerve terminal
triggers catecholamine release from adrenal chromaffin
cells. What happens, then, at the chromaffin cell
membrane when ACh molecules bind to the receptor? This
question has been answered by studying membrane
electrical properties of adrenal chromaffin cells in
culture during application of ACh. It has been reported
that chromaffin cells bathed in saline containing ACh
have reduced resting membrane potentials (Douglas, Kanno,
Sampson 1967). Iontophoretic application of ACh also
caused transient depolarization of cultured rat adrenal
chromaffin cells (Brandt, Hagiwara, Kidokoro, Miyazaki
1976; Biales, Dichter, Tischler 1976). Thus it appears
that ACh depolarizes the chromaffin cell. The precise
relationship between ACh concentration and depolarization
was studied by applying ACh by the "puffer" method
(Kidokoro, Miyazaki, Ozawa 1982). In this technique a
known concentration of ACh in a glass pipette with an
inner tip diameter of about 3½m is ejected by a pulse of
gas pressure. When the tip of the ACh pipette was placed
close to the cell (60-80½m) and sufficiently high gas
pressure (about 0.1 Kg/cm^2) was applied, the ACh
concentration surrounding the cell approximated that
inside the pipette. Application of 1½M ACh depolarized
the cell by a peak of approximately 10mV within 0.5 sec
and in some cases slowly declined during application for
5 sec. The membrane potential fluctuated prominently

during the ACh-induced depolarization and action potentials were superimposing (Fig. 1).

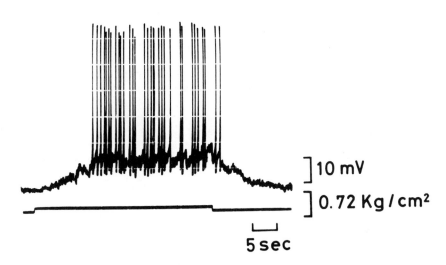

]10 mV

]0.72 Kg / cm²

5 sec

Figure 1. ACh-induced depolarization and action potentials. The oscilloscope traces represent the membrane potential recorded intracellularly and a pressure record indicate delivery of ΛCh from a pipette. The concentration of ACh was 10½M. The membrane potential was -59mV hyperpolarized from a resting level of -39mV with d.c. current (Kidokoro, Miyazaki, Ozawa 1982; reprinted with permission).

The depolarization became detectable at 0.1½M ACh and saturated at around 100½M. The half maximal concentration was around 10½M. This range of effective ACh concentrations is similar to that necessary to stimulate adrenaline secretion from the perfused rat adrenal medulla (Ishikawa, Harada, Kanno 1977; Kidokoro, Ritchie 1981). Therefore, it is reasonable to assume that there is causal relationship between ACh-induced depolarization and subsequent adrenaline secretion. Ca ions which enter the cell through ACh receptor channels as well as through voltage-dependent Ca channels during action potentials may contribute to catecholamine secretion.

ACh-induced potential fluctuations are probably due to stochastic opening and closing of ACh receptor channels

Upon application of ACh the adrenal chromaffin cell depolarized with characteristic potential fluctuations. It is, therefore, worthwhile to determine the nature of this potential noise (Kidokoro, Miyazaki, Ozawa 1982). It was analyzed in cells in which action potentials were no longer generated upon ACh induced depolarization probably due to damage by microelectrode penetration and which were hyperpolarized to improve the signal-to-noise ratio. A plot of the mean amplitude of the ACh-induced depolarizations vs the variance of the potential fluctuations in a given cell was approximately linear with a slope of 0.2mV (Fig. 2). If we assume that these potential fluctuations are due to summation of rectangular

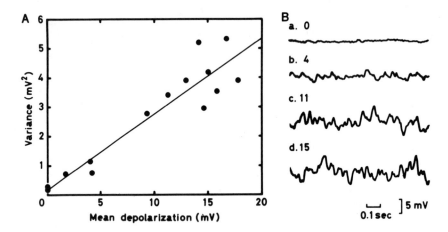

Figure 2. ACh-induced noise. A, relationship between the mean depolarization and the variance. Points on the ordinate are control without ACh application. ACh concentration inside the capillary was 10½M. The membrane potential was hyperpolarized to -83mV by injecting current through the recording electrode. The straight line is a least-square fit. Note that correction for changes in the electromotive force is not done in this plot. B, sample recordings. Various amounts of ACh were delivered through the pipette by changing pressure. The numbers to the left of each trace are the mean depolarization in mV (reprinted with permission from Kidokoro, Miyazaki, Ozawa 1982).

voltage pulses generated from random opening and closing of ACh receptor channels at a low frequency, then this value (0.2mV) estimates the amplitude of the pulse. The input resistances of the cells in which ACh-induced potential fluctuations were measured and, assuming ohmic behavior of the membrane, yielded a value of about 0.8pA for single channel current. By changing the membrane potential level, the reversal potential of the ACh induced depolarization was estimated to be -24mV. Hence an elementary single channel conductance change can be calculated to be 20pS, which is similar to that found for ACh receptor channels in skeletal muscle cells (Katz and Miledi, 1972; Anderson and Stevens, 1973; Neher and Sakmann, 1976).

Temporal characteristics of the potential fluctuations were studied by calculating an autocorrelation function. The autocorrelation function in a given cell declined exponentially with a mean time constant of about 40 msec at 30°C and at a membrane potential of -60 mV. Under the same assumptions described above, this value equals the mean duration of the elemental noise pulses. This value is extremely large compared with that for muscle ACh receptor channels (a few msec).

In these experiments membrane potential fluctuations were analyzed. Even if the underlying unitary current were rectangular, the resulting potential change would be distorted by the membrane time constant of 13 msec. Therefore, the assumption that the potential fluctuations were the result of summation of randomly occurring rectangular voltage pulses is not valid. Fortunately, since the estimated open time, 40 msec, is much longer than the membrane time constant the unitary potential change, although somewhat distorted, can be approximated with a rectangular pulse. Thus these estimated values for the unitary current and the mean open time are probably close to the real values.

Single ACh receptor channel current recording from the adrenal chromaffin cell

The above mentioned analyses were performed under the assumption that these potential fluctuations are due to summation of rectangular voltage pulses resulting from random opening and closing of ACh receptor channels at a low frequency. In order to test the validity of this assumption, I recorded single ACh receptor channel currents using patch clamp techniques (Hamill, Marty, Neher, Sakmann, Sigworth 1981). When the electrode contained 1½M ACh rectangular pulses of inward current were observed as shown in Figure 3A.

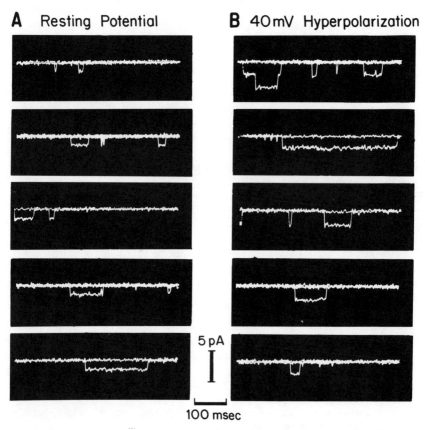

Figure 3. Single channel currents from ACh receptor channels in a rat adrenal chromaffin cell. The electrode contained 1½M ACh. A, at the resting potential. B, at 40mV hyperpolarization. Hyperpolarization was achieved by applying positive potentials inside the electrode. The experiment was carried out at room temperature.

The amplitudes of these events are similar (1.6pA) but the duration varied over a wide range. When the duration histogram was constructed, it was fitted with multiple exponentials (at least three). The longest time constant was about 30 msec. This value is the estimate of the mean duration of long individual current pulses and fits reasonably well. The values estimated from potential fluctuation analysis described above. Other components with short durations may not have been observed in the potential recording because of filtering by the membrane time constant. These events are most likely due to activation of ACh receptor channels, because when no ACh was included in the electrode virtually no events were observed. Previously we have demonstrated that 100½M hexamethonium blocks depolarization induced by 10½M ACh (Kidokoro, Miyazaki, Ozawa 1982). Therefore, I further tested the effect of hexamethonium on ACh-induced single channel currents. When 100½M hexamethonium was included in the recording electrode together with 10½ACh rectangular events with short durations (a few msec) were sporadically observed. These were probably too short in duration to affect the membrane potential. Although further investigation is required to determine the mode of hexamethonium action the observation is consistent with the previous results.

The mean open time of ACh receptor channels in adrenal chromaffin cells is characteristically long. Long open time is also reported for frog parasympathetic ganglion cells (about 30 msec, Ascher, Large, Rang 1979). There are other peculiarities in ACh receptor channels in chromaffin cells. Thus even when the membrane potential was altered the mean open time did not change appreciably (Fig. 3B). This is in contrast to the finding that, in skeletal muscle cells, the mean open time is prolonged by hyperpolarization (Anderson, Stevens 1972). On the other hand, in bullfrog sympathetic ganglion cells the mean open time is relatively unaffected by the membrane potential (MacDermott, Conner, Dionne, Parsons 1980).

DISCUSSION

ACh depolarizes the adrenal chromaffin cell with prominent membrane potential fluctuations. These fluctuations are most likely the result of stochastic

opening and closing of ACh receptor channels. These
potential fluctuations were particularly prominent in
chromaffin cells for two reasons. Since adrenal chromaffin cells are small (16½m in diameter, Brandt et al.
1976), the input resistance is extremely large (380 Mohm,
Kidokoro et al. 1982; around 6 Gohm, Fenwick et al.
1982). Consequently minute current flowing through an
individual channel can generate significant potential
fluctuations. The other factor which contributes to this
prominent potential fluctuation is the long open time of
ACh receptor channels in chromaffin cells. If 1.5pA of
current flows through an ACh receptor channel and the
cell has five Gohm input resistance, 7.5 mV
depolarization will result, provided the channel open
time is long compared with the membrane time constant.
In fact, in many cases action potentials were generated
during the time when one channel was open. This occurred
much more frequently than one can expect from
coincidental overlap of spontaneous action potentials and
open channels (Fig. 4). Similar observation has been
reported recently by Fenwick et al. (1982).

In vivo, however, significantly smaller input
resistances (25-40 Mohm) were observed in rat adrenal
chromaffin cells although the resting membrane potential
was comparable to that observed in culture (-50mV,
Ishikawa, Kanno 1978). They argued that cells may be
electrically coupled in situ. Dissociation of cells for
cell culture could have disrupted these couplings and
have resulted in artifically high input resistances.
While there is no electrophysiological study on
electrical coupling among chromaffin cells in situ, both
thin section and freeze-fracture electron microscopy did
not reveal any intercellular junctions in the rat
(Grynszpan-Winograd, 1975; Grynszpan-Winograd, Nicholas
1980). Furthermore, in culture we did not find
electrical coupling between two contacting chromaffin
cells (Brandt and Kidokoro, unpublished observation).
There is an alternative way to explain observed low input
resistance in chromaffin cells in situ. Undoubtedly,
penetration of small chromaffin cells with a conventional
microelectrode is difficult, particularly in situ where
cells are loosely supported in a spongework of venous
sinuses. Therefore, it is conceivable that upon
penetration a certain amount of Ca enters the cell
through leakage around the microelectrode and activates a

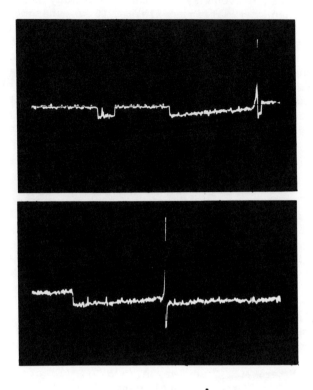

— $\quad$ | 2 pA

50 msec

Figure 4. Action potentials superimposing on unitary current through the ACh receptor channel. Decline of current amplitude during a long opening is presumably due to depolarization.

Ca-dependent K conductance, which has been demonstrated in chromaffin cells (Marty, 1981). This conductance increase can result in a small input resistance accompanied with a relatively large resting membrane potential. Thus it is still possible that intact chromaffin cells in situ have an input resistance as large as observed in culture.

In the central nervous system there are many neurons even smaller than chromaffin cells. If a phenomenon similar to that observed in cultured chromaffin cells occurs in these neurons, that is, if the opening of a single receptor channel can trigger an action potential, then the stochastic nature of these events will severely disturb signal transmission. This could be particularly problematic when a synapse is terminating on a thin dendrite where input resistance must be extremely high. There must be some mechanism to suppress inadvertent generation of action potentials.

REFERENCES

Adams, D.J., Dwyer, T.M., Hille, B. (1980). The permeability of endplate channels to monovalent and divalent metal cations. J Gen Physiol 75:493.

Anderson, C.R., Stevens, C.F. (1973). Voltage clamp analysis of acetylcholine produced endplate current fluctuations at frog neuromuscular junction. J Physiol 235:655.

Ascher, P., Large, W.A., Rang, H.P. (1979). Studies on the mechanism of action of acetylcholine antagonists on rat parasympathetic ganglion cells. J Physiol. 295:139.

Biales, B., Dichter, M., Tischler, A. (1976). Electrical excitability of cultured adrenal chromaffin cells. J Physiol 262:743.

Brandt, B.L., Hagiwara, S., Kidokoro, Y., Miyazaki, S. (1976). Action potentials in the rat chromaffin cell and effects of acetylcholine. J Physiol 263:427.

Bregestovsky, P.D., Miledi, R., Parker, I. (1979). Calcium conductance of acetylcholine induced endplate channels. Nature (Lond.) 279:638.

Douglas, W.W., Kanno, T., Sampson, S.R. (1967). Effects of acetylcholine and other medullary secretagogues and antiagonists on the membrane potential of adrenal chromaffin cells: an analysis employing techniques of tissue culture. J Physiol 118:107.

Douglas, W.W., Rubin, R.P. (1961). The role of calcium in the secretory response of the adrenal medulla to acetylcholine. J Physiol 159:40.

Douglas, W.W., Rubin, R.P. (1963). The mechanism of catecholamine release from the adrenal medulla and the role of calcium in stimulus-secretion coupling. J Physiol 167:288.

Fenwick, E.M., Marty, A., Neher, E. (1982). A patch-clamp study of bovine chromaffin cells and of their sensitivity to acetylcholine. J Physiol 331:577.

Grynszpan-Wynograd, O. (1975). Ultrastructure of the chromaffin cell. In Handbook of Physiology, Section 7, Endocrinology, Vol. VI, Adrenal Gland. p 295.

Grynszpan-Wynograd, O., Nicolas, G. (1980). Intercellular junctions in the adrenal medulla: a comparative freeze-fracture study. Tissue and Cell 12:661.

Hagiwara, S., Byerly, L. (1981). Calcium channel. Ann Rev Neurosci 4:69.

Hamill, O.P., Marty, A., Neher, E., Sakmann, B., Sigworth, F.J. (1981). Improved patch-clamp techniques for high-resolution current recording from cells and cell-free membrane patches. Pflügers Arch 391:85.

Ishikawa, K., Harada, E., Kanno, T. (1977). A quantitative analysis of the influence of external calcium and magnesium concentrations on acetylcholine-induced adrenaline release in rat adrenal gland perfused in an antidromic or orthodromic direction. Jap J Physiol 27:251.

Katz, B., Miledi, R. (1972). The statistical nature of the acetylcholine potential and its molecular components. J Physiol 224:665.

Kidokoro, Y., Miyazaki, S., Ozawa, S. (1982). Acetylcholine-induced membrane depolarization and potential fluctuations in the rat adrenal chromaffin cell. J Physiol 324:203.

Kidokoro, Y., Ritchie, A.K. (1980). Chromaffin cell action potentials and their possible role in adrenaline secretion from rat adrenal medulla. J Physiol 307:199.

Knight, D.E., Baker, P.F. (1982). Calcium-dependence of catecholamine release from bovine adrenal medullary cells after exposure to intense electric fields. J Membrane Biol 68:107.

MacDermott, A.B., Connor, E.A., Dionne, V.E., Parsons, R.L. (1980). Voltage clamp study of fast excitatory synaptic currents in bullfrog sympathetic ganglion cells. J Gen Physiol 75:39.

Marty, A. (1981). Ca-dependent K channels with large unitary conductance in chromaffin cell membranes. Nature 291:497.

Neher, E., Sakmann, B. (1976).   Noise analysis of drug
    induced voltge clamp currents in denervated frog muscle
    fibres.   J Physiol 258:705.
Takeuchi, N.   (1963).   Effects   of   calcium   on   the
    conductance change of the endplate membrane during the
    action of transmission.   J Physiol 167:141.

The Physiology of Excitable Cells, pages 139–147
© 1983 Alan R. Liss, Inc., 150 Fifth Avenue, New York, NY 10011

EFFECTS OF A GENERAL ANESTHETIC ON THE ACETYLCHOLINE RECEPTOR
CHANNEL PROPERTIES IN CULTURED XENOPUS MYOCYTES

James Lechleiter and Raphael Gruener

Department of Physiology, University of
Arizona, College of Medicine.
Tucson, AZ 85724.

When acetylcholine (ACh) binds to its nicotinic
receptor (AChR), the resultant permeability change is due
to the opening of ion channels associated with the
receptor. The period of time that such channels remain
open is referred to as the opentime. Modulation of
channel properties occurs during development (Michler and
Sakmann, 1980) and in response to pharmacologic agents
(Gage and Hamill, 1975), possibly as a result of changes in
the microenvironment of the receptor. Thus, changes in
membrane fluidity, for example, may be expected to affect
channel kinetics.

General anesthetics, regardless of their molecular
structure, increase membrane fluidity (Pang et al. 1979;
Lenaz et al. 1979; Shieh et al. 1975) and reduce nerve-
evoked postsynaptic depolarizations (Gage and Hamill,
1976). The latter effect results from an increase in the
decay rate of endplate currents, presumably due to the
shortening of AChR channel opentimes (Katz and Miledi,
1973). We investigated this possibility directly by
measuring single channel currents using the extracellular
patch clamp (Hamill et al. 1981). We propose that changes
in receptor channel properties may result, under a variety
of circumstances, from alterations in the membrane
environment in which the receptor is embedded. Thus, in
accord with Gage and Hamill (1975) we suggest that general
anesthetics affect receptor (synaptic) function by
fluidizing the lipid microenvironment near the receptor.
Such fluidization or disordering of membrane structure may
permit a faster relaxation (closure) of receptor channels
opened in response to the conformational change induced by

the interaction of the receptor with the transmitter ligand. Similarly, during embryonic development, alterations in membrane lipids (Boland and Martonosi, 1974; Kutchai et al. 1978; Nagatomo et al. 1980) may account for the transition from slow to fast channels associated with the process of innervation (Sakmann and Brenner, 1978; Fischbach and Schuetze, 1980; Kullberg et al. 1981). The opposite effect, of increasing the AChR channel opentime, as measured by the prolongation of the endplate current after denervation, was reported by Argentieri and McArdle (1981). These authors also reported a simultaneous change in membrane lipid composition consistent with a decrease in membrane fluidity.

MATERIALS AND METHODS

    Uninnervated muscle cells from Xenopus laevis embryos (stages 19-22) were grown in culture as previously described (Kidokoro et al. 1980). Experiments were carried out at room temperature. Cultures were continuously superfused with recording medium (in mM: 120 NaCl, 1.6 KCl, 1 $CaCl_2$, 8 Hepes, pH 7.4; for cell-free patches, 5 EGTA was added and $CaCl_2$ omitted) bubbled with air (control) or air with vaporized (Forreger vaporizer) halothane (2 and 3%). The flow rate was 1.1 ml/min and the chamber volume was 0.8 ml. Patch clamp electrodes were pulled from glass pipettes (Drummond, 50 µl) to resistances of 3-10 M Ohms. Electrodes were lightly heat-polished, coated with sylgard, then re-polished just prior to the experiment. On several occasions, the sylgard coating was omitted. This resulted in a lower signal/noise ratio but otherwise did not affect the results . Electrodes were filled with recording medium (diluted 10% with distilled water) and 0.2 µM ACh. All solutions were filtered (0.2µm, Millipore). Single channel currents were recorded and stored on FM magnetic tape (Racal Recorder). A 3 to 5 minute control record was taken before switching to halothane. We began to observe the effects of halothane within the first two minutes of exposure; therefore, we report here data obtained at least 3 minutes after initiation of halothane exposure to insure equilibration (by a 4-fold replacement of the chamber volume). Washout began 3-5 minutes after recording in halothane. Recovery was noted within 10-20 minutes. With this protocol we collected at least 200 events per recording period.

Our analysis of the effects of halothane on AChR channel properties is based primarily on cell-attached (n=2) and cell-free (n=2) patches. Each membrane patch was taken from a different cell and from a different one day old culture. All records were low-pass filtered (corner frequency=2.5 kHz ; Frequency Devices, model 901F) and digitized at 100 μsec intervals using a microcomputer (Dynabyte). The amplitudes, durations and inter-event intervals were computed from these digitized records. Channel events and closures composed of only one digitized point (=100 μsec), were ignored because of the limit of resolution of our patch-clamp circuit (time constant=120 μsec).

RESULTS

Figure 1 (top) shows a representative control record from a cell-attached patch [holding potential was -60 mV from resting potential (RP)]. Single channel events fell into 2 distinct "types" on the basis of current amplitudes in agreement with previous reports (Clark and Adams, 1981; Kidokoro et al. 1982). A histogram (figure 1; bottom) of the channel event amplitudes, from this patch, showed the characteristic bimodal distribution with means ($\pm$ SEM) of 5.7$\pm$0.58 pA and 8.33$\pm$0.51 pA (a separation amplitude of 7 pA was chosen by eye). The corresponding mean opentimes were 2.19$\pm$0.34 msec and 1.46$\pm$0.14 msec. When this patch was exposed to 2% halothane (figure 2), the mean channel opentimes were significantly reduced to 1.72$\pm$0.42 msec and 1.10$\pm$0.10 msec, respectively.

Under the same experimental conditions, cell-free patches (inside-out; holding potential -120 mV) also showed two "types" of channel event amplitudes in control records. Exposure of these patches to halothane also resulted in a significant shift to shorter opentimes for all channel events; however, it was no longer possible to clearly distinguish two populations. After washout, channel opentimes returned to control values including the bimodal distribution. When the same current amplitude chosen to separate large from small control channels was used to separate channel types during halothane exposure, the opentime of both channel populations was reduced. These data are shown in Table 1. Reductions in channel opentimes are expressed as a percentage of the

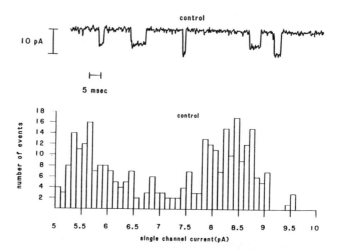

Figure 1. A control AChR single channel current record (top panel) from a cell-attached patch of a Xenopus myocyte in culture. The holding potential was -60 mV from resting potential (RP). Two "types", large and small, of channels can be distinguished. Amplitude distribution of channel events from this patch (bin width is 0.1 pA; the number of events was 345). The means of each "type" of channel correspond to 5.7±0.58 pA and 8.33±0.51 pA, based on a separating current of of 7 pA.

corresponding control mean opentime.

In addition to the reduction in channel opentime, halothane appears to affect the relative percentage of large-to-small current channels. Exposure of a cell-attached patch to halothane produced a shift from the control distribution (see fig. 1) such that the number of small current channels was markedly decreased, while the number of large current channels was increased (figure 3). The percentage of large to small current channels increased from 56% (control) to 80% (in 2% halothane).

DISCUSSION

In an attempt to test the hypothesis that the physical properties of the lipid environment influence the behavior

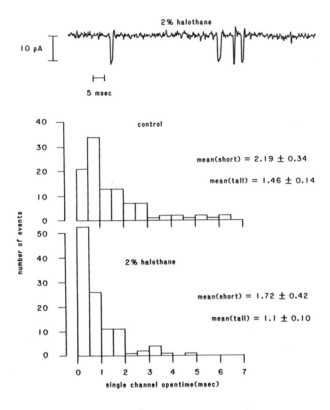

Figure 2. A current record (top) from a cell-attached patch (same as in fig. 1) after exposure to 2% halothane. Note that all channel events now appear to be similar (compare with control in fig. 1). Opentime histograms of control and halothane-treated patches. The mean channel opentime of the large (tall) current channels was reduced from 1.46+0.14 msec to 1.1+0.1 msec. Small (short) current opentimes were reduced from 2.19+0.34 msec to 1.72+0.42 msec.

of the AChR, we examined the effects of a general anesthetic, halothane, on the receptor channel opentime. It is known that the anesthetic potency of an agent is best predicted by its lipid solubility in comparison to other general anesthetics and therefore, is essentially

Table 1. Effects of Halothane on Mean Opentimes for Cell-
attached and Cell-free Patches.

type of patch	channel type	2% halothane*	3% halothane*
cell-attached	small	75.7% (2,61)	——
	large	82.5% (2,119)	67.0% (1,94)
cell-free	small	59.8% (1,128)	65.0% (1,117)
	large	56.9% (1,122)	38.5% (1,83)

* values are expressed as % of control mean opentimes (# of patches, # of events)

independent of its detailed size, shape and chemical nature (Franks and Lieb, 1982). This suggests that the mechanism of anesthetic action is not directly related to a specific ligand-protein interaction but rather to a more general effect on membrane structure. It is therefore highly probable that the anesthetic effect on receptor proteins is indirect in nature. It has been suggested that general anesthetics act by disordering or fluidizing (via volume

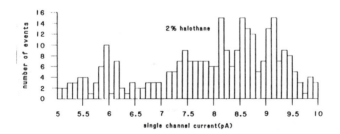

Figure 3. A histogram of current amplitudes (same as in fig. 1) after exposure to 2% halothane. Using a separating amplitude of 7 pA, the population means correspond to 5.89±0.45 pA and 8.55±0.72 pA. The large current channels now represent 80% of the total, in comparison to 56% under control conditions.

perturbations) the membrane lipids in which the receptor protein is embedded (Metcalfe et al. 1968; Gage and Hamill, 1975; Lenaz et al. 1979). In agreement with this hypothesis, Gage and Hamill (1976) showed that many general anesthetics [which increase membrane fluidity (Pang et al. 1979; Lenaz et al. 1979; Shieh et al. 1975)] reduce the amplitudes of nerve-evoked postsynaptic depolarizations. They argued that an increase in membrane fluidity allows an open AChR channel to close (relax) more rapidly. Our results show that single AChR channels close more rapidly in the presence of halothane. These results are therefore in agreement with Gage and Hamill (1975) who obtained similar findings using endplate current measurements. Furthermore, by using the patch clamp technique, we were able to determine that the opentime of both channel types (large and small current) was reduced and that the relative percentage of large-to-small current channels was increased by halothane. The latter result is also consistent with the idea that an increase in membrane fluidity during development (Kutchai et al. 1976) may account for the transition from slow to fast channels, a process associated with synaptogenesis (Sakmann and Brenner, 1978; Fischbach and Schuetze, 1980; Kullberg et al. 1981).

Finally, a reduction in channel amplitudes (due to sealing of patches and/or the limitation on the response time of our patch clamp) may have affected the bimodal distribution of channel events. However, the maintenance of this distribution in halothane-exposed intact patches and during recovery suggests that the channel "types" may be differentially affected by halothane. That is, halothane causes a reduction in the number of small current channels while increasing the number of large current channels.

We wish to acknowledge the generosity of Paul Brehm in helping us set up the patch clamp; Vince Dionne for his suggestions for our computer analysis; Steve Moffett for writing the programs and Yoshi Kidokoro, Joe Henry Steinbach and Chuck Stevens for their helpful discussions and technical suggestions. Thymol-free halothane was a gift from Louis Ferstandig, Halocarbon Labs, Inc.

JL was supported by an NIH training grant (842122) to the Department of Physiology.

REFERENCES

Argentieri TM, McArdle JJ, (1981). Endplate currents are prolonged at reinnervating neuromuscular junctions.
Boland R, Martonosi A, (1974). Developmental changes in the composition and function of sarcoplasmic reticulum. J Biol Chem 249(2):612.
Clark RB, Adams PR, (1981). ACh receptor channel populations in cultured Xenopus myocyte membranes are non-homogenous. Soc Neurosci Abstr 7:838.
Fischbach GD, Schuetze SM, (1980). A post-natal decrease in ACh channel opentime at rat end-plates. J Physiol 303:125.
Franks NP, Lieb WR, (1982). Molecular mechanisms of general anesthesia. Nature 300:487.
Gage PW, Hamill OP, (1975). General anesthetics: synaptic depression consistent with increased membrane fluidity. Neurosci. Let. 1:61.
Gage PW, Hamill OP, (1976). Effects of several inhalation anaesthetics on the kinetics of postsynaptic conductance changes in mouse diaphragm. Br J Pharmac 57:263.
Gage PW, Hamill OP, (1981). Effects of anesthetics on ion channels in synapses. In Porter R (ed): "International Reviews in Physiology," Baltimore: p.1.
Hamill OP, Marty A, Neher E, Sakmann B, Sigworth FJ, (1981). Improved patch-clamp techniques for high resolution current recording from cells and cell-free membrane patches. Pflügers Arch 391:85.
Katz B, Miledi R, (1973). The binding of acetylcholine to receptors and its removal from the synaptic cleft. J Physiol 231:549.
Kidokoro Y, Anderson MJ, Gruener R, (1980). Changes in synaptic potential properties during acetylcholine receptor accummulation and neurospecific interactions in Xenopus nerve-muscle cell culture. Dev Biol 78:464.
Kidokoro Y, Brehm P, Gruener R, (1982). Developmental changes in acetylcholine receptor distribution and channel properties in Xenopus nerve-muscle cultures. In press. Cold Spring Harbor Symposium.
Kullberg RW, Brehm P, Steinbach JH, (1981). Nonjunctional acetylcholine receptor channel opentime decreases during development of Xenopus muscle. Nature 289:411.
Kutchai H, Barenhotz Y, Ross TF, Wermer DE, (1976). Developmental changes in plasma membrane fluidity in chick embryo heart. Biochim Biophys Acta 436:101.
Kutchai H, Ross TF, Dunning DM, Martin M, King SL, (1978).

Developmental changes in the fatty acid composition and cholesterol content of chicken heart plasma membrane. J Comp Physiol 125:151.

Lenaz G, Curatola G, Mazzanti L, Bertoli E, Pastuszko A (1979). Spin label studies on the effects of anesthetics in synaptic membranes. J Neurochem 32:1689.

Metcalfe JC, Seeman P, Burgen ASV, (1968). Receptor stability and channel conversion in the subsynaptic membrane of the developing mammalian neuromuscular junction. Mol Pharmacol 4:87.

Michler A, Sakmann B (1980). Receptor stability and channel conversion in the subsynaptic membrane of the developing mammalian neuromuscular junction. Dev Biol 80:1.

Nagatomo T, Hattori K, Ideda M, Shimada K, (1980). Lipid composition of sarcolemmal, mitochondria and sarcoplasmic reticulum from newborn and adult rabbit cardiac muscle. Biochem Med 23:108.

Pang KY, Chang TL, Miller KW (1979). On the coupling between anesthetic induced membrane fluidization and cation permeability in lipid vesicles. Mol Pharmacol 15:729.

Sakmann B, Brenner HR, (1978). Change in synaptic channel gating during neuromuscular development. Nature 276:401.

Shieh DP, Ueda I, Eyring H (1975). NMR study of the interaction of general anesthetics with beta, gamma-depalmitoyl-L-alpha-lecithin bilayers. In Fink BR (ed): "Molecular Mechanisms of Anesthesiology," Raven Press, p. 307.

The Physiology of Excitable Cells, pages 149–163
© 1983 Alan R. Liss, Inc., 150 Fifth Avenue, New York, NY 10011

CHEMICAL MODIFICATION OF THE Na^+ CHANNEL: SPECIFIC AMINO
ACID RESIDUES

Douglas C. Eaton and Malcolm S. Brodwick

Department of Physiology and Biophysics
University of Texas Medical Branch
Galveston, TX   77550

Since the initial work which quantitatively described
the characteristics of Na^+ conductance in axonal membrane
[16], there have been increasingly more precise mathematical
and physical descriptions of the Na^+ conductance process
couched in terms of that original model proposed in 1952.
Unfortunately, even though it has always been clear that Na^+
channel activation and inactivation are fundamentally a
chemical process, the original model and subsequent elabor-
ations of the Hodgkin-Huxley model shed little light on the
actual chemical structures of the Na^+ channel. Recently,
chemical kinetic models of the Na^+ channel [2] have rein-
forced the concept of the Na^+ conductance as a chemical
event; however, these models also do not provide much
specific information about the chemical components of the
Na^+ channel.

Another related approach utilizes many of the same
methods developed by protein chemists to probe the chemical
structure of the active sites of enzymes: that is, obser-
ving the biological activity of the protein of interest,
before and after attacking the protein with reagents which
react with the side chains of specific amino acids [13, 21].
In the case of the Na^+ channel, the assay for activity is a
simple examination of the electrical characteristics of the
Na^+ current before and after exposure to the group-specific
reagents.

Several investigators have used chemical modification,
as described above, to gain specific chemical information
about the Na^+ channel (for a review of this work, see [6]

and [8]). In particular, the proteinaceous component of the Na[+] channel which is responsible for Na[+] channel inactivation has been particularly susceptible to attack from the internal surface of the axonal membrane, initially by proteolytic enzymes [3, 23], and also by group-specific reagents which attack tyrosine and arginine residues [5, 11, 22]. These results suggested a specific model of the Na[+] inactivation process in which a positively-charged arginine residue could act as a coulombic blocker at the inner mouth of the Na[+] channel. The positive charge of the arginine residue was stabilized by a hydrophobic or hydrogen-bond interaction of a spatially nearby tyrosine. This notion was lent additional support by the observation that small molecules that mimic the arginine residue, also block the Na[+] channel in a manner reminiscent of inactivation [11, 18, 20, 21].

The activity of all of the tyrosine-and arginine-reactive agents, as well as the proteolytic enzymes, is strictly limited to the inner surface of the axonal membrane. These results were originally thought to imply that the inactivation mechanism was located only on the inner surface of the axonal membrane, and that it, in fact, might not even lie within the membrane potential field [2, 4]. The idea was suggested that the potential dependence of the inactivation process arose because of a coupling to the activation process [4]. The strict interpretation of this coupling has subsequently been questioned [12]; however, in this paper we report on experiments with several negatively charged amino-reactive reagents whose actions imply an indirect coupling between Na[+] activation and inactivation, as well as a component of the inactivation molecule which is accessible from the exterior surface of the axonal membrane.

METHODS

All of the methods used in these experiments have been previously described [10, 11]. Since Na[+] currents were the object of study, K[+] currents were always eliminated by perfusing with K[+]-free perfusate, in which the K[+] was replaced with Cs[+].

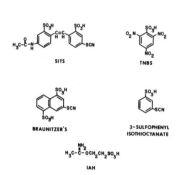

Fig. 1. Structure of re-agents. The anionic amino group reagents used in this study are all sulfonic acid derivatives. All were obtained from Pierce Chemical Company.

RESULTS

Effect of Internal Amino Group Reagents

There are a large variety of reagents that react rather specifically with the amino groups of proteins [13, 21]. The free amino groups are either the ε-amino group of lysine residues, or the amino group at the N-terminal end of the protein chain. Most of these reagents do not affect Na$^+$ channel function (see Table 1 and [11]). However, one class of these reagents is negatively charged (primarily sulfonic acid derivatives), and produces substantial effects on the Na$^+$ channel. Several agents of this class have become well known for their interaction with the anion exchange mechanism of red blood cells (for a review, see [7]). The most well known representative of this class is SITS (4-aceta-mido,4'-isothiocyano-stilbene-2,2'-disulfonic acid). The effect on the Na$^+$ currents of 0.1 mM SITS perfused through squid axon, is shown in Figure 2. The action of SITS appears somewhat complicated. At first glance, SITS clearly removes a component of inactivation, since the inward current no longer declines to near the baseline level. However, SITS could also be slowing the activation, and/or reducing the peak inward current. To investigate the effect of SITS on the activation process alone, it was necessary to study Na$^+$ current activation in isolation. We did this by first essentially removing all Na$^+$ channel inactivation with protease [3, 23], and then applying internal SITS. In Figure 3, we can clearly see that low concentrations of internal SITS are capable of substantially slowing the

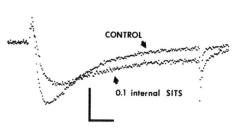

Fig. 2. The effect of 0.1 mM SITS applied to the inner surface of the squid axon membrane. The current responses to two voltage steps to zero mV, from a holding potential of -70 mV. The control record has an external solution with 1/8 normal Na^+ (replaced with TRIS), and an internal solution with 10 mM Na^+, 50 mM CsF, and 200 mM Cs glutamate. The SITS record is obtained under similar conditions except that 0.1 mM SITS has been added to the internal perfusate. Gating currents are noticeable at the beginning of the record. The calibration bars are 0.1 mA/cm^2 vertically, and 0.5 msec horizontally.

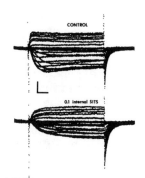

Fig. 3. The effect of internal SITS on Na^+ channel activation. The control records are a family of current responses to voltages from -70 to +100 in steps of 10 mV. The external solution is artificial sea water (ASW), and the internal solution is 50 mM NaF, 250 mM Cs glutamate. Inactivation has been removed by brief exposure of the axon to an internal solution containing 1 mg/ml alkaline protease. The lower record is the same as the control, except 0.1 mM SITS has been added to the internal solution. Calibration bars are 0.5 mA/cm^2 vertically, and 0.5 msec horizontally.

activation process, even though the reagent only slightly reduces the peak inward current. To obtain a more quantitative estimate of the effect of SITS on the Na^+ channel activation process, we adopted the formalism of Hodgkin and Huxley [16], and represent the activation as a third order reaction process, where the fractional occupation of any of the three states is given by m, with the time constant for transitions from the quiescent or closed state is τ_m. For steady-state activation (m_∞), application of SITS internally produces no change in magnitude or voltage dependence; however, the time required to produce the maximal

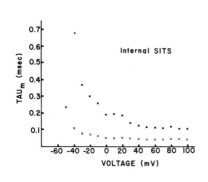

Fig. 4. The effect of internal SITS on the time constant of activation. Internal SITS (filled squares) slows opening of the Na⁺ channel compared to control (open squares) at all potentials tested, with the most profound effect at negative potentials. The effect of SITS might be interpreted as both a shift to more positive potentials plus a decrease of τ_m; however, the absence of any shift of the steady-state activation curve argues against this interpretation.

conductance is significantly longer, as demonstrated by the substantial increase in τ_m shown in Figure 4. Although SITS is the most potent of the anionic amino group reagents, several other anionic reagents produce similar effects on Na⁺ channel activation. Besides the effects of SITS (and other anionic amino reagents) on the Na⁺ channel activation process, Figure 2 clearly demonstrated an effect of internal SITS on the Na⁺ current inactivation process. Interestingly, the time constant for inactivation is virtually unaltered by treatment with internal SITS; however, the steady-state level of inactivation is dramatically altered. Again using the Hodgkin-Huxley formalism [16], we can represent the level of inactivation by the parameter h, with the steady-state level being h_∞. In Figure 5, the voltage-dependence of inactivation before and after a 5 minute SITS treatment demonstrates the removal of inactivation by SITS. Longer exposure to SITS removed proportionally more inactivation. The removal of inactivation by TNBS (trinitro-benzene-sulfonic acid) is as effective as SITS, while the other anionic amino reagents are less potent, but still effective when used at concentrations 2 to 5 times greater than SITS.

Effect of External Application of Reagents

In some respects, the external application of SITS and other anionic amino group reagents produces effects which

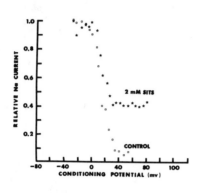

Fig. 5. The effect of SITS on steady-state levels of inactivation. Exposure of an axon with initially normal steadystate inactivation (open squares) to 2 mM SITS internally will remove additional inactivation until the current responses of the axons are similar to pronase-treated fibers (see Figure 3). Additional exposure after all the inactivation is removed, only reduces peak Na^+ conductance.

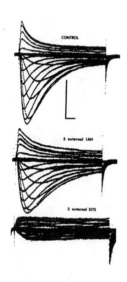

Fig. 6. Effect of external SITS and IAH. The current records at the top of the figure are the current responses of an axon to voltage steps from -70 to +100, in steps of 10 mV. The records are leakage corrected. The external solution is artificial sea water (buffered with HEPES), and the internal solution is 50 mM NaF plus 250 mM Cs glutamate. The middle and lower records are the same, except that 5 mM IAH and 3 mM SITS have been added to the external solution to produce the middle and lower records respectively. The calibration bars are 1 mA/cm^2 vertically, and 0.5 msec horizontally.

are similar to those produced by internal application. But there are also several additional features which serve to distinguish internal from external application. In Figure 6, the effect the external application of IAH produces a small but significant slowing of the inactivation. Subsequent application of SITS reduces peak inward current, slows activation, and slows inactivation to an extent where there is almost no reduction in Na^+ current within the time period

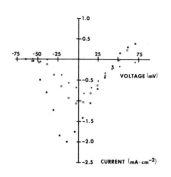

Fig. 7. The current-voltage relationship after external treatment with SITS. The filled squares represent peak inward current in a control axon similar to that of Figure 8. The unfilled squares and filled stars are the peak currents after the same axon has been treated with 1 and 3 mM external SITS.

of the stimulus pulse. In addition to the obvious effects of SITS on the current-time records, both TNBS and SITS produce a shift of the current-voltage relationship towards more positive potentials (Figure 7). This shift is at least partially responsible for the reduction in peak inward current.

The effect of SITS is more than merely a simple shift of the activation parameters. When we examined the voltage-dependence of activation (Figure 8), the activation process is shifted, but the shape of the relationship has changed. This implied to us a modification of the charge structures which were responsible for the voltage-sensitivity of Na$^+$ channel activation. The time constant for activation reflects both the shift in activation as well as the alteration in shape of the activation-voltage curve (Figure 9).

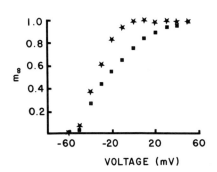

Fig. 8. The effect of external SITS on the voltage-dependence of Na$^+$ channel activation. Filled stars represent activation at various voltages in a control axon. Filled squares are activation in the same axon after treatment with 3 mM SITS externally.

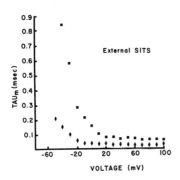

Fig. 9. External SITS effects the time constant for activation. In a manner similar to its effect internally, SITS slows Na^+ channel opening. Filled diamonds are control, and filled squares are after exposure to 3 mM SITS.

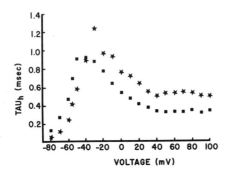

Fig. 10. External SITS slows Na^+ channel inactivation. Filled squares are control values, and filled stars are after exposure to 3 mM SITS.

The effectiveness of SITS in slowing the Na^+ activation is much greater for external application than for internal application (compare Figure 8). Also, unlike the case for internal application, there is an effect on the time constant of the inactivation process (Figure 10) which consists of a slight shift in a depolarizing direction, coupled with a slowing at high stimulus potentials.

The most interesting effect of SITS applied externally is an unusual effect on the steady-state levels of inactivation (h_∞). Figure 11 shows that the steady-state inactivation is substantially removed at high potentials, but there is a minimum in the h_∞ curve at intermediate potentials. To a certain extent, the presence of the minimum in the h_∞ vs voltage-curve appears to depend upon the value of the

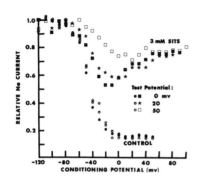

Fig. 11. External SITS removes Na$^+$ channel inactivation in a voltage-dependent manner. Fifteen minute exposure to 3 mM SITS in the external solution removed a substantial fraction of the Na$^+$ channel inactivation. However, the removal was almost complete at potentials above +40 mV, while only about 50% of the inactivation was removed at -10 mV. The extent of the voltage-dependence may have depended upon the test potential.

conditioning potential.

The effect of SITS on the time constant of activation suggested that the reagent might be affecting the gating of the Na$^+$ channel. To test this hypothesis, we examined the gating charge movement according to the protocol of Armstrong and Bezanilla [1]. As expected, the magnitude of the gating charge movement, and the total amount of charge moved, are both reduced. In addition, the time constant for charge movement is also slowed.

DISCUSSION

The results of our experiments are summarized in Table 1 for internal application of amino group reagents, and in Table 2 for the external application.

As specifically demonstrated in Table 1 for internal application of reagents, the ability to react with amino groups is not sufficient to produce the effects observed in this paper: the reagents also require a substantial anionic character to be effective. For both internal and external application, SITS, a doubly charged anion, is the most potent, while uncharged or partially-charged amino reagents have little or no effect. In red blood cells, the ability of amino reagents to block the anion exchanger is

Table 1. Internally Applied Reagents

Reagent	m_∞	τ_m	h_∞	τ_h	g_{Na}	Voltage Shift
SITS	no effect	slows	↑	slows	↓	slight
BRAUNITZER'S	no effect	slows slightly	slight increase	slows slightly	↓	none
SULFOPHENYL ISOTHIOCYANATE	no effect	slows slightly	slight increase	slows slightly	↓	none
TNBS	no effect	slows slightly	↑	slows	↓	10 - 20 mV
IAH	no effect	no effect	↑	no effect	little effect	none
5-MNT						
MALEIC ANHYDRIDE						
DIKETENE	NO EFFECTS					
ETHYL TRIFLUOROACETATE						
SUCCINIC ANHYDRIDE						

Table 2. Externally Applied Reagents

Reagent	m_∞	τ_m	h_∞	τ_h	g_{Na}	Voltage Shifts
SITS	alters shape	slows	alters shape	slows	↓	20 - 30 mV
TNBS	slight	slows	alters shape	slows	↓	slight
IAH	no effect	no effect	↑	slows slightly	no effect	none

similar to our results: only anionic agents have substantial access to the amino group of the anion binding site [7]. For the case of the RBC anion exchanger, the binding site is an area with several positive groups, but also an area of membrane which is substantially lipophilic in character. It seems likely that the interaction site in squid axon, both internally and externally, is of a similar nature.

The proximity of a lipophilic area to the inner mouth of the Na $^+$ channel is also supported by experiments in which alkyl-guanidinium compounds were applied to the internal surface of axons which had previously had their normal inactivation removed with protease [18, 20, 21]. The guanidinium derivatives were able to mimic normal inactivation if the alkyl chains were long enough to allow substantial hydrophobic interaction. This implied a lipophilic site within 5 to 10 nm of the guanidinium blocking site. Also suggestive of such as lipophilic region is the ability of lipophilic local anaesthetics, such a benzocaine, to block the Na $^+$ channel from the inside of the axon [15].

Indeed, the fact that the time constant of activation is slowed in a similar manner both internally and externally, suggests a generalized effect of SITS on the membrane aside from its specific sided actions on inactivation. In many respects, the effect of SITS on activation suggests the action of various crosslinking agents [17].

The sided effects (summarized in Tables 1 and 2) suggest a relatively simple site for interaction of SITS internally. In many respects, the action of SITS (and the other anionic amino reagents) in blocking inactivation, parallels that of arginine-and tyrosine-reactive reagents [5, 6, 11, 22]. These agents simply remove inactivation with little or no effect on the time course of inactivation. The voltage-dependence of any inactivation remaining after treatment is unaffected, suggesting that there are only two types of channels: those with inactivation completely removed, and those which have, as yet, not reacted with the reagent.

The implication of these results is that one (or more) lysine amino groups is in a portion of the Na $^+$ channel responsible for inactivation. When these amino groups are modified, the inactivation mechanism is blocked either due to steric hindrance of the bulky added groups, or due to unfavourable coulombic interaction (when, for the case of SITS, the positive amino group is replaced with a doubly-charged anion). A coulombic effect cannot be completely ruled out, but seems less likely than steric effects, since TNBS also is very effective in removing inactivation when applied internally, even though the adduct of TNBS with amino groups is uncharged.

The external site is more interesting. Merely the fact that a small, hydrophilic reagent can affect inactivation from the exterior surface of the axon, suggests that the inactivation component of the Na^+ channel is a more compli-cated entity than that which was suggested by previous experiments with reagents and enzymes [3, 5, 11, 17, 22, 23].

Also, the alteration by SITS of the time constant for inactivation, and the shape of the voltage-dependence for inactivation, suggest that the reaction site for SITS is within, and is sensitive to, the membrane potential field. That is, SITS reacts with a positively-charged group which would normally move toward the exterior surface of the membrane in response to depolarization of the axon. After reaction with SITS however, the anionic adduct will tend to move in the opposite direction in response to the membrane field, i.e., inward. This idea is supported by examining the relative effectiveness of the different reagents in altering the voltage-dependence of inactivation. SITS, which replaces a positive amino group with a doubly-charged anion, is most effective, followed by 3-sulfophenylisothio-cyanate (one cationic charge to one anionic), TNBS (one cation to no charge), and IAH (cationic charge unaltered), which is without effect. It seemed possible to us that SITS might be interacting with a site close to, or identical to, the site with which the toxin from the scorpion, Leiurus, interacts. The action of Leiurus toxin, in many respects, is very similar to the external action of SITS. The inter-action site for scorpion toxin appears to be within the membrane field, and its primary effect is to slow and remove inactivation [8, 9]. We are interested in examining the possibility that SITS can interfere with the specific binding of radio-labeled scorpion toxin to the Na^+ channel.

Finally, in summary, Figure 12 shows one very schema-tized drawing of the inactivation mechanism that is both consistent with previous results obtained by chemical modification experiments reviewed in [6] and [8], as well as the results presented here.

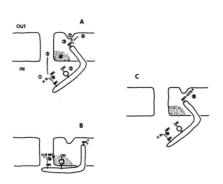

Fig. 12. A schematic representation of the Na$^+$ channel inactivation mechanism and its interaction with external SITS. Previous work has suggested that some portion of the Na$^+$ channel inactivation mechanism is freely exposed on the inner surface of the axon. Associated with this exposed area, there are arginine and tyrosine residues (1 and 2 in A) necessary for inactivation. The results presented here also suggest the presence of an amino group on the same mechanism which is near the external surface of the axon (3). After the channel is open (as in A), inactivation might occur by the guanidino group of arginine (1), moving as indicated into the inner mouth of the channel (5). This movement would be favoured by the interaction of the tyrosine residue (2) with a hydrophobic region of the inner membrane surface (6), and also by the membrane depolarization favouring the movement of the amino group (3) outward toward the external surface (4). The amino group (3) might also be affected by interaction of scorpion toxin at its surface receptor (7). The final state after inactivation has occured, is depicted in B. When SITS is present externally, it can react with the available amine and produce a double-charged anion in its place (8 in C). Thus, the tendency for movement in the membrane is no longer outward, but is now inward, which tends to inhibit inactivation.

1. Armstrong CM, Bezanilla F (1975). Currents associated with the ionic gating structures in nerve membrane. Ann N Y Acad Sci 264:265-277.
2. Armstrong CM, Bezanilla F (1977). Inactivation of the sodium channel. II. Gating current experiments. J Gen Physiol 70:567-579.

3. Armstrong CM, Bezanilla F, Rojas E (1973). Destruction of sodium conductance inactivation in squid axons perfused with pronase. J Gen Physiol 62:375-391.

4. Bezanilla F, Armstrong CM (1977). Inactivation of the sodium channel. I. Sodium current experiments. J Gen Physiol 70:549-566.

5. Brodwick MS, Eaton DC (1978). Sodium channel inactivation in squid axon is removed by high internal pH or tyrosine specific reagents. Science 200:1494-1496.

6. Brodwick MS, Eaton DC (1982). Chemical modification of excitable membranes. In Haber B, Perez-Polo JR, Coulter JD (eds): "Proteins in the Nervous System," New York: Alan R. Liss, pp. 51-72.

7. Cabantchik AI, Knauf PA, Rothstein A (1978). The anion transport system of the red blood cell. The role of membrane protein evaluated by the use of 'probes'. Biochim Biophys Acta 515:239-302.

8. Cahalan MD (1980). Molecular properties of sodium channels in excitable membranes. In Cotman CW, Poste G, Nicolson GL (eds): "The Cell Surface and Neuronal Function," New York: Elsevier North Holland Biomedical Press, pp 1-47.

9. Catterall WA (1977). Activation of the action potential Na^+ ionophore by neurotoxins. J Biol Chem 252:8669-8676.

10. Eaton DC, Brodwick MS (1980). Effects of barium on the potassium conductance of squid axon. J Gen Physiol 75:727-750.

11. Eaton DC, Brodwick MS, Oxford G, Rudy B (1978). Arginine specific reagents remove sodium channel inactivation. Nature 271:473-476.

12. Gillespie JI, Meves H (1980). The time course of sodium inactivation in squid giant axons. J Physiol 299:289-307.

13. Glazer AN, Delange RJ, Sigman DS (1975). "Chemical Modification of Proteins," New York: American Elsevier, Inc.

14. Hille B (1972). The permeability of the sodium channel to metal cations in myelinated nerve. J Gen Physiol 59:637-658.

15. Hille B (1977). Local anaesthetics: Hydrophilic and hydrophobic pathways for the drug receptor reaction. J Gen Physiol 69:497-515.

16. Hodgkin AL, Huxley AF (1952). A quantitative description of membrane current and its application to conduction and excitation in nerve. J Physiol 177:500-544.
17. Horn R, Brodwick MS, Eaton DC (1980). Effects of protein crosslinking reagents on membrane currents of squid axon. Am J Physiol 238:C127-C132.
18. Kirsch GE, Yeh JZ, Farley JM, Narahashi T (1980). Interaction of n-alkyl-guanidines with the sodium channel of squid axon membrane. J Gen Physiol 76:315-335.
19. Lo M-VC, Shrager P (1981). Block and inactivation of sodium channels in nerve by amino acid derivatives. I. Dependence on voltage and sodium concentration. Biophys J 35:31-43.
20. Lo M-VC, Shrager P (1981). Block and inactivation of sodium channels in nerve by amino acid derivatives. II. Dependence on temperature and drug concentration. Biophys J 35:45-57.
21. Means G, Feeney RE (1971). "Chemical Modification of Proteins," San Francisco: Holden Day.
22. Oxford G, WuC, Narahashi T (1978). Removal of sodium channel inactivation in squid giant axons by N-bromoacetamide. J Gen Physiol 71:227-248.
23. Rojas E, Rudy B (1976). Destruction of the sodium conductance inactivation by a specific protease in perfused nerve fibres from Loligo. J Physiol 262:501-531.

Top: M. Henkart; R. Eckert; D. Eaton; Mrs. Eaton
Bottom: G. Eisenman; N. Eisenman; N. Standen; J. Deitmer;
J. Patlak; R. Horn; G. Wooden

The Physiology of Excitable Cells, pages 165–180

CONDITIONAL PROBABILITY MEASUREMENTS ON TWO MODELS OF
NA CHANNEL KINETICS

Joseph B. Patlak
Department of Physiology and Biophysics
University of Vermont
Given Building
Burlington, VT 05405

INTRODUCTION

Many models have been proposed to explain the molecular
mechanisms that underlie the activation and inactivation of
Na currents observed in voltage clamped nerve and muscle
(see French and Horn, 1983: Armstrong, 1982: Bezanilla, 1982
for reviews). An important aspect of such models is the
extent to which the processes of activation and inactivation
interact with one another. Hypotheses range from those
which hold that the two reactions are always independent
(Hodgkin and Huxley, 1952) to those which hold that the Na
channel must open before it can inactivate (Goldman and
Hahin, 1978: Keynes and Rojas, 1974: Moore and Jacobsson,
1971). A number of authors have proposed models in which
the two reactions have some of their states in common, and
are therefore referred to as partially coupled (Armstrong
and Gilly, 1978: Bean, 1981: Nonner, 1980: Armstrong and
Bezanilla, 1977).

Although many lines of evidence support the
intermediate view of partial coupling rather than either of
the two extremes, full or no coupling, it has been difficult
to prove conclusively that any one theory is correct (French
and Horn, 1983). During the past several years, currents
through single Na channels have been measured (Sigworth and
Neher, 1980). Since single channel recordings provide
information about channel kinetics that cannot be obtained
by macroscopic recordings (Patlak, 1981), this method has
the potential to further elucidate Na channel kinetics.

Horn, Patlak, and Stevens (1980) have used single channel measurements to examine the question of coupling between activation and inactivation. Their results showed a high proportion of pulses which had no channel openings in response to a depolarizing pulse, even under conditions where it was unlikely that channels would be inactivated in the resting state or would fail to open their activation gates. This evidence alone pointed strongly to a pathway for inactivation which does not involve channel opening.

In order to further quantify this conclusion, Horn, et. al. presented a method of data analysis which could distinguish between different classes of inactivation models. This method used conditional probability. It can be used to distinguish between coupled or independent inactivation as follows:

A channel responds differently to each pulse that it experiences because it is controlled by stochastic processes. Over the course of many pulses, its activity will average out to a consistent pattern, the channel's probability of being open, P(t). (This probability corresponds directly to the macroscopic currents observed when many channels function simultaneously.) The time it takes for a channel to open for the first time after the start of a pulse will be different for each pulse, but there will always be a finite probability that a channel will first open only after some arbitrary interval, Δt has elapsed. For channels that <u>must</u> open before they inactivate, there can be no inactivation during the interval Δt. The conditional probability, $W(t,\Delta t)$, (defined as the probability that a channel is open after Δt, given that it did not open before this time) will be insensitive to the duration of Δt in this case. $W(t,\Delta t)$ will be very similar to P(t) but will be shifted by the interval Δt. On the other hand, if inactivation proceeds independently of activation, then some of the channels that do not open during Δt will inactivate and never open. In this case, $W(t,\Delta t)$ will never exceed P(t) at any time.

These two contrasting predictions allowed Horn et. al. to test theories of inactivation by measuring the conditional probability, $W(t,\Delta t)$, on single channel data. They found that the data fulfilled their predictions for the independent model, and not those for the coupled model. However, there were several limitations to their presentation of this new method. The amount of data available for analysis was very limited, their patches contained multiple channels, and no explicit theory for the conditional probability expected for various models was given. Therefore, the ability of the technique to resolve differences between various kinetic schemes remained unclear.

In order to test the predictions of this form of data analysis, and to determine the sensitivity of the results to different channel activation models, I have measured the conditional probability, $W(t,\Delta t)$, of channel data that were generated by a computer using several standard models that have been proposed to explain channel kinetics. The conditional probability for channels that operate according to Hodgkin and Huxley's independent inactivation model, Armstrong and Gilly's partially coupled, and a fully coupled model are presented in this paper, along with the methods that were used to generate the single channel data. Furthermore, I present an alternative formulation of the conditional probability and discuss its relevance to physiological measurements.

METHODS

Kinetic models from two references were examined in depth for this analysis: Hodgkin and Huxley (1952), and Armstrong and Gilly (1979). Both models were represented by a series of Markovian states that had the following properties (Colquhoun and Hawkes, 1981):

1) The probability, P_i, of leaving a state during any short time interval was constant with time.
2) The probability of a transition occurring was independent of the past history of the channel.
3) The mean lifetime of any state was equal to the inverse of the sum of the rates leaving the state.

The Hodgkin-Huxley model was formulated as a series of three closed states leading to an open state, and an independent set of states representing activation and inactivation, as illustrated in Figure 1A. This scheme is equivalent to that expected from three independent 'm' particles and an 'h' particle. The α's and β's were calculated according to the equations given in Hodgkin and

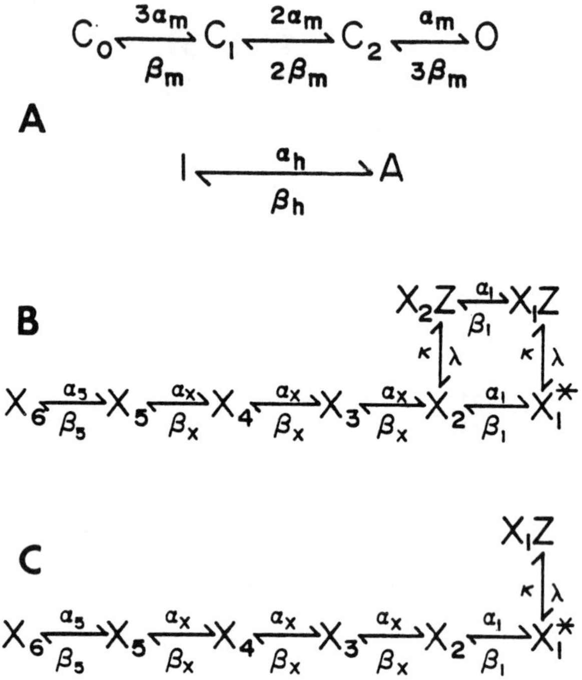

$$C_0 \underset{\beta_m}{\overset{3\alpha_m}{\rightleftharpoons}} C_1 \underset{2\beta_m}{\overset{2\alpha_m}{\rightleftharpoons}} C_2 \underset{3\beta_m}{\overset{\alpha_m}{\rightleftharpoons}} O$$

A

$$I \underset{\beta_h}{\overset{\alpha_h}{\rightleftharpoons}} A$$

B

$$X_2Z \underset{\beta_1}{\overset{\alpha_1}{\rightleftharpoons}} X_1Z$$

$$X_6 \underset{\beta_5}{\overset{\alpha_5}{\rightleftharpoons}} X_5 \underset{\beta_x}{\overset{\alpha_x}{\rightleftharpoons}} X_4 \underset{\beta_x}{\overset{\alpha_x}{\rightleftharpoons}} X_3 \underset{\beta_x}{\overset{\alpha_x}{\rightleftharpoons}} X_2 \underset{\beta_1}{\overset{\alpha_1}{\rightleftharpoons}} X_1^*$$

with vertical transitions $\kappa \downarrow \lambda$

C

$$X_1Z$$

$$X_6 \underset{\beta_5}{\overset{\alpha_5}{\rightleftharpoons}} X_5 \underset{\beta_x}{\overset{\alpha_x}{\rightleftharpoons}} X_4 \underset{\beta_x}{\overset{\alpha_x}{\rightleftharpoons}} X_3 \underset{\beta_x}{\overset{\alpha_x}{\rightleftharpoons}} X_2 \underset{\beta_1}{\overset{\alpha_1}{\rightleftharpoons}} X_1^*$$

with vertical transition $\kappa \downarrow \lambda$

Figure 1. State diagrams for the models analyzed. (A) The Hodgkin-Huxley model, shown as two independent reactions. At the start of each pulse the channel was in state C_0 and state A. A channel was open only if it was in states O and A simultaneously. (B) The Armstrong-Gilly model. The activation sequence is a series of closed states (X_6 through X_2) and the open state, X_1^*. The states X_2Z and X_1Z are the inactivated states. (C) Fully coupled activation and inactivation, generated by removal of the state X_2Z from the Armstrong-Gilly model.

Huxley, 1952. Conversion between the relative potentials used by these authors and the absolute membrane potentials used here was made by assuming a resting potential of −60 mV for Hodgkin and Huxley's work.

The Armstrong-Gilly model (1979) was formulated as shown in Fig. 1B. The rate constants for the three voltages that were used for this analysis were taken from Table II of their reference. The Armstrong-Gilly model could also be modified to give a fully coupled model by removing state X_2Z and its associated transitions. This is illustrated in Fig. 1C.

For computer simulation, individual rate constants were translated into the probabilities that a jump would occur during a discrete interval, i, equal to the time represented by each displayed point. These probabilities were calculated according to the equation:

$$P_i = (1 - \exp(-R*i))$$

where R is the appropriate rate constant. Many states had more than one pathway leaving them. In these cases, P_i was determined for all rates together by using the approximation:

$$P_{ia} + P_{ib} \cong P_{ia+b} = (1 - \exp(-(a+b)*i))$$

where a and b are individual rate constants leaving a single state. The error from such an approximation is less than 5% when the individual P_i values are less than 0.1. This condition was fulfilled for all but the fastest of rate constants in the above models, and was never significant for the processes examined here.

The probability, P_i, could be simulated for each point by determining if a given random number fell within a predetermined subset of all possible random numbers. In order to maintain true randomness for long series of numbers, and to improve speed, random numbers were sampled from a white noise source rather than being generated arithmetically. The peak-to-peak amplitude of the noise was adjusted so that its amplitude was approximately 1/4 of the dynamic range of the computer's analog-to-digital converter. The least significant eight bits from the 12 bit sample were used as a random number. Histograms of the random numbers thus generated were flat for all 256 possible values. Samples were taken at intervals with a minimum of about 50 μs. Since the maximum frequency of the noise signal was between 10 and 20 kHz, the correlation between sequential random numbers was negligible. Two random numbers were combined to give a final 16 bit number, giving a range of 65,536 possible numbers.

Transitions between states were defined as having occurred if the random number was within the interval whose width was defined by $P_i*65,535$. If transitions to more than one state were possible, several nonoverlapping intervals were used to determine which transitions, if any, occurred. Transitions were determined for sequential time points by a program which assumed that the channel always started in the left-most closed state. For each time point, the computer stored either a 0 or a 1, representing the closed and open states, respectively. Sets of sequential points (traces) were averaged with each other on a point by point basis. Programs to determine conditional probability measured the time of the first opening and averaged the trace if this time was greater than a preset threshold.

The programs were run on a microcomputer. Approximately 10 pulses could be generated and analyzed per second. All data shown here are from data sets which included 1000 pulses. This number was chosen because it represents a practical upper limit to the number of pulses that could be expected from an excised patch of cell membrane in a real experiment.

RESULTS

The modeling procedure that was used assumed that at time zero the left-most closed state was always occupied, and that the rate constants were those appropriate for the given voltage. Therefore, the single channel responses thus produced were equivalent to the currents from an idealized patch clamp recording in which the voltage was changed at time t=0 to the new value. The single channel signals thereby generated are shown in Figs. 3A and 4A. The open channel is represented by a downward deflection with instantaneous rise and fall.

When such traces are averaged together, the ensemble average is the probability that the channel will be open at any time after the start of the pulse, P(t). Figs. 3B and 4B show P(t) for ensembles of 1000 individual pulses to different voltages. The absolute membrane potential during the pulse is shown to the right of each trace.

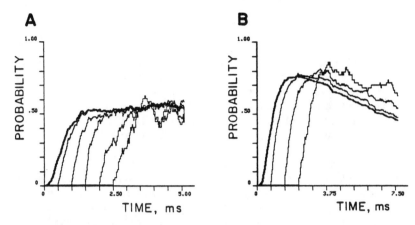

Figure 2. (A) Conditional probabilities for a channel with no inactivation. The dark trace is the probability of being open for this set of 1000 generations based on a Hodgkin-Huxley model at a membrane potential of −20 mV where β_h was set equal to zero. The number of entries into the five conditional probabilities shown were 798, 435, 219, 98, and 55. (B) Conditional probability for a completely coupled model of inactivation. The dark trace is the probability of being open for 1000 generations of a channel's activity based on an Armstrong-Gilly model which had been modified to prevent transitions to the inactivated state from all except the open state. The membrane potential was +10 mV. The number of entries into the conditional probabilities was 594, 199, and 56.

Channels With No Inactivation

Channel data were simulated for models which have no inactivation, as shown in Figure 2A. The parameters used here were for the Hodgkin-Huxley activation sequence without the separate inactivation reaction. The probability of being open, P(t), rises after a delay and reaches a steady level (illustrated by the dark line in Fig. 2A), as is observed for Na channels that have had their inactivation removed by Pronase or N-Bromoacetamide (Patlak and Horn, 1982). In this case, the conditional probability W(t,Δt), should superimpose for all values of the delay time (after W(t,Δt) has had a chance to rise to its steady state probability). This is illustrated in Fig. 2A which shows

$W(t,\Delta t)$ and $P(t)$ for pulses to -20 mV without inactivation. After about 2 ms the conditional probability is indistinguishable from the $P(t)$. Note also that the conditional probability rises to its steady value with a curve that is heavily dominated by a single exponential, because the channel has a high probability of being in the C_2 state after the Δt interval. The opening rate is therefore dominated by the relaxation to the steady state distribution between occupancy of state 0 and C_2.

The conditional probability curves become noisier as the interval, Δt, is increased because a smaller subset of all the pulses fulfills the conditions. Fewer individual responses are averaged together, giving an increase in the curve's variance. For very long Δt intervals, the number of instances where the condition of no opening is fulfilled approaches zero.

Channels With Coupled Inactivation

As stated earlier, the conditional probability, $W(t,\Delta t)$, for channels that must open before they inactivate should be a time-shifted replica of the overall probability. In this case, $W(t,\Delta t)$ reaches the same maximum probability as the $P(t)$ because the channel has never inactivated during Δt for those responses that satisfy the conditions of $W(t,\Delta t)$. $W(t,\Delta t)$ should cross $P(t)$ to reach approximately the same peak probability of opening, and it should subsequently decline with the same time constant as $P(t)$. These predictions were confirmed by my simulations, as shown in Figure 2B, which shows the result of channel simulations using the model shown in Figure 1C. $W(t,\Delta t)$ clearly crosses $P(t)$ for all values of Δt, and it reaches peak values similar to, or slightly greater than, the maximum of $P(t)$.

Independent Inactivation -- The Hodgkin-Huxley Model

Figure 3A shows the responses of a single Hodgkin-Huxley type channel to sequential pulses to -20 mV. The traces represent about 5 ms of activity following the start of a pulse. Note the high percentage of traces with no opening. These are instances when the channel inactivated before its first opening.

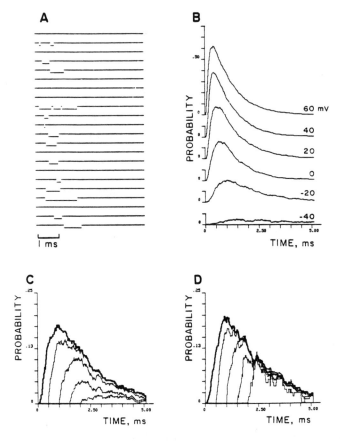

Figure 3. Conditional probability analysis of the Hodgkin and Huxley model. (A) A single channel's responses for 22 sequential computer simulations based on kinetic parameters as given by Hodgkin and Huxley (see text). The membrane potential was -20 mV. The channel's openings are indicated by downward deflections. (B) Probability that the channel is open as a function of time, P(t), for several voltages. The curves were calculated by averaging 1000 individual pulses such as in (A). (C) The conditional probabilities, W(t,Δt), are the subset of pulses where the channel did not open during Δt. These were calculated for four different Δt intervals from the set of 1000 generations whose overall probability is shown by the dark line. (D) The modified conditional probability, W*(t,Δt), where entries to the conditional probability were determined only by the movement of the 'm-gate'. The dark trace is the overall probability.

The curves in Figure 3B are the average of 1000 responses of this channel to pulses of the voltages given by each trace. These curves are the equivalent of the $g_{Na}(t)$ curves calculated by Hodgkin and Huxley. As expected, their time course and amplitude reproduce the original data, except that these curves have more noise due to the limited number of channel events which contributed to them.

Note that even for large depolarizations, the maximum P(t) reaches only about 0.6. This low probability is due primarily to instances where the channel inactivated before it opened, as would be expected from a gating theory in which the two processes were independent. Since the chance that the inactivated channel will return to the active state is quite low, a significant fraction of the single channel traces had no opening, as can be seen in Fig. 3A. Overall, the probability that the channel never opened during a pulse to -20 mV was almost 0.5.

The conditional probability curves for various times after the start of the pulse, W(t,Δt), are always less than or equal to P(t). This is illustrated in Fig. 3C, where P(t) is shown as a dark line for pulse height -20 mV. The four lighter curves are the conditional probabilities that no channel has opened before 0.5, 1, 1.5 and 2 ms for this set of 1000 pulses. The diminution of W(t,Δt) with increasing Δt is clearly apparent.

This diminution is due to the nature of the definition of conditional probability: All traces which have no event before a specified time satisfy the condition. This condition is always satisfied by traces in which no opening occurs. As such, the conditional probability will have a substantial minimum number of entries, even for very long delay intervals. This minimum number is exactly the number of pulses in which the channel fails to respond.

The traces where the channel never opens have a greater effect on the amplitude of W(t,Δt) than on P(t) because the number of entries in W(t,Δt) in which the channel actually opens declines rapidly to zero as the delay interval increases. These openings are averaged with a constant number of failure traces. W(t,Δt) therefore becomes disproportionatley smaller. In the example of Fig. 3C, the number of entries to the conditional probability curves was 823, 636, 552, and 516. The number of failures was 502.

Although the overall probabilities are also diminished by the failures, the extent is not as large, because pulses with early openings are not excluded, giving a much higher level of activity for these later times.

Horn, Patlak, and Stevens predicted that $W(t,\Delta t)$ would superimpose on the falling phase of $P(t)$ for the independent model. (This would correspond to all of the traces in Fig. 2A being multiplied by the single declining exponential that is the independent process of inactivation.) Although this simple interpretation is appealing, it is clearly not the case. It is instructive to examine why the conditional probability curves diminish for the independent model.

The difference between the results presented above and the earlier prediction of Horn, et. al. is due to the exact definition of $W(t,\Delta t)$: Events which have no channel openings during Δt satisfy the condition. As stated above, all instances in which the channel inactivates before it opens satisfy this condition. However, not all of these failures 'belong' in $W(t,\Delta t)$, because in many of the events the channel does indeed activate during Δt, but this activation is masked by a previous inactivation. Such traces are 'erroneously' included in $W(t,\Delta t)$. An alternative statement of the conditions for acceptance corrects this problem. $W^*(t,\Delta t)$ is defined as the conditional probability that the channel is open, given that its <u>activation gates</u> do not open during the interval Δt. This means that some, but not all of the failures will be included in $W^*(t,\Delta t)$. As the delay interval increases, the number of entries in $W^*(t,\Delta t)$ declines to zero, as expected.

Fig. 3D shows the result of such an analysis. It gives the result that was predicted for independent gating in Horn, Patlak, and Stevens, i.e. the $W^*(t,\Delta t)$ traces rise quickly and reach the same level as $P(t)$. The number of entries for the four conditional probability traces shown was 796, 485, 232, and 130.

Partial Coupling –– The Armstrong–Gilly Model

One additional question concerning the conditional probability method is its ability to resolve the differences between independent and partially coupled models of inactivation. One such partially coupled model is that of

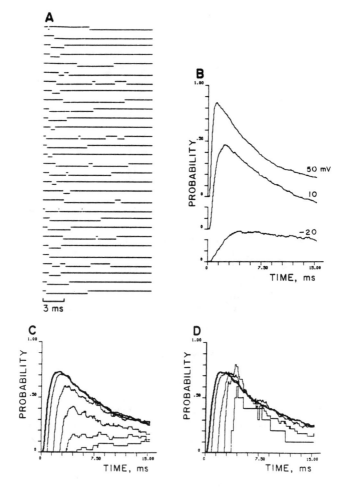

Figure 4. Conditional probability analysis of the Armstrong and Gilly model. (A) 30 sequential simulations of the single channel responses to pulses to +10 mV. Opening of the channel is indicated by a downward deflection. (B) The probability of being open curves determined from 1000 pulses each at three different potentials. (C) 5 conditional probability measurements on a set of 1000 pulses whose overall probability is shown as the dark curve. The potential was +10 mV. (D) Modified conditional probability measurements on a set of 1000 pulses. Entry into the conditional subset was determined only by movement of the channel's 'activation gate', as in Fig. 3D.

Armstrong and Gilly, which allows for inactivation only from the open state and the last closed state. Fig. 4 shows the results of a conditional probability analysis on data generated by this model.

Fig. 4A shows the single channel signals produced by pulses to +10 mV (absolute membrane potential), and Fig. 4B shows the probability of being open curves for the three different potentials for which rate constants were given in the original model. Although these curves are quantitatively different from those produced by the Hodgkin-Huxley model, they reproduce the data and theoretical curves from Armstrong and Gilly quite well. The curves differ largely because the experimental conditions in the two original papers were different.

Since this was a partially coupled model, it was still possible for the channel to inactivate without opening. This occurred at all three voltages, although much less frequently than for the Hodgkin-Huxley model. At +10 mV, approximately 5% of pulses had no opening events. These failures caused the conditional probability curves to fall below the $P(t)$ curves for reasons identical to those described above for the H-H model. This is illustrated in Fig. 4C, which shows 5 $W(t,\Delta t)$ traces for a data ensemble of 1000 pulses to 10 mV. The number of entries in the $W(t,\Delta t)$ curves was 532, 199, 91, 61, and 50. The number of failures was 44.

As with the Hodgkin-Huxley model, the diminution of the $W(t,\Delta t)$ curves with increasing delay times can be reversed with an alternate statement of the conditions, $W^*(t,\Delta t)$. Fig. 4D shows the result of an analysis where only channels that had first reached the right-most states (X_1 or X_1Z in the Armstrong-Gilly formulation) after the specified delay interval were included in the analysis. The number of entries in the four curves shown was 531, 153, 45, and 10. Neither the original conditional probability nor the modified version, as illustrated in Fig.'s 4C and D, are qualitatively different from those produced by a completely independent model of inactivation.

DISCUSSION

Horn, Patlak, and Stevens (1981), presented a new method of analysis of single channel data with the object of distinguishing between models of inactivation that were coupled to, or independent of activation for Na channels. For coupled inactivation, the conditional probability should rise to a level equal to the peak of the overall probability, and fall parallel to this curve. The specific prediction was that $W(t,\Delta t)$ was a time shifted replica of $P(t)$. In the case of independent activation, they stated that the conditional probability should rise rapidly and meet the overall probability, then coincide with it during the falling phase. Since the data presented in that paper behaved in the latter way, they concluded that inactivation was most likely independent of activation.

I have tested these predictions by performing a conditional probability analysis on single channel data generated by a computer model which simulates independent, partially coupled, or completely coupled models of inactivation. This analysis showed that the interpretation of the conditional probability expected for independent inactivation in Horn, et. al. is incorrect because all pulses which fail to elicit a channel opening are included in the conditional probability. In order for the original interpretation to be strictly correct, the conditional probability must be determined by the activity of the activation gates alone. Unfortunately, this analysis method is not possible for real single channel data.

However, this mistaken interpretation does not invalidate the usefulness of conditional probability measurements for examining the kinetic behavior of data from real channels. The interpretation of the conditional probability for channels with coupled activation and inactivation was essentially correct, and the method does have sufficient resolution to distinguish clearly between the two models of inactivation. Furthermore, quantitative analysis of the $W(t,\Delta t)$ curves for independent inactivation would be possible by fitting to the diminishing function $W(t,\Delta t)$. The data give information about the channel's kinetics that is independent of other types of measurements. Furthermore, any model of channel kinetics must yield $W(t,\Delta t)$ curves which are identical to those produced by the data.

Between the two extremes of completely coupled, or completely independent inactivation is a continuum of possibilities for partially coupled inactivation. The Armstrong-Gilly model is only one example. In these cases the conditional probability measurements of single channel data do not produce curves which are qualitatively different from those of completely independent inactivation. In order to further distinguish between such models it is necessary to perform quantitative analysis of the actual data and compare them to the predictions of specific models, for both the P(t) and W(t,Δt) curves. Only then would it be possible to make judgements as to which model is the most appropriate.

It is interesting to note that the data from Horn, et. al. do not fit the prediction of the completely independent model, as originally stated in their paper. These data also do not fit the expectations of a completely coupled model, leaving us with the conclusion that the actual behavior of the Na channel lies somewhere between these two extremes. However, the data in their presentation are very limited, and a final interpretation of the channels kinetics will have to await a more extensive analysis.

The author wishes to thank Dr. Michael Berman for many helpful discussions, and Drs. Richard Horn and Ray Gibbons for comments on the manuscript. The work was supported by National Institutes of Health grant HL-28192.

REFERENCES

Armstrong CM   (1981).   Sodium channels and gating current. Physiol Rev   61:644-683.
Armstrong CM, Bezanilla F   (1977).   Inactivation of the sodium channel. II. Gating current experiments.   J Gen Physiol   70:567-590.
Armstrong CM, Gilly WF   (1979).   Fast and slow steps in the activation of Na channels.   J Gen Physiol   74:691-711.
Bean BP   (1981).   Sodium channel inactivation in the crayfish giant axon: Must channels open before inactivating.   Biophys J   35:595-614.
Bezanilla F   (1982).   'Proteins in the Nervous System: Structure and Function'. Haber B, Perez-Polo R (eds).   New York: Alan R. Liss, pp. 51-72.

Colquhoun D, Hawkes AG   (1981).   On the stochastic properties of single ion channels. Proc Royal Soc Lond B. 211:205-235.

French RJ, Horn R   (1983).   Sodium channel gating: Models, Mimics, and Modifiers.   Ann Rev Biophys Bioeng   (in press).

Goldman LG, Hahin R   (1978).   Initial conditions and the kinetics of the sodium conductance in Myxicola giant axons. J Gen Physiol   72:879-898.

Hodgkin AL, Huxley AF   (1952). A quantitative description of the membrane current and its applications to conduction and excitation in nerve.   J Physiol (Lond)   117:500-544.

Horn R, Patlak J, Stevens CF   (1980).   Sodium channels need not open before they inactivate.   Nature   291:426-427.

Keynes RD, Rojas D   (1974).   Kinetics and steady-state properties of the charged system controlling sodium conductance in the squid giant axon. J Physiol (Lond) 239:393-434.

Moore LE, Jacobsson E   (1971).   Interpretation of sodium permeability changes of myelinated nerve in terms of linear relaxation theory.   J Theor Biol   33:77-89.

Nonner W   (1980).   Relation between the inactivation of sodium channels and the immobilization of gating charge in frog myelinated nerve.   J Physiol (Lond)   299:573-603.

Patlak JB   (1981).   Single channel recording.   Biophys J 33:267a.

Patlak J, Horn R   (1982).   Effect of N-bromoacetamide on single sodium channel currents in excised membrane patches.   J Gen Physiol   79:333-351.

Sigworth FJ, Neher E   (1980).   Single Na channel currents observed in cultured rat muscle cells.   Nature   287:447-449.

The Physiology of Excitable Cells, pages 181–189

COUNTING KINETIC STATES: THE SINGLE CHANNEL APPROACH

Richard Horn and Nicholas B. Standen

Department of Physiology, UCLA Medical School,
L.A., Ca. 90024, and Department of Physiology,
University of Leicester, Leicester LE1 7RH,
England.

The advent of single channel recording (Bean et al 1969; Neher, Sakmann 1976) has greatly enhanced the possibility of understanding the molecular properties of ionic channels. The experimenter can now measure both the amplitude of the single channel current and the switching between open and closed states. We will concentrate on the latter process, which is usually called "gating". How the analysis of single channel records significantly improves our knowledge of gating we hope will become apparent in the course of this chapter.

The description of gating involves many aspects of the process by which an ionic channel regulates current flow across the membrane. For example one can ask what makes a channel open or close, and the answer may be one or a combination of such factors as the membrane potential, agonist and/or antagonist concentration, temperature, lipid environment, and permeant ion concentration. If these factors can be varied systematically, then we can describe, at least qualitatively, the way in which they affect the opening and closing of ionic channels.

In order to make these descriptions more precise, most physiologists resort to kinetic schemes, in which transitions occur between two or more states. Typically the states are "Markovian", and the transitions between states obey the properties of a true Markov process. These are technical terms derived from the theory of stochastic processes (e.g., see Cox, Miller 1965; Feller 1968,1971). The relationships between the mathematical theory and kinetic schemes has been treated previously (e.g., Conti, Wanke 1975; Colquhoun, Hawkes, 1977,1981,1982), and will not be discussed here. Suffice it to say that the dwell time, T_i, in a given state, S_i, is an exponentially distributed random variable. In other words Prob{channel remains in state S_i for $T_i < t$} = $1 - \exp(-\lambda_i t)$, where λ_i is the "hazard rate" for leaving state S_i. In kinetic schemes λ_i equals the sum of all rate constants leaving state S_i.

For example, suppose a single channel obeys the kinetic scheme

$$S_1 \underset{k_{-1}}{\overset{k_1}{\rightleftarrows}} S_2 \underset{k_{-2}}{\overset{k_2}{\rightleftarrows}} S_3 \underset{k_{-3}}{\overset{k_3}{\rightleftarrows}} S_4$$

where S_1, S_2, and S_4 are closed states, and S_3 is an open state. By inspection we can write

$$\lambda_1 = k_1$$
$$\lambda_2 = k_{-1} + k_2$$
$$\lambda_3 = k_{-2} + k_3$$
$$\lambda_4 = k_{-3}$$

and

The theory for Markov processes insures that the mean dwell time in state S_i is $\lambda_i^{-1} = \tau_i$.

When we consider the kinetic process as inherently random, we can begin to make sense of the apparently chaotic openings and closings of channels. Kinetic schemes, such as the one above can also describe the fluctuations, or "noise", of a population of many independent channels opening and closing at random (Stevens 1972; Conti, Wanke 1975; Colquhoun, Hawkes 1977). Finally, such schemes yield differential equations which describe the "macroscopic currents" of a great many channels acting more or less in concert (Conti, Wanke 1975; Colquhoun, Hawkes 1977). For example, if the resting state of all channels in the above scheme is S_1, and depolarization causes an instantaneous increase in k_1, k_2, and k_3 and decrease in k_{-1}, k_{-2}, and k_{-3}, then a step depolarization will cause a macroscopic current which increases and then decreases as channels randomly pass through S_3, the open state, into S_4.

One of the main difficulties in kinetic analysis is determining just how many open and closed states there are in a scheme. This can be especially difficult if one has to resort to macroscopic currents and fluctuation analysis of the averaged behavior of many channels. Suppose, for example, that we want to know the number of open states in a kinetic scheme. If the actual process is well described by an N-state scheme, the macroscopic current will have the time course of a sum of N-1 exponentially decaying components, regardless of the number of open states. The same is true of fluctuation analysis, which will yield an auto-correlation function with N-1 exponentially decaying components, or a spectral density function with N-1 Lorentzian components. On the other hand, if we can observe the behavior of a single channel, we can directly measure the duration of the openings. When these are collected and plotted as a histogram (see below), it will have M exponentially decaying components for a kinetic scheme with M open states. In general M is less than or equal to N, and is often much less than N.

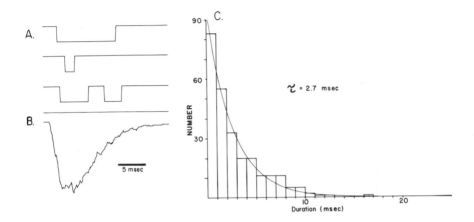

Fig. 1. A. Simulated single channel currents in a patch with one channel for the scheme in the text. B. Averaged currents from 200 simulated traces. C. Open time histogram for the single channel records. $k_1 = 1080$, $k_2 = 540$, $k_3 = 270$, $k_{-1} = 45$, $k_{-2} = 90$, $k_{-3} = 0$ sec^{-1}.

As an example we have simulated single channel currents for the above kinetic scheme on a digital computer using an exponentially distributed random number to determine dwell time in the various states. Fig. 1 shows simulated currents of a single channel for a series of independent trials in which the channel begins in S_1. Setting $k_{-3} = 0$ insures that in each trial the channel will eventually reach S_4 and stay there. Note that we can always tell when a channel reaches S_3, the open state, and when it leaves it. But we can never know for certain which state it is in when the channel is closed. By indicating an open channel as a downward deflection, we have used the usual convention for an inward current, as observed in patch recording of sodium current (Sigworth, Neher 1980; Horn et al 1981) or calcium current (Brown et al 1982; Fenwick et al 1982; Hagiwara, Ohmori 1982; Reuter et al 1982). Furthermore we can average the simulated traces to give a simulated macroscopic current (Fig. 1; see also the chapter by Patlak in this volume).

We have plotted in Fig. 1 the histogram of measured open times, which when scaled properly estimates the probability density of dwell time in the open state, $p(t) = \lambda_3 e^{-\lambda_3 t}$. The scaling simply insures that the area under the theoretical curve equals the area occupied by the histogram itself. We judged the best fit by eye, and a chi-square goodness-of-fit test shows that this is a good fit (P ~ 0.4). The scheme used to simulate the data had $\tau_3 = 1/\lambda_3 = 2.8$ msec, which agrees well with the value of 2.7 msec estimated from the histogram.

With data like these we might safely conclude that the kinetic scheme has one open state, a fact that would be very difficult to determine from macroscopic data. A multi-exponential histogram, on the other hand, would signify more than one open state. Recent studies, using single channel data, suggest that some types of channels have just one open state (e.g. calcium channel: Fenwick et al 1982; Hagiwara, Ohmori 1982; sarcoplasmic reticulum potassium channel: Labarca et al 1980; anomalous rectifier potassium channel: Fukushima 1981a,1982), whereas others may have more (e.g. calcium-activated potassium channel: Barrett et al 1982; endplate channel: Colquhoun, Sakmann 1981; Jackson et al 1983; glutamate-activated channel: Cull-Candy, Parker 1982).

The sodium channel, however, has been somewhat elusive to this form of analysis. A variety of evidence (discussed in French, Horn 1983) suggests that sodium channels have more-than-one kinetically distinct open state, each with the same conductance. For example, histograms of open time were markedly non-exponential in one study (Patlak, Horn 1982). In another report the histograms were fit by single exponential curves, but the data systematically deviated from the theoretical curves (Quandt, Narahashi 1982). Fukushima (1981b) also fit his histograms by single exponentials. However, the paucity of his data precludes a definitive statement as to the quality of the fit.

The problem with accepting the above as evidence for multiple open states is that the histograms in the first two studies cited were obtained from membrane patches containing more than one channel. In such cases openings of different channels may overlap, and these "events" are discarded when forming the histogram of open times. This is illustrated in Fig. 2, which shows 3 "events" which may have occured in a patch with more than one channel. Only the first, non-overlapping, event can be used for the histogram. The second event contains the overlapping openings of two channels. The last event, although non-overlapping, is terminated by the period of observation, i.e. the channel did not close. Therefore this information is also discarded. Histograms formed in this way are, in general, non-exponential, even if the kinetic scheme for each channel has only one open state. We will demonstrate this by simulation.

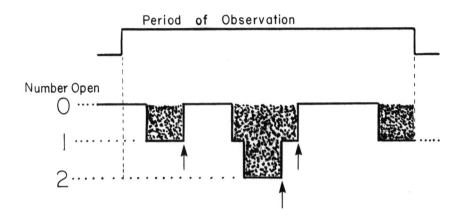

Fig. 2. Schematic record from a patch with more-than-one channel. See text for details.

Using the same scheme and rate constants as above, we have simulated the current response to 400 trials in a patch with four channels. Some of these responses are shown in Fig. 3, along with the average response, which (as it should) has the same time course as that in Fig. 1. The histogram for these data is also shown in Fig. 3. It is clearly non-exponential. Our best fit single exponential (the solid line) has a time constant of 1.8 msec, which underestimates the theoretically expected mean dwell time of 2.8 msec. Furthermore a chi-square test indicates that a single exponential is a poor fit (P < .001). The theoretical histogram for a single channel is superimposed for comparison (dashed line), and clearly does not fit the observed histogram. Similarly, the average open time for non-overlapping events (1.65 msec) underestimates the expected open time for a single channel. The discrepancy can be predicted if one considers that longer openings will tend to be preferentially removed from the histogram by overlapping openings from other channels.

The ambiguities in interpreting the histogram do not, however, preclude an unbiased estimate of open time from data such as those in Fig. 3. A maximum likelihood estimate for open time is simply obtained by summing all of the time for which channels are open and dividing by the number of closings observed in all of the records (Fenwick et al 1982; Horn, Lange 1983; Horn, unpublished calculations). This can be shown to hold rigorously for any kinetic scheme with one open state. For example, in the record of Fig. 2 the maximum likelihood estimate of $1/\lambda_3$ is the shaded area divided by three, the number of closings (indicated by arrows). For the 400 traces used in Fig. 3, the maximum likelihood estimate for τ_3 is 2.9 msec, which is obviously a much better estimator of the theoretical lifetime than anything given by the histogram.

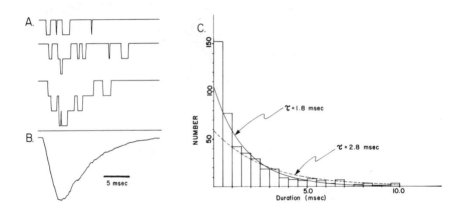

Fig. 3. A. Sample records of simulated currents from a patch with 4 channels. Same rate constants as in Fig. 1. B. Averaged currents from 400 traces. C. Open time histogram. Solid line is best fit single exponential. Dashed line is theoretical curve for single channel.

With this in mind it is easy to devise a strategy for determining the number of open states of sodium channels. One can search for a membrane patch with just one channel, or else create conditions that reduce the likelihood of overlapping events. We have chosen the latter approach, using a combination of fast and slow inactivation in a patch (with at least three channels) to reduce the number of overlaps.

We recorded sodium currents for 450 voltage pulses to approximately -35 mV from a holding potential of -75 mV in a membrane patch from a neuroblastoma of the cell line NlE-115 (cf. Quandt, Narahashi 1982). Since this was an "on-cell"

recording rather than an excised patch, we estimated the
resting potential of the cell by comparing the amplitudes of
single channel currents with those from excised patches
(e.g., see Horn et al 1981). The activating depolarization
was preceded by a 20 msec step to -95 mV, and the interpulse
interval was 300 msec. The combination of a somewhat positive holding potential and pre-pulse voltage, as well as the
rapidity of stimulation reduced the frequency of overlapping
events to 6.4% of the total number of openings.

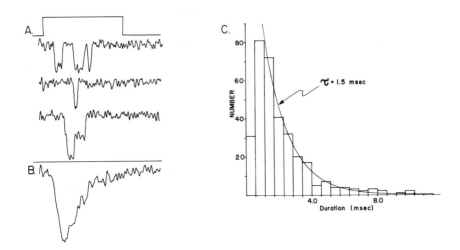

Fig. 4. A. Single channel currents from a neuroblastoma
cell, as described in the text. The lowest trace shows an
overlapping event. Uppermost trace represents the depolarizing pulse of 30 msec duration. Its amplitude is scaled at
1 pA for the single channel records. B. Averaged response
for the 450 current records. C. Open time histogram for
these currents and best-fit single exponential. Pipette
solution (in mM): 150 NaCl, 2 $CaCl_2$, 0.5 $MgCl_2$, 10 HEPES
(pH 7.4). Temp: 11 °C.

　　Examples of single channel records and the averaged record are shown in Fig. 4, as well as the histogram of open
time. Except for the first bin, which is reduced by sam-
pling errors, the histogram is reasonably well-fit by a single exponential (solid line), predicting a mean lifetime of
1.5 msec. The chi-square test was applied to all the data
beyond the first bin and showed a good fit at the 10% level.
Since the data contained a small fraction of overlapping
events, we also calculated the maximum likelihood estimate
for open time. It was 2.2 msec, somewhat larger than the
value predicted by the histogram, suggesting that the small

percentage of overlapping events produced a modest bias in our histogram.

Thus our preliminary data support a model in which the sodium channel has a single open state, rather than multiple open states as suggested in previous studies. There are a number of pitfalls in accepting this conclusion, however. First of all the data were obtained at a single test voltage; the experiment should be repeated over as wide a voltage range as possible. Second, longer duration pulses should be used. A short pulse duration tends to interrupt long openings, and thus exludes them from the histogram. This is not likely to be a problem in our experiment, however, since the averaged current had completely inactivated by the end of the pulse. Finally, the experiment should be repeated in the absence of inactivation (see Patlak, Horn 1982). The rapid closing of inactivation gates may limit the access of individual channels to other open states.

Supported by NSF grant PCM 76-20605, NIH grants NS 703-01 and NS 186-08, and a Wellcome Trust travel grant to N.B.S.

Barrett JN, Magleby KL, Pallotta BS (1982). Properties of single calcium-activated potassium channels in cultured rat muscle. J Physiol 331:211.
Bean RC, Shepherd WC, Chan H, Eichner JT (1969). Discrete conductance fluctuations in lipid bilayer protein membranes. J gen Physiol 53:541.
Brown AM, Camerer H, Kunze DL, Lux HD (1982). Similarity of unitary Ca^{2+} currents in three different species. Nature 299:156.
Colquhoun D, Hawkes AG (1977). Relaxation and fluctuations of membrane currents that flow through drug-operated channels. Proc R Soc Lond B 199:231.
Colquhoun D, Hawkes AG (1981). On the stochastic properties of single ion channels. Proc R Soc Lond B 211:205.
Colquhoun D, Hawkes AG (1982). On the stochastic properties of bursts of single ion channel openings and of clusters of bursts. Phil Trans R Soc Lond B 300:1.
Colquhoun D, Sakmann B (1981). Fluctuations in the microsecond time range of the current through single acetylcholine receptor ion channels. Nature 294:464.
Conti F, Wanke E (1975). Channel noise in nerve membranes and lipid bilayers. Quart Rev Biophys 8:451.
Cox DR, Miller HD (1965). The Theory of Stochastic Processes. Chapman and Hall, London.
Cull-Candy SG, Parker I (1982). Rapid kinetics of single glutamate-receptor channels. Nature 295:410.
Feller W (1968). An Introduction to Probability Theory and

Its Applications, Volume I. New York: Wiley.

Feller W (1971). An Introduction to Probability Theory and Its Applications, Volume II. 2nd Ed. New York: Wiley.

Fenwick EM, Marty A, Neher E (1982). Sodium and calcium channels in bovine chromaffin cells. J Physiol 331:599.

French RJ, Horn R (1983). Sodium channel gating: models, mimics, and modifiers. Ann Rev Biophys Bioeng 12:319.

Fukushima Y (1981a). Single channel potassium currents of the anomalous rectifier. Nature 294:368.

Fukushima Y (1981b). Identification and kinetic properties of the current through a single Na^+ channel. Proc Natl Acad Sci 78:1274.

Fukushima Y (1982). Blocking kinetics of the anomalous potassium rectifier of tunicate egg studied by single channel recording. J Physiol 331:311.

Hagiwara S, Ohmori H (1982). Studies of calcium channels in rat clonal pituitary cells with patch electrode voltage clamp. J Physiol 331:231.

Horn R, Lange K (1983). Estimating kinetic constants from single channel data. Biophys J in press.

Horn R, Patlak J, Stevens CJ (1981). Effect of tetramethylammonium on single sodium channel currents. Biophys J 36:321.

Jackson MB, Wong BS, Morris CE, Lecar H, Christian CN (1983). Successive openings of the same acetylcholine receptor-channel are correlated in their open-times. Biophys J in press.

Labarca P, Coronado R, Miller C (1980). Thermodynamic and kinetic studies of the gating behavior of a K+-selective channel from the sarcoplasmic reticulum membrane. J gen Physiol 76:397.

Neher E, Sakmann B (1976). Single channel currents recorded from membrane of denervated frog muscle fibres. Nature 260:799.

Patlak J, Horn R (1982). The effect of N-bromoacetamide on single sodium channel currents in excised membrane patches. J gen Physiol 79:333.

Quandt FN, Narahashi T (1982). Modification of single Na^+ channels by batrachotoxin. Proc Natl Acad Sci 79:6732.

Reuter H, Stevens CF, Tsien RW, Yellen G (1982). Properties of single calcium channels in cardiac cell culture. Nature 297:501.

Sigworth FJ, Neher E (1980). Single Na^+ channel currents observed in cultured rat muscle cells. Nature 287:447.

Stevens, CF (1972). Inferences about membrane properties from electrical noise measurements. Biophys J 22:1028.

Top: D. Eaton; M. Henkart
Bottom: R. Horn; H. Ohmori

The Physiology of Excitable Cells, pages 191–204
© 1983 Alan R. Liss, Inc., 150 Fifth Avenue, New York, NY 10011

TEA AND TMA ALTER THE I-V CHARACTERISTICS OF THE GRAMICIDIN
CHANNEL AS IF THEY BIND TO CHANNEL SITES AT DIFFERING DEPTHS
IN THE POTENTIAL FIELD BUT DO NOT CROSS.

George Eisenman and John Sandblom.

Department of Physiology, UCLA Medical School,
L.A., Ca. 90024, and Department of Medical
Biophysics, University of Uppsala, Sweden.

In view of Susumu Hagiwara´s early interest in the effects of the tetraethylammonium (TEA) ion, dating back to
the finding (Tasaki, Hagiwara 1957) that it prolonged the
action potential of the squid giant axon, it seems appropriate to describe here an heretofore undetected ability of
this impermeant ion, and its relatives, to bind competitively to cation binding sites in the simple peptide channel
formed in bilayer membranes by gramicidin A. Our purpose is
to demonstrate the existence of such binding for the gramicidin channel and to extend the 3 barrier 4 site model for
this channel (Sandblom, Eisenman, Hagglund 1983) to include
the effects of such impermeant species, which will be shown
to be more complex than simply "blocking" in such a
multiply-occupiable channel. Indeed, we will show that an
impermeant species which blocks by its ability to compete
for a site in a permeation pathway in which multiple occupancy is possible can not only appear to be inert under certain experimental conditions but can even actually increase
the conductance of a channel under appropriate circumstances. We hope that this analysis will be useful as a
prototype for analogous phenomena in biological channels.

Alkylated ammonium ions like TEA and TMA have long been
regarded as being inert for the gramicidin channel (Urban
1978; Andersen 1982) and have been used as supporting elec-
trolytes in several studies (Andersen 1982; Dani, Levitt
1981). However, an indication that things are not so simple
is contained in the finding both by Andersen (1982) for TEA
and by Urban (1978) for the related diethylammonium ion that
these ions actually increased the conductance significantly.
This observation, far from indicating inertness of these
species, will be shown below to be expected theoretically as
a direct consequence of competitive binding in a multi-ion
channel. For, although an impermeant, "blocking", species
which competes for a site in such a channel will, indeed,
decrease conductance, this expectation holds only at low

permeant ion occupancies. Indeed, we will demonstrate that the "blocker" should actually increase the conductance of a channel at higher occupancies for the simple reason that repulsions between the blocking ion and the permeant ion actually speed up the exit rate from the channel sufficiently to override the decreased availability of sites occupied by the blocker.

In this paper we will not only describe the effects of TEA on the gramicidin A channel but we will also compare these with the effects of the smaller TMA species and show that, although these species behave similarly for low applied voltages, they behave differently for high voltages in a way consistent with what would be expected if TMA could actually penetrate more deeply into the channel. Our primary experimental measurements are of the single channel I-V shape based upon a triangular wave, many-channel technique, whose justification has been given elsewhere (Eisenman et al 1982). Such measurements (made in GMO/Decane or GMO/Hexadecane bilayers) in the presence of TEA or TMA, when compared to those made in the absence of these species, will show that these impermeant species alter the concentration dependence of the I-V shape in a way that can be interpreted quantitatively as a direct consequence of competition for binding sites at the channel mouth in the case of TEA, as well as for binding sites further into the channel in the case of TMA. In addition, combining these measurements with the published values of the single channel conductances in the presence and absence of these species, we are able to produce a completely satisfactory description of the total electrical behavior for Cs in the presence and absence of TEA with only a single additional parameter, the binding constant of TEA to the outermost sites of the channel. This will be seen from the good agreement between all data points and the corresponding theoretical curves in all the figures which represent data obtained not only in the presence of TEA but also in its absence. It is to be emphasized that only a single additional parameter, a TEA binding constant of 20 M^{-1}, has been needed, together with the parameters published elsewhere for Cs (Eisenman, Sandblom 1983) to represent all of the data of the present paper.

THE EFFECTS OF TETRAETHYLAMMONIUM (TEA)

We restrict considerations here to highly permeant species, typified by Cs, for which it is easy to establish that currents due to permeant contaminants of the TEA and TMA solutions are totally negligible. For less permeant species, such as Li, it is sufficiently difficult to assess the low concentration limiting shape that quantitative treatment will require better data than we presently have. For this reason, our conclusions may apply strictly only to species which, like Cs (Eisenman, Sandblom 1983), have both a low voltage dependence of the entry step and a relatively insignificant central barrier.

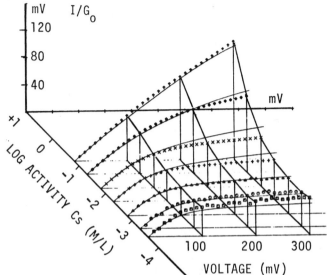

Fig. 1 Three dimensional plot of current, voltage, and logarithm of Cs activity in the presence of 1.0 M TEACl.

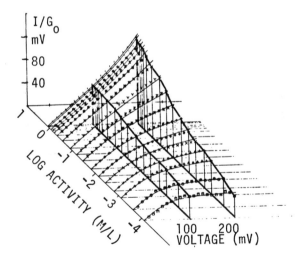

Fig. 2. Three dimensional plot of current, voltage, and logarithm of Cs activity in the absence of TEA and in the presence of 9 mM $MgCl_2$.

Figs. 1-3 enable comparison of the I-V shapes in the presence of 1.0M TEACl (Fig. 1) and in its absence (Fig. 2). Figs. 4-5 extend this comparison to the single channel conductances at zero voltage (G_o). All points are experimental, all curves theoretical. We will confine comments now to comparison of the experimental data; the theoretical curves and their interpretation will be discussed later. The effect of TEA on the I-V shape for the permeant species, Cs, can be seen by comparing Figs. 1 and 2. These figures plot 3-dimensional surfaces of "normalized currents", I/G$_o$, as functions of voltage and of log Cs ion activity in the presence of 1.0 M TEACl (Fig. 1) and in its absence (Fig. 2). (The "normalized currents" are simply the "normalized chord conductances" times the voltage (cf Eisenman, Sandblom 1983)). The points represent experimental data, the curves are theoretically calculated (see later). Planes drawn at 100 mV, 200 mV, and 300 mV have also been constructed intersecting these surfaces; and the currents corresponding to these are shown in Fig. 3 to make it easier to see the dependence of current on the activity of Cs.

The first thing that will be noticed on comparing Figs. 1 and 2 is that the shape appears to be independent of the presence of TEA at the lowest and highest Cs ion concentrations. This does not signify an absence of an effect of TEA but rather that the effect of TEA is observable only at intermediate Cs concentrations, as can be seen on closer inspection of the concentration dependence with the aid of Fig. 3, which plots the dependence of I/G$_o$ on Cs activity for 3 selected voltages (corresponding to the planes at 100, 200 and 300 mV on the current surfaces of Figs. 1-2. The data in the presence of 1.0 M TEACl are shown by squares and dashed curves (labelled "1.0 M TEA") and in its absence by dots and solid curves (labelled "pure Cs").

The following conclusions can be reached from the data of Figs. 1-3: 1. The I-V shape is the same in the limit of low permeant ion concentration whether TEA is present or absent (the identity in limiting shape for Cs in the presence of TEA and in its absence can be seen directly in Eisenman and Sandblom (1983, Fig. 8)). 2. The I-V shape is also the same at high permeant ion concentrations (cf. Fig. 3). This is one of the findings that has led to the belief that TEA was inert. 3. However, at intermediate concentrations there is a dramatic effect of TEA on the I-V shape (seen most clearly in Fig. 3). An additional detail observable in Fig. 3 is that the first change of I-V shape on increasing Cs activity (which may be taken as a sign of binding of Cs to the first binding site (Eisenman et al 1982)) is seen to occur at a much lower Cs activity in pure Cs than when 1.0 M TEACl is present. Indeed, if one views the change of shape in Fig. 3 as a "titration" whose midpoint should occur at half occupancy of the first binding site, then the formal

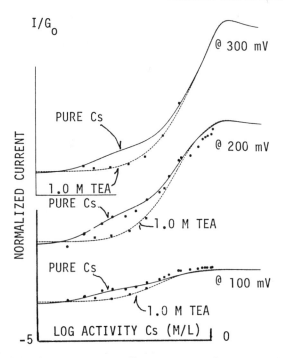

Fig. 3. "Normalized currents" at the indicated voltages as a function of Log Cs activity in the presence of 1.0 M TEACl and in its absence.

effect of TEA is as if it markedly decreased the strength of the first binding site. This, of course, is intuitively what would be expected for a competitive blocker; and the quantitative basis for this effect will be developed in a later section. 4. The limiting shape at low permeant ion concentration is so strongly sublinear that it implies that the major barrier to permeation is at the channel mouth both in the presence of TEA and in its absence and that this barrier, moreover, exhibits virtually no voltage dependence for the entry step. This will be recognized as exactly what Bamberg and Lauger (1977) concluded to be the situation in the presence of divalent blockers; although they did not realize that this was also the limiting shape for Cs because their data for pure Cs solutions were not obtained at sufficiently low Cs concentrations. Because Cs in pure solution also is characterized by a relatively negligible central barrier relative to the barrier for exit, as well as by virtually no voltage dependence of the entry step (Eisenman, Sandblom 1983), it is not surprising that there is no change of limiting shape seen for Cs on adding TEA.

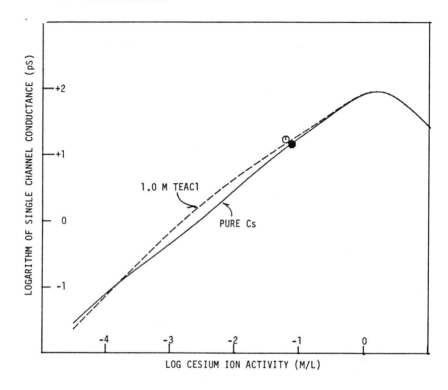

LOG CESIUM ION ACTIVITY (M/L)

Fig. 4. Predicted effect of 1.0 M TEACl and single channel conductances in the absence of TEACl (solid line) and in the presence of 1.0M TEACl (dashed line). Notice that a pure "blocking" effect is expected only at the lowest Cs activity and that an enhancement of conductance is expected over most of the experimental data range. Andersen's (1982) data in the absence of TEA (solid circle) and in the presence of 400 mM TEACl (open circle) are shown for comparison.

The effect of TEA on the single channel conductance for Cs is illustrated in Fig. 4 which presents Andersen's measurements at 25 mV (made at 25°C and corrected to 22°C) in 0.1 M CsCl (filled circle) and in 0.1 M CsCl and 0.4 M TEA (open circle) for comparison. Note that experimentally the conductance is <u>higher</u> in the presence of TEA than in its absence! This is as described by Andersen (1982) and also as found by Urban (1978) for a related molecule, diethyl ammonium. The effect may look small on a log-log plot, but the increase of 19% observed by Andersen on adding 400 mM TEACl actually corresponds to the plotted increase of 48% when an activity coefficient decrease from 0.756 to 0.606 on

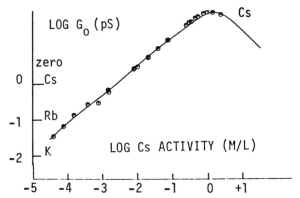

Fig. 5. Single channel conductances in pure CsCl solutions (after data of Eisenman Neher and Sandblom from Eisenman and Sandblom Fig. 2) for comparison with Fig. 4.

adding TEACl is taken into account. The explanation of this heretofore puzzling finding will come from the comparison of the theoretical curves to be discussed later. (Incidentally, Andersen´s data point in the absence of TEA can be seen to agree well with our data in Fig. 5, as seen by its falling precisely on the solid theoretical curve in Fig. 4 which has been transposed from Fig. 5).

THE EFFECTS OF TETRAMETHYLAMMONIUM (TMA)

For applied voltages less than 200 mV the effects of TMA closely resemble those of TEA, as shown in Figs. 6-7. Fig. 6 shows the I-V shapes in 1.0M TMACl for the indicated concentrations of the permeant species Li, Na, K, and Cs; and Fig. 7 compares the differences in I-V shape for Cs produced by TEA vs. TMA for a series of comparable Cs concentrations. The similarity in the effects of 1.0 M TEA and TMA up to 200 mV are most clearly seen in this figure. However, for potentials higher than 200 mV a clear downturn in I/G_o is present for TMA for the lowest concentrations of Cs. This is not seen for TEA. Following the precedent of Horn et al (1981) for TMA in the Na channel, we interpret this as a sign of a voltage-dependent block by TMA on the reasoning that, although TMA cannot cross, it can enter into the channel to a sufficient depth to sense some of the applied potential (cf. Horn et al, 1981; Eisenman et al, 1982, for further details). The absence of such an effect for TEA indicates that this larger species cannot penetrate into any portion of the channel where it can sense the applied potential. We conclude therefore that TEA can only compete for the outermost sites, which makes possible the surprisingly simple analysis given below.

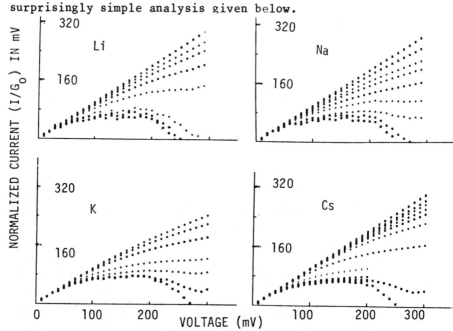

Fig. 6. Normalized currents as a function of voltage in the presence of 1.0 M TMACl. From the lowest curves to the highest curves the permeant ion concentrations are as follows (in mM/L): Li (1,3,12,32,100,200,300,500); Na (1,3, 10,30,100,200,300,500,1000); K (1,3,10,30,100,200,300); Cs (0.5,1,3,10,30,100,300,500,700,900,1100,1500).

A detail worth noting in Fig. 6 is the finding that that the voltage-dependent "blocking" by TMA indicated by the downturn of the I-V curves, is seen only at the very lowest permeant ion concentrations, well below those studied heretofore. Presumably, this is why it has not been noticed previously. Also apparent in this figure is the detail that the concentrations at which the I-V curves turn down correlate, as expected for competitive binding, with the sequence of binding affinities. Thus, the effects of TMA are seen to higher concentrations with just those species (e.g. Li, Na) which are thought (Eisenman, et al, 1978; Dani, Levitt 1981; Eisenman, Sandblom 1983) to bind the least strongly. For example, although a strong voltage-dependent block is still seen in the presence of 12 mM of the weakly binding Li ion, almost none is apparent at this concentration for the more strongly binding Na and K ions; and for the most strongly binding Cs ion the effect is almost undetectable at concentrations higher than 1 mM.

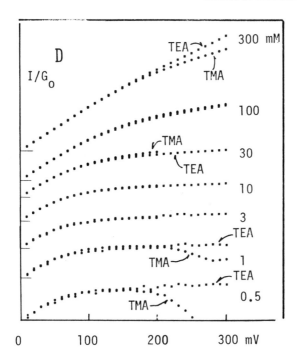

Fig. 7. Comparison of I-V shapes in 1.0 M TEAC1 vs. 1.0 M TMAC1 for the indicated concentrations of Cs.

A COMPARISON AND SUMMARY OF TEA AND TMA EFFECTS

We may now summarize and compare the effects of TEA and TMA as follows. The impermeant ion TEA alters the concentration dependence of the I-V shape of the gramicidin A channel in the following way. For permeant species (such as Cs) which have almost no voltage dependence of entry (<1.4%) and a much slower rate of exit than crossing the chief effect observed is to increase by a factor of 30 the activity of permeant ion at which the first shape change occurs while leaving unaltered the I-V shapes in the high and low concentration limits. The effects of TMA are indistinguishable from those of TEA up to 200 mV, but for higher voltages an additional effect is seen at low permeant ion concentrations, namely a downturn in the I-V curve. These findings are consistent with the postulate that TEA and TMA both bind to a site external to the potential field with a dissociation constant of 50 mM and that TMA can also penetrate sufficiently deeply into the channel to sense the applied potential, thereby producing a voltage-dependent block.

Table 1. Extension to blocking species of the Peak & Well functions of Sandblom et al 1983

$$P_{ox}^{oo} = P_{:\bar{x}} + C_x P_{x:x_o}(1+A)$$

$$P_{xx}^{xo} = P_{x:x_i} + C_x P_{x:x_i x_o}(1+A)$$

$$P_{ox}^{xo} = P_{\bar{x}}^- + 2C_x P_{x x_o}^-(1+A) + C_x^2 P_{x_o \bar{x} x_o}(1+A)^2$$

$$k_{oo} = 1 + 2C_x k_{x x_o}(1+A) + C_x^2 k_{x_o x_o}(1+A)^2$$

$$k_{ox} = k_{x_i} + 2C_x k_{x_i x_o}(1+A) + C_x^2 k_{x_o x_i x_o}(1+A)^2$$

$$k_{xx} = k_{x_i x_i} + 2C_x k_{x_i x_i x_o}(1+A) + C_x^2 k_{x_o x_i x_i x_o}(1+A)^2$$

WHERE: $A = k_{T_o} C_T / k_{x_o} C_x$

$$P_{ox}^{oo} = P_{:\bar{x}} + C_x P_{x:x_o} + P_{\bar{x}:T_o}^- C_T$$

$$P_{xx}^{xo} = P_{x:x_i}^- + C_x P_{x:x_i x_o}^- + P_{\bar{x}:x_i T_o}^- C_T$$

$$P_{ox}^{xo} = P_{\bar{x}}^- + 2C_x P_{x x_o}^- + C_x^2 P_{x_o \bar{x} x_o} + 2C_T P_{\bar{x} T}^- + 2C_x C_T P_{x_o \bar{x} T_o} + C_T^2 P_{T_o \bar{x} T_o}$$

$$k_{oo} = 1 + 2C_x k_{x x_o} + C_x^2 k_{x_o x_o} + 2C_T k_{T_o} + 2C_x C_T k_{x_o T_o} + C_T^2 k_{T_o T_o}$$

$$k_{ox} = k_{x_i} + 2C_x k_{x_i x_o} + C_x^2 k_{x_o x_i x_o} + 2C_T k_{x_i T_o} + 2C_x C_T k_{x_o x_i T_o} + C_T^2 k_{T_o x_i T_o}$$

$$k_{xx} = k_{x_i x_i} + 2C_x k_{x_i x_i x_o} + C_x^2 k_{x_o x_i x_i x_o} + 2C_T k_{x_i x_i T_o} + 2C_x C_T k_{x_o x_i x_i T_o} + C_T^2 k_{T_o x_i x_i T_o}$$

A QUANTITATIVE MODEL FOR THE EFFECTS OF TEA.

The 3 barrier 4 site model of Sandblom, Eisenman, Hag-glund (1983) has a site at each end of the channel external to the barrier at the channel mouth. This site is freely accessable to the facing solution and lies external to the transmembrane potential field. It therefore has exactly the properties required as a binding site for a large, non per-meant species like TEA, which we have tentatively concluded (from the comparison with TMA) does not penetrate into the transmembrane potential field. It is thus natural to extend this model to include the effects of a non permeant species which can bind only to an outer site. For reasons of space the derivation is formulated with the aid of Table 1, which simply extends the definitions of the "peak and well func-tions" of Sandblom et al (1983, Table 1) to include the ef-fect of a non-permeant species which can competitively bind to an outer site.

Table 1 introduces terms that depend upon the concen-tration of the impermeant as well as of the permeant species and is presented for explictness. It, and its discussion in the rest of this paragraph, can be skipped by the reader who wishes to pass over the theoretical details and who only needs to know that this table, together with Eqs. 36 and 39 of Sandblom et al (1983), has been used in drawing all the theoretical curves in the presence of TEA. The left column of Table 1 defines the "peak and well variables" in the presence of an impermeant competitor quite generally (for the case of a symmetrical channel in symmetrical solutions) and holds for any charge or sign. The right column of Table 1 is specialized to apply under the assumption, adequate for our present purposes in the case of TEA and Cs, that the binding of TEA to the outer sites has the same effects on the energy profile as the binding of Cs to these sites. Under this assumption all concentration terms, C_X, in the p's and k's on the left are replaced by $(1 + A)C_X$, where

$$A = k_{To} \, C_T \, / \, C_X \, k_{Xo}$$

in which k_{To} and k_{Xo} are the binding constants of TEA and Cs to the outer site and C_T and C_X are the concentrations of TEA and Cs, respectively. Note that when k_{To} is zero, the resulting equations, of course, reduce to the form in the absence of blocker.

Consequences of the extended model

Only one new parameter has been introduced to charac-terize all the TEA effects reported here, namely the binding constant of TEA to an outer site, k_{To}. This has the value of 20 M^{-1}, which is 35 times weaker than the corresponding binding constant for Cs, whose value is 700 M^{-1}. All the other parameters for the theoretical curves presented here were taken for pure Cs from Table 1 of Eisenman and Sandblom (1983). In view of the drastic simplifying assumption that the effects of TEA are exactly the same as those produced by Cs, it is remarkable how satisfactory the agreement is between theoretical expectations and experimental data. This should be apparant from the close correspondence between all the theoretical curves shown and their corresponding data points. This means that the 3B4S" model holds for the total electrical behavior, both in the presence of TEA and in its absence. The effects of TEA therefore provide an independent verification of the reality of the outer sites postulated in this model to exist in addition to those internal to the barriers at the channel mouth in the 3B2S model (Urban, Hladky 1979).

To recapitulate, in the case of the I-V behavior, Figs. 1-3, and particularly Fig. 3, show how TEA alters the I-V shape, and especially its concentration dependence. On the other hand, Fig. 4 shows a startling expectation of this model, namely, an enhancement, instead of a block, of the single channel conductance over a significant range of experimental conditions (e.g., between 0.0003 and 0.3 M concentration of permeant ion). This result, in fact, explains the previously puzzling findings by Urban (1978) and by Andersen (1982) that TEA and diethyl ammonium ions actually increased the conductance of the permeant species. Note that this effect depends upon concentration of the permeant species, being seen as an enhancement only in the concentration range where significant occupancy of the inner site occurs for the permeant ion. The intuitively expected "blocking" is, indeed, seen for sufficiently low permeant ion concentrations where the expected decrease in the entry rate is due simply to the decrease in availability of outer sites, owing to their occupancy by TEA. The less intuitively obvious enhanced conductance in the middle concentration range is, of course, a consequence of the effect of TEA to increase the exit rate of Cs from inner sites (by repulsion), which over a certain range of concentrations (and voltages) outweighs the decrease in availability of external sites owing to their occupancy by TEA. Finally, the lack of effect of TEA seen at high permeant ion concentrations is because TEA is driven off the sites competitively by the permeant species; and it is only under this condition that it is, in fact, "inert."

Although an alternative possibility that the TEA binds to a separate site, not in the conduction path, instead of to the external site of the channel has not been excluded, this possibility is unattractive on several counts. First, it is unnecessary. Second, any site which can bind a group Ia cation should also be available as a binding site for a larger cation, provided it can reach it. Since TMA can apparently actually penetrate into the inner portions of the channel and since TMA and TEA closely resemble each other as far as the outer site effects are concerned, it seems reasonable to assume that the site to which they bind outside the potential field is indeed the outermost site of the permeation pathway.

CONCLUSIONS

It has been possible to describe quantitatively all the effects of TEA on the I-V shape for Cs by extending the 3 barrier 4 site model of Sandblom et al (1983) to include the effects of an impermeant binding species and by also assuming that although TEA cannot enter the channel, it can compete for a site external to the channel mouth and, when bound, has the same effects as Cs on the energy profile. One implication of this work is that one cannot conclude that an impermeant species is inert unless the tests for its effects have been performed over a sufficiently wide range of permeant ion concentrations and of voltage. In particular, the model predicts that although TEA should be "inert" at sufficiently high permeant ion concentrations, it will indeed block at low permeant ion concentrations and produce complex effects on I-V shape as well as on single channel conductance for intermediate concentrations. Indeed, the model predicts that the conductance should actually be increased for Cs activities between 0.3 to 200 mM, explaining this previously puzzling finding of Andersen for TEA and Urban for diethylammonium. The explanation is simply that TEA speeds up the exit rate of Cs from an occupied channel more than it decreases the availability of sites for Cs to jump into.

We thank the NSF (PCM 76-20605) and the USPHS (GM 24749) for support, Chris Clausen for setting up the data analysis procedures, and Olaf Andersen and Dick Horn for valuable discussions.

Andersen OS (1982). Ion movement through gramicidin A channels. Interfacial polarization effects on single-channel current measurements. Biophysical J in press.
Bamberg E, Lauger P (1977). Blocking of the gramicidin A channel by divalent cations. J Membrane Biol 35:351.
Dani JA, Levitt DG (1981). Binding constants of Li, K, and Tl in the gramicidin channel determined from water permeability measurements. Biophys J 35:485.
Eisenman G, Sandblom J, Neher E (1978). Interactions in cation permeation through the gramicidin channel Cs, Rb, K,

Na, Li, Tl, H, and effects of anion binding. Biophys J 22:307.

Eisenman G, Sandblom J, Hagglund J (1982). Electrical behavior of single-filing channels. In Adelman W, Chang W, Leuchtag R, Tasaki I (eds): "Structure and Function in Excitable Cells", New York: Plenum Press. in press.

Eisenman G, Sandblom J (1983). Energy barriers in ionic channels: Data for gramicidin A interpreted using a single-file (3B4S") model having 3 barriers separating 4 sites. Proc Int Meeting on Physical Chemistry of Transmembrane Ion Motions, Paris 1982. Bioelectrochem Bioenergetics. in Press.

Horn R, Patlak J, Stevens CJ (1981). The effect of tetramethylammonium on single sodium channel currents. Biophys J 36:321.

Sandblom J, Eisenman G, Hagglund JV (1983). Multi-occupancy models for single-filing ionic channels. Theoretical behavior of a 4-site channel with 3 barriers separating the sites. J Membrane Biol 71:61.

Tasaki I, Hagiwara S (1957). Demonstration of two stable potential states in the squid giant axon under tetraethylammonium chloride. J Gen Physiol 40:859.

Urban BW (1978). The kinetics of ion movements in the gramicidin channel. Doctoral thesis, University of Cambridge, England.

Urban BW, Hladky SB (1979). Ion transport in the simplest single-file pore. Biochim Biophys A 554:410.

SECTION II
STUDIES ON OOCYTES, EGGS, AND EMBRYOS

Banquet speakers (clockwise from upper left): T. Wiesel,
T.H. Bullock; A. Watanabe; A. Grinnell; W. Mommaerts; M.
Cohen; C. Edwards

The Physiology of Excitable Cells, pages 207–209
© 1983 Alan R. Liss, Inc., 150 Fifth Avenue, New York, NY 10011

INTRODUCTION

A major effort of the Hagiwara laboratory in recent
years has been the study of the ion channels of oocytes, eggs
and embryos and the roles these channels might play in development and differentiation. Many laboratories working in
this area can trace their intellectual lineage back to Hagi,
and he has published a series of reviews (Hagiwara, Miyazaki
1977a; Hagiwara, Jaffe 1979). The papers in this section
address various aspects of embryology and the differentiation
of the cell membrane.

The observation that electrical events in the sea urchin
egg at fertilization could function as a fast block to polyspermy (Jaffe 1976) has lent importance to the study of the
properties of oocytes of other species. JAFFE here describes fertilization potentials from eggs of two marine
worms, one an annelid and the other a hemichordate, both of
which show depolarizations at fertilization. As summarized
in her paper, eggs of different species show a wide variety
of electrical responses at fertilization, and not all depolarizations are correlated with polyspermy blocks.
Mammalian eggs hyperpolarize in response to sperm, as described in the paper by MIYAZAKI. These periodic hyperpolarizations appear to be driven by the release and reuptake
of calcium from internal stores. It is suggested that the
series of hyperpolarizations and the cyclic increases in
internal free calcium may be related to the process of egg
activation. This series of papers suggests that the relative ubiquity of the alterations of ion permeability shown
by egg membranes at fertilization are reflections of large
reorganizations of the egg internal milieu, and that variations in the permeability of the plasma membrane are involved in these changes.

A major rearrangement of the surface membranes of immature oocytes appears to occur during the process of maturation, which releases oocytes from meiotic arrest and
prepares them for successful fertilization and activation.
Papers in this section discuss this process in echi-
nodermes, amphibians and mammals. LANSMAN, in his paper,
describes the ionic nature of the excitability in immature

starfish oocytes and some of the modifications to conductances that occur during maturation. Inward currents in
starfish oocytes were first studied by Hagi in the early
1970's (Hagiwara, Ozawa, Sand 1975). The work presented
here raises the possibility that the channel responsible for
the fertilization potential is also present in the immature
oocyte, and that its activation at fertilization underlies
the fertilization potential. KADO describes the fine structure of the *Xenopus* oocyte membrane and its reorganization
during maturation. He also discusses the possible significance of a sodium channel which shows an unusual voltage-
dependent "induction" as well as an "activation" which is
similar to that of other sodium channels. This channel is
further characterized in the paper by BAUD in which she describes the ability of lithium to both permeate and block the
sodium channel. YOSHIDA examines the membrane properties of
mammalian oocytes, which (as described by MIYAZAKI) differ
from starfish and amphibian oocytes in their responses to
fertilization. The mouse egg action potential is calcium-
dependent, a property it shares with many other oocytes, but
the calcium channel also appears to be permeable to sodium
and other monovalent cations. These papers point out the
large diversity of the channels found in oocytes; the differing character of these oocytes comprise a fascinating problem in comparative embryology.

Changes in the ion channels present in the membranes of
cells during development are being examined in increasing
detail. Unfortunately, this type of study is hampered by
the decrease in cell size that occurs during successive cell
divisions. To circumvent this problem, Hagiwara and
Miyazaki (1977b), in an early study, used high external
potassium to induce "differentiation without cleavage" in
a marine worm. They observed the initial presence of calcium channels and the progressive addition of potassium
and sodium channels during development. HIRANO and
TAKAHASHI, in their study, use the drug cytochalasin B to
arrest the cleavage of tunicate embryos at various stages,
and follow the differentiation of identified blastomerers
into neural, muscular, or epidermal types with accompanying
changes in the ionic dependence of their membrane currents.
During differentiation, they see calcium currents disappear
from presumptive neural cells, and change their properties
in presumptive muscle and epidemal cells. Experiments
suggest that an "egg-type" calcium channel disappears and a
"differentiated-type" calcium channel appears during

development.

The papers in this section thus encompass a large breadth of topics which relate to the structure of ion channels, their role in fertilization and activation, and their plasticity during development.

References

Hagiwara S, Jaffe LA (1979). Electrical properties of egg cell membranes. Ann Rev Biophys Bioeng 8:385.

Hagiwara S, Miyazaki S (1977a). Ca and Na spikes in egg cell membranes. Prog Clin Biol Res 15:147.

Hagiwara S, Miyazaki S (1977b). Changes in excitability of the cell membrane during 'differentiation without cleavage' in the egg of the annelid, *Chaetopterus pergamentaceus*. J Physiol 272:197.

Hagiwara S, Ozawa S, Sand O (1975). Voltage clamp of two inward current mechanisms of the egg cell membrane of a starfish. J Gen Physiol 65:617.

Jaffe LA (1976). Fast block to polyspermy in sea urchin egg is electrically mediated. Nature 261:68.

Top: L.A. Jaffe; J. Deitmer; M. Barish
Bottom: R. Kado; J. Wallace

The Physiology of Excitable Cells, pages 211–218

FERTILIZATION POTENTIALS FROM EGGS OF THE MARINE
WORMS CHAETOPTERUS AND SACCOGLOSSUS

Laurinda A. Jaffe

Physiology Department
University of Connecticut Health Center
Farmington, CT 06032

ABSTRACT

Fertilization potentials of an annelid (Chaetopterus) and
a hemichordate (Saccoglossus) are described. In Chaetopterus,
the potential of the egg membrane shifts upon fertilization
from −58 to +40 mV; the potential remains positive for 65 ± 33
min. In Saccoglossus, the response is of similar magnitude,
but shorter duration.

INTRODUCTION

Fertilization potentials have been recorded from eggs of
a variety of organisms (Table 1; see also Hagiwara, Jaffe
1979). In many species, the response to sperm involves a
shift from a negative to a more positive membrane potential;
the potential remains more positive than the unfertilized
level for several minutes. In other species, the response
to fertilization involves recurring hyperpolarizations, an
oscillation, or no significant change in potential.

This paper will describe fertilization potentials from
eggs of two marine worms, Chaetopterus variopedatus (annelid)
and Saccoglossus kowalevskii (hemichordate). Fertilization
in both of these species occurs externally in sea water (Lillie
1906: Colwin, Colwin 1954a, b; Costello, Henley 1971).

MATERIALS AND METHODS

Chaetopterus variopedatus were obtained from Pacific
Biomarine (Venice, California); experiments were done in
October. Eggs (~80 μm diameter) were obtained by clipping
the parapodia of the adult worms. Experiments were done
in natural sea water, at about 22°C. Eggs were always in
the sea water for at least 30 minutes before electrical
recording was started. This should be sufficient time to
allow nuclear maturation to proceed to the first meiotic
metaphase, where it pauses until fertilization (Lillie
1906; Allyn 1912); however, sometimes nuclear maturation
did not occur, and eggs retained their germinal vesicles.
For electrical recording, I chose eggs in which no germinal
vesicle was visible when observed at 50X magnification with
a stereomicroscope, but it is possible that some eggs were
oriented in such a way that a germinal vesicle was present,
but not visible. Eggs were dejellied by mechanical agita-
tion, after which some of them adhered to the bottom of
plastic petri dishes. Cleavage of such eggs sometimes
proceeded normally, but in many experiments, cleavage was
abnormal.

Saccoglossus kowalevskii were collected from mudflats
in Chappaquiddick, Massachusetts, in August. Eggs (~400 μm
diameter) were obtained by putting the female worm at 27°C
for about 12 hours, then removing it to 22°C; spawing occurred at 27°C and continued at 22°C. Electrical recordings
were made from eggs that lacked germinal vesicles, as observed
with a darkfield compound microscope at 30X magnification.
Sperm was obtained by cutting the male gonad. Experiments
were done in natural sea water, at about 22°C.

A single microelectrode filled with 3 M KCl (30 to 60
MΩ) was used for both recording voltage and passing current,
using an AM-1 amplifier from Biodyne Electronics (Santa
Monica, California). Data was recorded on a Gould 220 chart
recorder (Cleveland, Ohio). After the electrode was inserted
in the egg, sperm were introduced near the egg with a Pasteur
pipet.

RESULTS

The electrical properties of the unfertilized Chaetopterus
egg after germinal vesicle breakdown have been described by

Hagiwara and Miyazaki (1977). The membrane potential observed in the present study was -58 ± 5 mV (SD, n = 16),
similar to that reported by Hagiwara and Miyazaki. Action
potentials like those reported by Hagiwara and Miyazaki
were also observed.

Figure 1 shows a fertilization potential from a Chaetopterus egg; it occurs 2 to 60 seconds after insemination.
The potential rises to a peak, then falls back to a plateau.
The peak amplitude of the fertilization potential (+40 ± 4 mV,
SD, n = 14) is similar to the peak amplitude of the action
potential in the unfertilized egg (+37 ± 5, SD, n =14). In 6
eggs, recording was continued until the potential returned to
a negative level; the duration of the positive phase was 65 ±
33 min (SD, n= 6). Figure 2C of Hagiwara and Jaffe (1979)
shows the full time course of a fertilization potential from a
Chaetopterus egg.

Fig. 1. Fertilization potential from a Chaetopterus egg.

Fertilization potentials were recorded from two Saccoglossus
eggs; one such record is shown in Figure 2. The unfertilized
egg had a membrane potential of -70 mV. The membrane began to
depolarize 6 sec after adding sperm, and reached a peak of +2
mV, with three superimposed spikes to +17 to +22 mV. Four min
after the rise began, the potential had returned to -50 mV. A
fertilization envelope (FE) had elevated by 1 1/2 min post-
insemination. The response of the second egg was similar.

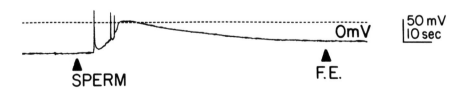

Fig. 2. Fertilization potential from a Saccoglossus egg.

TABLE 1. Fertilization Potentials

Organism	Transient shift to more positive V	Electrical polyspermy block	References*
Echinoderm			
starfish	yes	yes	Miyazaki (1979), Miyazaki and Hirai (1979), Dale et al. (1981), Lansman (1983)
sea urchin	yes	yes	Chambers and de Armendi (1979), DeFelice and Dale (1979), Jaffe (1980), Jaffe et al. (1982), Whitaker and Steinhardt (1983)
Hemichordate	yes	?	this paper
Tunicate	yes	?	Lambert and Lambert (1981) Kozuka and Takahashi (1982),
Fish (medaka)	yes	no	Nuccitelli (1980a,b)
Amphibian			
anurans	yes	yes	Cross and Elinson (1980), Cross (1981), Schlicter and Elinson (1981), Grey et al. (1982)
urodeles	no	no	Charbonneau et al. (1983)
Mammal			
hamster	no	no	Miyazaki and Igusa (1981,1982), Igusa and Miyazaki (1983), Igusa et al. (1983)
mouse	no	no	Igusa et al. (1983), Jaffe et al. (1983)
rabbit	yes	?	McCulloh et al. (1983)
Annelid	yes	?	this paper
Mollusc	yes	?	Finkel and Wolf (1980)
Echiuroid	yes	yes	Gould-Somero et al. (1979), Jaffe et al. (1979), Gould-Somero (1981)
Brown alga	yes	?	Robinson et al. (1981)

*Selected references. For papers before 1979, see Hagiwara and Jaffe (1979).

DISCUSSION

The positive-going fertilization potentials reported here in Chaetopterus and Saccoglossus might serve to prevent polyspermy or to help activate the egg. Table 1 lists organisms in which a fertilization potential has been shown to provide an electrically-mediated polyspermy block. In those studies, it was demonstrated that 1) fertilization could be prevented by holding the potential of the unfertilized egg membrane positive during insemination, and 2) polyspermy could be induced by holding the potential of the just-fertilized egg membrane negative. It will be interesting to try such experiments with Chaetopterus and Saccoglossus.

The relationship between the fertilization potential and activation of development is less clear (see Hagiwara and Jaffe 1979). However, artifically depolarizing the membrane of the unfertilized Chaetopterus egg with high-K^+ sea water can elicit "development without cleavage" (Lillie 1902, 1906; Hagiwara, Miyazaki 1977). Perhaps high-K^+ sea water activates the egg by mimicking the natural positive shift in membrane potential which occurs at fertilization.

ACKNOWLEDGEMENTS

The experiments with Chaetopterus were done in the laboratory of Dr. S. Hagiwara at UCLA. The experiments with Saccoglossus were done in the laboratory of Dr. L. G. Tilney at the Marine Biological Lab, Woods Hole. I would also like to thank Drs. L. H. Colwin and A. L. Colwin for their advice on working with Saccoglossus.

REFERENCES

Allyn HM (1912). The initiation of development in Chaetopterus. Biol Bull 24:21.
Chambers EL, de Armendi J (1979). Membrane potential, action potential and activation potential of eggs of the sea urchin, Lytechinus variegatus. Exp Cell Res 122:203.
Charbonneau M, Moreau M, Picheral B, Vilain JP, Guerrier P (1983). Fertilization of amphibian eggs: a comparison of electrical responses between anurans and urodeles. Dev Biol (in press).

Colwin LH, Colwin AL (1954a). Fertilization change in the membranes and cortical granule layer of the egg of Saccoglossus kowalevskii (Enteropneusta). J Morph 95:1.

Colwin LH, Colwin AL (1954b). Sperm penetration and the fertilization cone in the egg of Saccoglossus kowalevskii (Enteropneusta). J Morph 95:351.

Costello DP, Henley C (1971). "Methods for Obtaining and Handling Marine Eggs and Embryos" Woods Hole, Massachusetts: Marine Biological Laboratory.

Cross NL (1981). Initiation of the activation potential by an increase in intracellular calcium in eggs of the frog, Rana pipiens. Dev Biol 85:380.

Cross NL, Elinson RP (1980). A fast block to polyspermy in frogs mediated by changes in the membrane potential. Dev Biol 75:187.

Dale B, Dan-Sohkawa M, DeSantis A, Hoshi M (1981). Fertilization of the starfish Astropecten aurantiacus. Exp Cell Res 132:505.

DeFelice LJ and Dale B (1979). Voltage response to fertilization and polyspermy in sea urchin eggs and oocytes. Dev Biol 72:327.

Finkel T, Wolf DP (1980). Membrane potential, pH and the activation of surf clam oocytes. Gam Res 3:299.

Gould-Somero M (1981). Localized gating of egg Na^+ channels by sperm. Nature 291:254.

Gould-Somero M, Jaffe LA, Holland LZ (1979). Electrically mediated fast polyspermy block in eggs of the marine worm, Urechis caupo. J Cell Biol 82:426.

Grey RD, Bastiani MJ, Webb DJ, Schertel ER (1982). An electrical block is required to prevent polyspermy in eggs fertilized by natural mating of Xenopus laevis. Dev Biol 89:475.

Hagiwara S, Jaffe LA (1979). Electrical properties of egg cell membranes. Ann Rev Biophys Bioeng 8:385.

Hagiwara S Miyazaki S (1977). Changes in excitability of the cell membrane during 'differentiation without cleavage' in the egg of the annelid, Chaetopterus pergamentaceus. J Phyiol 272:197.

Igusa Y, Miyazaki S (1983). Mechanism of periodic increase in cytoplasmic Ca^{2+} reflected in hyperpolarizing responses during fertilization of the hamster egg. J. Physiol (in press).

Igusa Y, Miyazaki S, Yamashita N (1983). Periodic increase in cytoplasmic Ca^{2+} reflected in hyperpolarizing responses of the egg during cross-species fertilization between hamster and mouse. J Physiol (in press).

Jaffe LA (1980). Electrical polyspermy block in sea urchins: nicotine and low sodium experiments. Dev Growth, Diff 22:503.

Jaffe LA, Gould-Somero M, Holland L (1979). Ionic mechanism of the fertilization potenital of the marine worm Urechis caupo (Echiura). J Gen Physiol 73:469.

Jaffe LA, Gould-Somero M, Holland LZ (1982). Studies of the mechanism of the electrical polyspermy block using voltage clamp during cross-species fertilization. J Cell Biol 92:616.

Jaffe LA, Sharp AP, Wolf DP (1983). Absence of an electrical polyspermy block in the mouse. Dev Biol 96:(in press).

Kozuka M, Takahashi K (1982). Changes in holding and ion-channel currents during activation of an ascidian egg under voltage clamp. J Physiol 323:267.

Lambert CG, Lambert G (1981). Formation of the block to polyspermy in ascidian eggs: time course, ion requirements, and role of the accessory cells. J Exp Zool 217:291.

Lansman JB (1983). Fertilization-induced conductance of the starfish egg membrane. J Physiol (submitted).

Lillie FR (1902). Differentiation without cleavage in the egg of the annelid, Chaetopterus pergamentaceus. Arch Entwmech Org 14:477.

Lillie FR (1906). Observations and experiments concerning the elementary phenomena of development in Chaetopterus. J Exp Zool 3:154.

McCulloh DH, Rexroad CE Jr, Levitan H (1983). Insemination of rabbit eggs is associated with slow depolarization and repetitive diphasic membrane potentials. Dev Biol 95:372.

Miyazaki S (1979). Fast polyspermy block and activation potential. Electrophysiological basis for their changes during oocyte maturation of a starfish. Dev Biol 70:341.

Miyazaki S, Hirai S (1979). Fast polyspermy block and activation potential. Correlated changes during oocyte maturation of a starfish. Dev Biol 70:327.

Miyazaki S, Igusa Y (1981). Fertilization potential in golden hamster eggs consists of recurring hyperpolarizations. Nature 290:702.

Miyazaki S, Igusa Y (1982). Ca-mediated activation of a K current at fertilization of golden hamster eggs. Proc Natl Acad Sci USA 79:931.

Nuccitelli R (1980a). The electrical changes accompanying fertilization and cortical vesicle secretion in the medaka egg. Dev Biol 76:483.

Nuccitelli R (1980b). The fertilization potential is not necessary for the block of polyspermy or the activation of development in the medaka egg. Dev Biol 76:499.

Robinson KR, Jaffe LA, Brawley SH (1981). Electrophysiological properties of fucoid algal eggs during fertilization. J Cell Biol 91:179a.
Schlichter LC, Elinson RP (1981). Electrical responses of immature and mature Rana pipiens oocytes to sperm and other activating stimuli. Dev Biol 83:33.
Whitaker MJ, Steinhardt RA (1983). Evidence in support of the hypothesis of an electrically mediated fast block to polyspermy in sea urchin eggs. Dev Biol 95:244.

The Physiology of Excitable Cells, pages 219–231
© 1983 Alan R. Liss, Inc., 150 Fifth Avenue, New York, NY 10011

PERIODIC HYPERPOLARIZATIONS IN FERTILIZED HAMSTER EGGS:
POSSIBLE LINKAGE OF Ca INFLUX TO INTRACELLULAR Ca RELEASE

Shun-ichi Miyazaki

Department of Physiology
Jichi Medical School
Tochigi-ken, 329-04, Japan

INTRODUCTION

During the past decade, extensive work has been done in
Dr. Susumu Hagiwara's laboratory on Ca and K channels of the
egg cell membrane, and on the fertilization potential of
various oocytes (Hagiwara, Jaffe 1979). At the earliest
stage of fertilization, eggs show a change in the membrane
potential, which is called the fertilization (or activation)
potential, and an intracellular release of Ca ions (Epel
1978). This work has been performed mainly with echinoderm
eggs, but no studies had been reported of electrical phenomena occurring during fertilization of mammalian eggs until
we first demonstrated a characteristic fertilization potential in golden hamster eggs (Miyazaki, Igusa 1981a). The
potential is composed of recurring hyperpolarizations.
Since each hyperpolarizing response (HR) is due to a Ca-
activated K conductance increase (Miyazaki, Igusa 1982),
the periodic HRs are likely to reflect a periodic increase
in intracellular Ca ion concentration ($[Ca]_i$). I describe
here the analyzed properties of the cyclic HRs and then
discuss possible mechanisms involved in the periodic increase
of $[Ca]_i$. This study seems to be of general importance,
since intracellular free Ca mediates a number of cellular
processes, such as muscle contraction (Ebashi, Endo 1968),
membrane permeabilities to ions (Hagiwara, Byerly 1981),
excocytotic secretion (Douglas 1968) and cell motility
(Hitchcock 1977).

Golden hamster eggs freed from the surrounding zona
pellucida have been used as the experimental system for

fertilization <u>in vitro</u>. The fairly large size of the egg
(72 μm in diameter) enables successful impalement by two or
three microelectrodes. After microelectrode impalement(s),
sperm suspension was added to the bath at 31°C. Zona-
removed eggs always allow multiple sperm entries. Mammalian
sperm are so large that attachment of each sperm to the egg
surface is visible under a microscope during electrical
recording. The fertilization potential consists of transient
hyperpolarizations (Fig. 1), quite different from the positive-going responses in oocytes of many other species
(Hagiwara, Jaffe 1979). The first 3 to 4 hyperpolarizing
responses (HRs) reach -80 to -75 mV from the resting potential of ca. -30 mV, and later HRs reach -65 mV, superimposed
on a gradual hyperpolarizing shift of the membrane potential
to -40 mV. The first HR appears at about the time when
active flagellar motion of the first sperm stops (1 "st" in
Fig. 1), when sperm and egg are likely to fuse (Yanagimachi
1978a). A single sperm can induce multiple HRs (Fig. 1A),
and later HRs do not always coincide with the cessation of
flagellar motion of additional sperm (Fig. 1B). The series

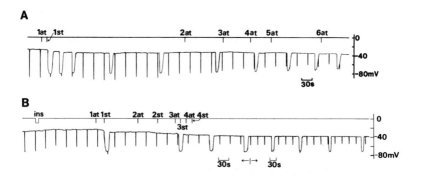

Fig. 1. Fertilization potential of zona-free hamster eggs.
Timing: ins, insemination; at, attachment of sperm to the
egg surface; st, cessation of flagellar motion. The numbers
indicate each sperm in the order of attachment. In B, no
further sperm attached. Constant current pulses are repetitively applied throughout. Normal medium (mM): NaCl, 94.6;
KCl, 4.8; CaCl$_2$, 1.7; MgSO$_4$, 1.2; KH$_2$PO$_4$, 1.2; Na lactate,
22; Na pyruvate, 0.5; glucose, 5.6; NaHCO$_3$, 25.1 (equilibrated with CO$_2$, pH 7.4); bovine serum albumin, 4 mg/mL.

of HRs is established if more than 4 - 5 sperm have fused to the egg, and it continues for longer than 1 hr, even when no further sperm attach to the egg. HRs occur and sperm can enter the egg, even when the membrane potential is held at any level between -150 and +50 mV (Miyazaki, Igusa 1982), indicating that the hamster egg lacks the voltage-dependent block of sperm entry which has been found in sea urchins and other species (Hagiwara, Jaffe 1979).

Each HR is based on a Ca-activated K current which has been found in a variety of cells (Meech 1978). The reversal potential of HRs shifts with a Nernstian slope for K ions when external K concentration is changed, whereas it is unaltered by the removal of Cl ions (Miyazaki, Igusa 1982). HRs can be blocked by the intracellular injection of EGTA. Injection of Ca ions into an egg induces a hyperpolarization similar to the HR. The gradual hyperpolarizing shift of the resting potential is also associated with a conductance increase, and it is suppressed by EGTA injection (Miyazaki, Igusa 1982).

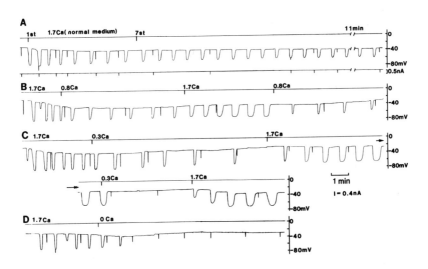

Fig. 2. Effect of lowered $[Ca]_o$ on HRs. Eggs were inseminated in normal medium and then perfused with lower Ca media (expressed in mM), starting at the mark. Constant current pulses of 0.4 nA were applied in all records, to monitor a change in the membrane conductance. The upper record of C is continuous with the one below (see arrow).

EFFECTS OF EXTERNAL DIVALENT CATIONS ON HRs

Periodic HRs are likely to reflect a periodic increase in $[Ca]_i$. In preliminary experiments using a Ca-sensitive microelectrode, Ca transients are confirmed. A question arises whether the increase in $[Ca]_i$ is due to a Ca influx across the plasma membrane or due to Ca release from intra-cellular stores. In normal medium (Ca = 1.7 mM), HRs occur in an all-or-none manner at fairly constant intervals of 45 - 60 sec, although the amplitude of each HR becomes smaller (Fig. 2A). The frequency of HRs is decreased reversibly by perfusing lower Ca media (Fig. 2B, C). The HR is abolished eventually in 0.3 mM Ca during sustained perfusion or within 2 min after the introduction of Ca-free medium (Fig. 2C, D). This is not due to a possible effect of Ca-free solution on sperm-egg interactions, because HRs are also abolished by replacement of Ca with Sr or Mn ions which are substitutable for Ca in the incorporation of sperm into the egg cytoplasm

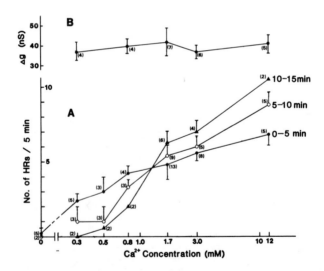

Fig. 3. A, the mean number of HRs occurring in every 5 min, counted between 0-5, 5-10 and 10-15 min after the introduc-tion of various Ca media, in eggs inseminated in normal medium or 3 mM Ca saline. B, K conductance at the peak of the HR which occurred between 4 and 6 min after the intro-duction of a given medium. Mean ± SD (n, in a parenthesis) is plotted against the logarithm of $[Ca]_o$. 3 or 12 mM Ca solution was made of Cl-salts only and buffered with Hepes.

(Yanagimachi 1978b). Thus, the occurrence of HRs requires the presence of external Ca ions, suggesting that they are based primarily on Ca influx. Figure 3A shows the averaged number of HRs occurring in every 5 min during perfusion of a given medium. The dependence of the HR frequency on $[Ca]_0$ becomes more pronounced with time. The frequency of HRs is remarkably reduced by adding 9 mM Mn or Co ions to 3 mM Ca solution (Igusa, Miyazaki 1983), suggesting that the pre-sumed Ca influx is inhibited by Mn or Co ions.

In contrast to the HR frequency, the amplitude (Fig. 2), conductance increase and reversal potential of each HR are little affected by the change of $[Ca]_0$, or addition of Mn or Co. Increased K conductance can be estimated by subtraction of the membrane conductance just before or after the HR from that at its peak, since current-voltage relations are almost linear (Miyazaki, Igusa 1982). Mean values ranged between 35 and 42 nS at $[Ca]_0$ between 0.3 and 12 mM (Fig. 3B). The reversal potentials of HRs also ranged only between -86 and -83 mV upon the change of $[Ca]_0$ from 0.8 to 12 mM. These results indicate that each HR is not a direct consequence of Ca influx. It should be noted that no depolarizing phase precedes each HR (Figs. 1 and 2). The HR seems to be based on intracellular events such as Ca release from stores, although its occurrence is dependent on extracellular Ca.

The HR frequency is decreased on sustained depolariza-tion with DC current, and increased on hyperpolarization. It seems that the greater electromotive force of Ca ions increases Ca influx and thereby facilitates the occurrence of HRs. The slow hyperpolarization may reflect continuous Ca influx stimulating a Ca-activated K conductance. The hyperpolarizing shift reaches -40 mV in 10 min after insem-ination in normal medium (Fig. 2A). It is suppressed by injected EGTA. The membrane potential can be made less negative by lowering the $[Ca]_0$ (Fig. 1B-D) or by adding Mn or Co, associated with the lowered HR frequency. On the other hand, the hyperpolarizing shift reaches -50 mV in 3 mM Ca (Igusa, Miyazaki 1983): the greater Ca influx upon raising the $[Ca]_0$ seems to cause the more negative resting potential, due to Ca-activated K conductance.

REGENERATIVE HR INDUCED BY Ca INJECTION

It has been shown in sea urchin eggs that Ca ions are

released from intracellular Ca stores at an early stage of
fertilization (Steinhardt et al. 1977). Increased free Ca
induces cortical exocytosis (Epel 1978). Gilkey and others
(1978) have demonstrated a wave of increased Ca traversing
through the cytoplasm in aequorin-loaded eggs of medaka fish.
They ascribed the propagation to a process of Ca-induced Ca
release, which has been found in the sarcoplasmic reticulum
(Endo 1977). With respect to the Ca-induced Ca release, a
regenerative HR is observed in the hamster egg in response
to intracellular injection of Ca ions. As shown in Fig. 4A,
a hyperpolarization is produced by ionotophoretic Ca injection

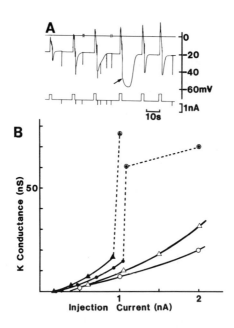

Fig. 4. A, a regenerative response induced by Ca injection
into an unfertilized egg with a single 2 sec pulse. The
initial depolarization is produced by the injection current
itself, and the succeeding hyperpolarization is due to the
injected Ca ions. Short, hyperpolarizing pulses are applied
through the $CaCl_2$ electrode, to monitor conductance changes.
B, relationship between K conductance at the peak of the
HR and injection current of Ca in two eggs (circles and tri-
angles). Filled symbols, subthreshold HRs before a regenerative HR was generated; filled symbols with a circle, regenerative HRs; open symbols, HRs at 5-10 sec after the end of
a regenerative HR.

with a single 2 sec pulse through a second electrode filled
with 0.5 M $CaCl_2$. Incremental injection of Ca ions into an
unfertilized egg reveals a regenerative HR with an apparent
threshold (Fig. 4A, arrow): a hyperpolarizing excitation is
induced by stimulation of injected Ca but not depolarizing
current pulse. The critical injection current of Ca is about
1 nA. This excitation is followed by a refractory period;
that is, a suprathreshold injection pulse fails to generate
another regenerative HR immediately after the previous one
(Figs. 4A and 5A). The recovery of the refractory period is
illustrated in Fig. 5B, where the ratio of the area in the
record of a regenerative HR and the succeeding responses with
the same pulses is plotted against the interval. The next
regenerative HR is obtained about 2 min later. When K conductance at the peak of induced HR is plotted as a function
of Ca injection current, a steep jump is demonstrated at the
critical injection generating the regenerative HR (Fig. 4B,
broken lines). In contrast, the relation is approximately
linear to 5-10 sec after the regenerative HR, at the early
phase of the refractory period (open symbols). Thus, the
hyperpolarizing excitation induced by Ca injection is

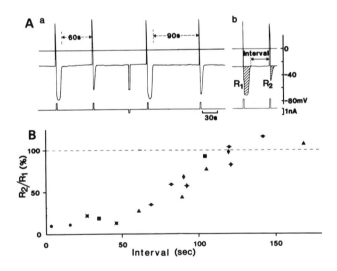

Fig. 5. Refractory period of the regenerative HR. A(a), a
sample record. B, the time course of the refractory period.
Measurements are shown in Ab. A simple circle was obtained
from one unfertilized egg. For other symbols, the same sym-
bol was obtained from the same egg.

considered to be due to a nonlinear increase in $[Ca]_i$ rather than a nonlinear dependence of the K conductance on a linear increase in $[Ca]_i$.

The regenerative HR in unfertilized eggs has the following additional properties (Igusa, Miyazaki 1983). It is produced even in Ca-free medium. It is induced by about twofold smaller injection current of Ca on lowering the temperature from 31 to 26°C, which should also lower the intracellular buffering power. Ca injections at two points about 50 μm apart in the egg of 70 μm diameter interfere with each other; the excitation induced by injection at a point causes a refractory period to the excitation by injection at the other point. Therefore, the increase in $[Ca]_i$ that produces the regenerative response seems to cover nearly the whole inner surface of the membrane.

Ca INJECTION INTO INSEMINATED EGGS

Ca injection into inseminated eggs provides more information on the mechanism of periodic HRs. In Fig. 6A and B, HR(s) induced by Ca injection are interposed between HRs occurring at a constant interval after application of sperm. The periodic HRs (indicated by dots) will be referred to as "sperm-mediated HRs" (S-HRs), although each HR may not be mediated by an action of sperm. S-HRs are interrupted by the interposed HR(s) (Fig. 6A, arrows 3 and 4): the periodicity of S-HRs is reset by a Ca-induced HR. The interference between S-HRs and the Ca-induced HR indicates that both HRs are based on process(es) at common sites.

A large HR is induced by Ca injection in an all-or-none manner (Fig. 6A, arrows 1-3). The size of the HR is little increased with further increase in the injected Ca (arrow 6). This all-or-none property is shown in Fig. 6C in terms of K conductance during HRs as a function of Ca injection current. The critical injection current is as small as 0.1 nA or less, tenfold smaller than necessary in unfertilized eggs. Such small Ca pulses produce no hyperpolarization at all immediately after the all-or-none HR (Fig. 6A, arrow 5), and thus an all-or-none like recovery of the refractory period is seen (Fig. 6D). All these findings strongly suggest that both S-HRs and artificially induced HRs are based on release of Ca from intracellular stores. The mechanism seems to be similar to the Ca-induced Ca release. The refractory period shortens

to about 30 sec in inseminated eggs (Fig. 6D) and is comparable to the period of S-HRs. One of the factors determining
the refractory period is presumably the re-uptake of Ca
stores, which seems to be augmented in inseminated eggs.

Oscillatory potential changes similar to S-HRs are produced when Ca is injected repetitively into inseminated eggs
with constant current pulses (0.5 sec, 1 Hz) (Fig. 6B). The
occurrence of S-HRs is prevented as long as Ca is injected.
The frequency of Ca-induced HRs is dependent on injected Ca
(Fig. 6B). The membrane potential at the pauses of these
HRs shifts gradually in the hyperpolarizing direction, similar to the hyperpolarizing shift. These results indicate

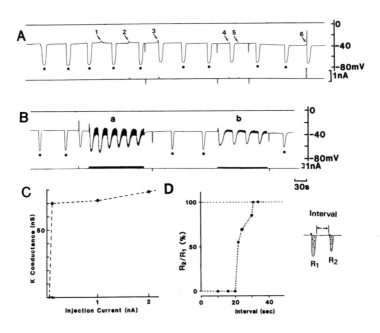

Fig. 6. HRs induced by Ca injection into inseminated eggs.
A, HRs induced with a single 2 sec pulse (arrows 1-6) at the
pause of periodic S-HRs (dots). B, repetitive Ca injections
at 1 Hz with constant pulses of 0.5 sec duration. C, all-or-
none property of Ca-induced HR in terms of the K conductance
as a function of Ca injection current. D, all-or-none like
recovery of the refractory period of the Ca-induced HRs with
a 0.1 nA, 2 sec pulse.

that an elevated continuous Ca influx, when present, can produce periodic HRs. Similar HRs are induced even in Ca-free medium by repetitive injections of Ca, but their frequency is lower than that in normal medium with comparable injection current (Igusa, Miyazaki 1983).

POSSIBLE MECHANISM

As a possible mechanism for periodic increase in $[Ca]_i$ the following model can be proposed, although other possibilities cannot be ruled out. The first sperm-egg fusion induces intracellular Ca release in the egg, resulting in the first HR. This process may be common to other kinds of eggs (Steinhardt et al. 1977; Gilkey et al. 1978). Sperm-egg interactions then cause remarkable changes in the egg, such as an increase in membrane permeability to Ca ions and a lowered threshold for Ca-induced Ca release. Transported Ca ions due to the elevated Ca influx are sequestered in intracellular stores adjacent to the plasma membrane, and accumulated Ca is released by an appropriate rise in $[Ca]_i$, possibly mediated by the Ca-induced Ca release. This process could be repeated in a cyclic manner with an interval corresponding to the refractory period shown by Ca injection. This model is based on the idea of a linkage of continuous Ca influx to intracellular Ca release. The pause of HRs is considered to be the time required to reload the Ca stores. The influx of Ca, therefore, is capable of determining the rhythm of HRs by changing the rate of Ca accumulation in the stores. This model is supported by the observation that the frequency of S-HRs is increased on raising the $[Ca]_o$ and decreased by lowering the $[Ca]_o$ or by adding Mn or Co, with little change in the amplitude of each S-HR. The Ca transport is probably not mediated by the voltage-gated Ca channel which has been found to be present in hamster eggs (Miyazaki, Igusa 1981b), because the frequency of S-HRs is increased by sustained hyperpolarization and S-HRs (with reversed polarity) can occur at the holding potential of -150 mV (Miyazaki, Igusa 1982). Henkart and Nelson (1979) have concluded that the endoplasmic reticulum apposed to the plasm membrane (subcellular cisterns) constitutes an intracellular Ca store releasable by surface stimuli in fibroblasts (L cells) which show "hyperpolarizing activation".

It is well known that the Ca ionophore A23187 is capable of inducing parthenogenetic activation in various kinds of eggs, including the hamster egg (Steinhardt et al. 1974). This drug causes only 1 or 2 large, transient HRs in Ca-free solution containing 5 mM EGTA as well as in normal medium (Miyazaki, Igusa 1981b). Zona-free hamster eggs allow multiple entries of heterologous sperm. When they are inseminated with mouse sperm, the start of the HR series is delayed about 15 min, in good agreement with a delay in egg activation (Igusa et al. 1983). Therefore, the early stage of the HR series is considered to be responsible for egg activation. However, the biological significance of such a series of HRs that persists for more than 1 hr is yet unknown. Ca-mediated hyperpolarizations have been found in other tissues and they are suggested to be related to cell motility, such as phagocytosis in fibroblastic L cells (Okada et al. 1981) or the activation of chemotactic migration in macrophages (Gallin, Gallin 1977). In fertilized hamster eggs the entire sperm, both its head and tail, is incorporated into the egg in a few hours, covered with wave-like protrusions of the egg cytoplasm, some parts in a phagocytic fashion (Yanagimachi 1978a). Periodic HRs may be related to this sperm incorporation. In the mouse egg, oscillatory $[Ca]_i$ transients of aequorin signals have been demonstrated during fertilization (Cuthbertson et al. 1981). We have recorded a series of Ca-mediated small HRs (3-4 mV) in inseminated mouse eggs (Igusa et al. 1983). Periodic HRs have been also reported in the rabbit egg (McCulloh et al. 1981). The HRs may be a common feature of the fertilization of mammalian eggs.

REFERENCES

Cuthbertson KSR, Whittingham DG, Cobbold PH (1981). Free Ca increases in exponential phases during mouse egg activation. Nature 294:754.

Douglas WW (1968). Stimulus-secretion coupling: the concept and clues from chromaffin and other cells. Br J Pharmacol 34:451.

Ebashi S, Endo M (1968). Calcium ion and muscle contraction. Prog Biophys Molec Biol 18:123.

Endo M (1977). Calcium release from the sacroplasmic reticulum. Physiol Rev 57:71.

Epel D (1978). Mechanisms of activation of sperm and egg during fertilization of sea urchin gametes. Curr Top Dev Biol 12:185.

Gallin EK, Gallin JI (1977). Interaction of chemotactic factors with human macrophages. J Cell Biol 75:277.

Gilkey JC, Jaffe LF, Ridgway EB, Reynolds GT (1978). A free calcium wave traverses the activating egg of the medaka, *Oryzias latipes*. J Cell Biol 76:448.

Hagiwara S, Byerly L (1981). Calcium channel. Ann Rev Neurosci 4:69.

Hagiwara S, Jaffe LA (1979). Electrical properties of egg cell membranes. Ann Rev Biophys Bioeng 8:385.

Henkart MP, Nelson PG (1979). Evidence for an intracellular calcium store releasable by surface stimuli in fibroblasts (L cells). J Gen Physiol 73:655.

Hitchcock SE (1977). Regulation of motility in non muscle cells. J Cell Biol 74:1.

Igusa Y, Miyazaki S (1983). Effects of altered extra cellular and intracellular calcium concentration on hyperpolarizing responses of the hamster egg. J Physiol in press.

Igusa Y, Miyazaki S, Yamashita N (1983). Periodic hyperpolarizing responses in hamster and mouse eggs fertilized with mouse sperm. J Physiol in press.

McCulloh DH, Rexroad CE, Levitan H (1981). Rabbit egg membrane potential's response at fertilization. J Cell Biol 91:176a.

Meech RW (1978). Calcium-dependent potassium activation in nervous tissues. Ann Rev Biophys Bioeng 7:1.

Miyazaki S, Igusa Y (1981 a). Fertilization potential in golden hamster eggs consists of recurring hyperpolarizations. Nature 290:702.

Miyazaki S, Igusa Y (1981 b). Ca-dependent action potential and Ca-induced fertilization potential in golden hamster eggs. *In The mechanism of gated calcium transport across biological membranes.* ed Ohnishi ST, Endo M 305, Academic Press NY.

Miyazaki S, Igusa Y (1982). Ca-mediated activation of a K current at fertilization of golden hamster eggs. Proc Natl Acad Sci USA 79:931.

Okada Y, Tsuchiya W, Yada Y, Yano J, Yawo H (1981). Phagocytic activity and hyperpolarizing responses in L-strain mouse fibroblasts. J Physiol 313:101.

Steinhardt R, Epel K, Carrol EJ, Yanagimachi R (1974). Is calcium ionophore a universal activator for unfertilized eggs? Nature 252:41.

Steinhardt RA, Zucker R, Schatten G (1977). Intracellular calcium release at fertilization in the sea urchin egg. Dev Biol 58:185.

Yanagimachi R (1978 a).   Sperm-egg association in mammals.
  Curr Top Dev Biol 12:83.
Yanagimachi R (1978 b).   Calcium requirement for sperm-egg
  fusion in mammals.   Biol Reprod 19:949.

Top: R. Meech; R. Kado; L.A. Jaffe
Bottom: C. Baud; H. Ohmori; O. Sand; S. Miyazaki; K.
Nicolaysen; S. Krasne

The Physiology of Excitable Cells, pages 233-246
© 1983 Alan R. Liss, Inc., 150 Fifth Avenue, New York, NY 10011

COMPONENTS OF THE STARFISH FERTILIZATION POTENTIAL:
ROLE OF CALCIUM AND CALCIUM-DEPENDENT INWARD CURRENT

Jeffry B. Lansman

Department of Physiology
University of California
Los Angeles, CA 90024

The eggs of many species undergo characteristic
changes in membrane potential as a consequence of the
interaction of the gamete membranes at fertilization. The
electrical response preceeds the changes in the egg's
metabolism collectively referred to as activation and
constitutes the first sign that the egg has initiated
development (Hagiwara, Jaffe 1979). It is becoming
increasingly clear that the electrical events associated
with activation are, like the subsequent steps of the
activation process, triggered by the transient rise in
intracellular free calcium that occurs following sperm
attachment (Whitaker, Steinhardt 1982). The fertilization
potential may, therefore, belong to a more general class of
excitation in which calcium ions couple cell surface
stimulation to a change in membrane permeability during
cell activation.

In this chapter, I will review some recent work on the
components of the starfish fertilization potential.
Evidence will be presented for the involvement of a
calcium-dependent inward current in the action potential
and in the activation potential triggered by sperm. In the
former case, inward sodium current is activated by calcium
ions entering the egg through voltage-gated calcium
channels. In the latter, inward sodium current appears to
be triggered by the rise in intracellular free calcium
associated with egg activation.

FERTILIZATION POTENTIAL

    Successful sperm-egg interaction at fertilization
initiates a two part depolarization of the egg membrane.
Figure 1 shows examples of fertilization potentials
recorded from the giant eggs (0.5-1.0 mm dia.) of the
starfish <u>Mediaster</u> <u>aequalis</u> (1A) and <u>Leptasterias</u> <u>hexactis</u>
(1B). Shortly after adding sperm to the artificial
seawater (ASW) bathing an egg, the egg membrane depolarizes
past threshold and an intracellular microelectrode records
a rapid depolarization due to the generation of an action
potential (phase 'a'). The egg membrane continues to
slowly depolarize reaching a peak of +10 to +15 mV (phase
'b'). Under normal conditions, phase 'b' requires the
interaction of the sperm and egg and is a response unique
to activation. For this reason it will be called the
activation potential to distinguish it from the action
potential which can be triggered by depolarization in the
unfertilized egg. Below, I will consider each part of the
fertilization potential.

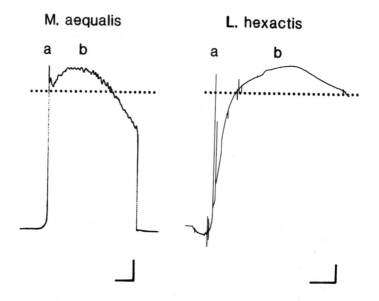

Figure 1. Fertilization potential of the starfish egg.
Eggs were matured <u>in vitro</u> with 1-methyladenine. Following
addition of sperm, the membrane potential depolarizes
triggering an action potential (phase 'a'). This is

followed by a sperm-dependent activation potential (phase 'b'). Note the second small action potential triggered in the <u>Leptasterias</u> egg after the membrane had partially repolarized. (Calibration bars, vertical = 10 mV, horizontal = 5 min. Dotted line = 0 mV.)

Action potential

The action potential of the starfish egg depends on Ca entry by all criteria used to distinguish Ca-dependent action potentials (Hagiwara, Byerly 1981). However, early work suggested the existence of an additional Na component (Miyazaki <u>et al.</u>, 1975a; Shen, Steinhardt 1976). Figure 2 shows the effects of Na removal on the action potential of the <u>Leptasterias</u> egg. The action potential duration is significantly shortened after replacing all extracellular Na with Tris. The effect of Na on inward current recorded under voltage clamp was investigated by Hagiwara, Ozawa, and Sand (1975). Na removal both reduced the peak amplitude of inward current and increased its rate of decay. In the absence of Ca, however, Na alone did not produce inward current suggesting that the pathway carrying Na was unlike the voltage-gated sodium channel of the squid axon.

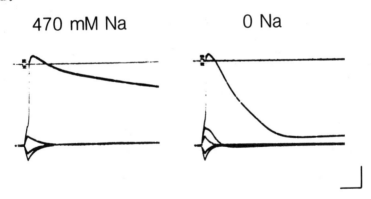

470 mM Na   0 Na

Figure 2. Action potential of the <u>Leptasterias</u> egg in 25 mM Ca ASW. (Calibration bars, vertical = 20 mV, horizontal = 0.2 sec.)

Recently, this problem has been reexamined in detail

and several lines of evidence suggested that Na carries current through cation channels that are activated as a result of Ca entering the cell through voltage-gated Ca channels (Lansman 1983). First, total inward current in normal ASW (470 mM Na) is the sum of a calcium component and a sodium component which can be separated by subtraction on a semilogarithmic scale. The slower component is absent when Na is replaced by Tris or when Ba substitutes for Ca. Secondly, inward tail currents measured after rapidly repolarizing the membrane during current flow is also composed of two components. The slower component of inward tail current is absent when Na is replaced by Tris. Finally, the reversal potential for the slow tail is near zero mV.

Separation of the calcium and sodium components of peak inward current is shown in Fig. 3. Figure 3A shows sample records of current responses to a 1 second voltage pulse to the potential indicated at the left of the figure. Inward currents in normal 25 Ca ASW (470 mM Na, Fig. 3Aa,c) do not decay to zero current by the end of the voltage step and are characterized by a slowly relaxing tail after the pulse is terminated. Removal of Na abolishes both the slowly decaying component during the pulse and the slow tail of inward current following cessation of the pulse. Peak inward current as a function of the pulse potential is plotted in Fig. 3B for the immature oocyte and Fig. 3C for the mature egg. The sodium current I-V relations (dotted lines) were obtained by subtracting the peak calcium current (measured in Na-free ASW) from the peak total current (measured in normal ASW). This method of current separation is justified since the peaks of the two currents occur at approximately the same time.

If the time courses of the underlying calcium and sodium permeability changes are sufficiently different, the total membrane conductance at any time can be split into its calcium and sodium components. Figure 4 shows the time course of the permeability change to Na during a voltage pulse to -30 mV. Pulses of increasing duration were applied to the egg membrane and current responses superimposed on the oscilloscope. Inward current transients following the termination of the voltage pulse consisted of two components, a fast inward relaxation of the Ca permeability and a slower inward relaxation of the Na permeability.

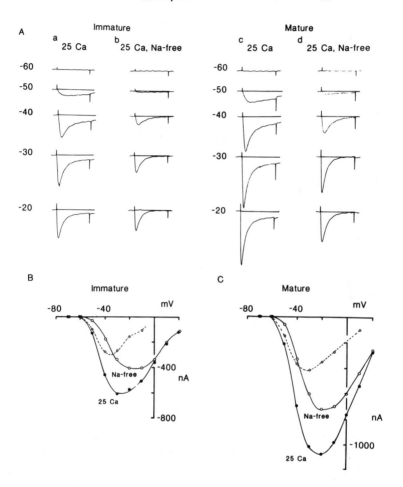

Figure 3. Inward currents recorded under voltage clamp in a <u>Leptasterias</u> egg before and 1 hour after exposure to 1-MA. In B and C, I–V relations plot peak inward current in 25 Ca ASW (●) and Tris-substituted Na-free 25 Ca ASW (O). The difference between these I–V relations (◇) represents the Na component.

Since the initial amplitude of each component of the tail current is proportional to its conductance, the time course of the underlying permeability change to Na is given by the envelope of the initial amplitudes of the slow inward relaxations (circled in Fig. 4) Inward current underlying

the starfish egg action potential, then, is due to current
through two channels of differing ion selectivities: one, a
voltage-gated Ca channel, the other, a cation channel
activated by Ca ions entering the cell.

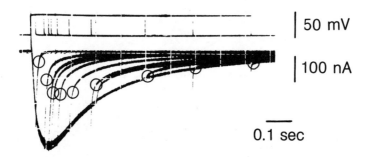

50 mV

100 nA

0.1 sec

Figure 4. Time course of the permeability change to
Na. The membrane was rapidly repolarized to -80 mV
following voltage steps to -30 mV of increasing duration.
The slow tail is absent when Na is replaced with Tris or
TEA.

Changes in inward current during oocyte maturation

During oogenesis, starfish oocytes develop until
reaching meiotic prophase. Under the influence of the
hormone 1-methyladenine (1-MA) which is released from
surrounding follicle cells, the oocyte completes meiosis
(Kanatani 1973). This can be seen as the breakdown of the
germinal vesicle and formation of polar bodies. Several
dramatic changes in action potential properties occur
during this process including a decrease in threshold, an
increase in rate of rise, and an increase in overshoot
(Miyazaki et al. 1975; Shen, Steinhardt 1976; Moody,
Lansman 1983).

Voltage clamp experiments show that maturation is
associated with an increase in the amplitude of inward
current (Fig. 3). Comparison of the I-V relations of the
immature oocyte (Fig. 3B) with those of the mature egg
(Fig. 3C) shows that the amplitudes of both the calcium and
sodium currents increase after maturation.

When the external calcium concentration is raised both

the calcium and sodium currents should increase. The calcium current increases because of the larger driving force for current flow while the sodium current increases because of its dependence on I_{Ca}. Figure 5 shows an experiment in which the peak calcium and sodium currents during a voltage pulse to -30 mV were plotted as a function of the calcium concentration in the external solution. Figure 5a shows the relationship between I_{Ca} and $[Ca]_o$ before and one hour after exposure to 1-MA. Because the calcium current saturates when the external calcium concentration is high, Ca ion permeation has been suggested to involve the binding of Ca to a channel site (Hagiwara, Takahashi 1967). The saturation of I_{Ca} at higher calcium concentrations in the mature egg could be due to a change in either the binding constant or the number of channels. The result, however, is a larger calcium current in the mature egg. Fig. 5b shows the relationship between sodium current and the external Ca concentration. It is unchanged except for a parallel shift of the relation along the voltage axis. In the mature egg, the Na current can be produced at a much lower external Ca concentration. This shift could be due simply to the larger Ca current in the mature egg; however, an increase in the number of Na channels or a change in the sensitivity of the sodium channel to Ca could also occur.

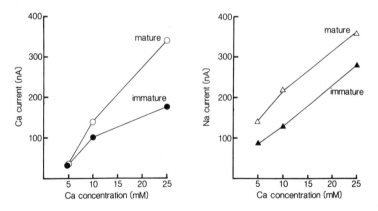

Figure 5. Peak calcium and sodium currents during a voltage pulse to -30 mV plotted as a function of the external Ca concentration. Data from a single Leptasterias egg before and 1 hour after exposure to 1-MA. Current expressed as absolute value.

Figure 6 shows an experiment designed to examine whether there is a change in the sensitivity of the sodium current to Ca. Data were obtained in a single cell for voltages (-50 to -30 mV) which activate a large inward calcium current. In Fig. 6, the relationship between Na conductance and calcium current before (●) and 1 hour after (○) exposure of the egg to 1-MA is shown. The sodium conductance is an approximately linear function of calcium current when I_{Ca} is small. The slope of the relation does not change suggesting that the sensitivity of the Na system to Ca is unaffected by the maturation process. There also appears to be little change in the saturation level of the Na conductance for Ca currents greater than about 300 nA. The number of Na channels therefore, may not change during maturation. These experiments show that calcium current activation of sodium influx is operative during the development of the egg prior to fertilization.

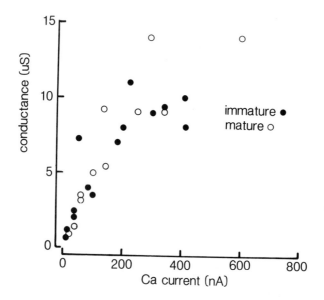

Figure 6. Na conductance calculated from $g_{Na} = I_{Na}/V-E_R$ where E_R was experimentally determined to be approx. 0 mV. Data from a single cell. Results for different voltage steps (-50 to -30 mV) were pooled.

Fertilization currents.

Depolarization of the egg by sperm is due to an inward current carried predominantly by Na which flows through channels having only moderate selectivity for monovalent cations (Lansman 1983). Figure 7 shows a record of the inward fertilization current recorded from a Mediaster egg. The action potential component is absent since the egg is held at its resting potential (-70mV) by the voltage clamp. In this record, short depolarizing pulses were applied to the egg membrane every 10 seconds and the corresponding current deflections are proportional to conductance. The conductance turns on, rises to a maximum at peak inward current, and then decays toward the pre-fertilization level.

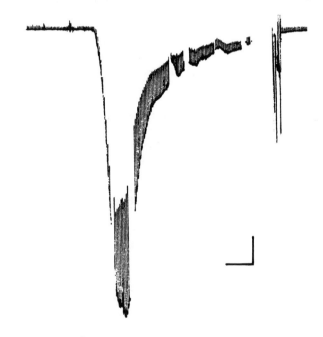

Figure 7. Fertilization current recorded under voltage clamp from a mature egg of the starfish, Mediaster aequalis. Holding potential, -70 mV. Short depolarizing pulses were applied to the egg membrane once every 10 seconds. (Calibration bars, vertical=50 nA, horizontal = 10 min.)

What mechanism controls the time course of the sperm-initiated conductance change? There are at least three possibilities: sperm may bind directly to a receptor which controls the opening of a channel, sperm may insert channels into the egg membrane, or sperm may induce the release of an intracellular second messenger which controls channel activation. The last possibility is supported by experiments in which eggs are artificially activated with Ca ionophore. In the sea urchin egg, a potential change similar to the fertilization potential occurs upon exposure to Ca ionophore (Steinhardt, Epel 1974). Similarly, the Ca ionophore is a sufficient stimulus for producing an activation current in the tunicate egg under voltage clamp (Kozuka, Takahashi 1981). Since a rise in the internal free calcium concentration of the egg is an almost universal feature of the program of egg activation (Epel 1978) it is reasonable to assume that the fertilization current is linked to changes in $[Ca]_i$. This suggests, by extension, that the inward current triggered by calcium influx during the action potential may represent a 'transient' activation of fertilization channels.

Although Ca ions may control the underlying permeability change, voltage also influences the fertilization current (Lansman 1983). Figure 8 shows an example of how voltage regulates the fraction of channels that are open at any time during fertilization. After insemination when the fertilization current had reached its peak inward level, the membrane potential was shifted to 0 mV. At this potential there is almost no inward current. Inward current is activated by hyperpolarizing the egg membrane in 10 mV steps as shown. The steady-state value of the time-dependent current is plotted in the lower half of Figure 8 for families of currents evoked 15 and 30 minutes following insemination. The net effect of voltage on the fertilization current is to produce a time-dependent inward rectification. At a fixed time during fertilization, the membrane slope conductance is large at negative potentials and decreases with depolarization. The same relationship holds 30 min. after insemination except that now the total conductance at the resting potential (-70 mV) is less. The total conductance (at a given voltage) available at different times during the fertilization current most likely represents the number of activated fertilization channels. This, in turn, presumably represents the level of intracellular free calcium.

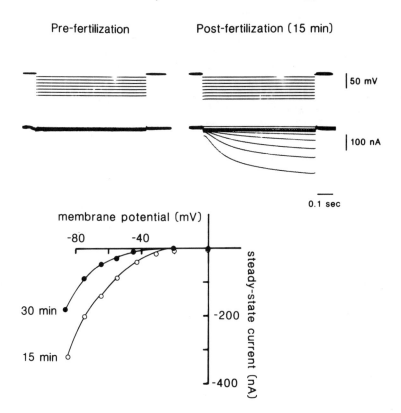

Figure 8. The effect of voltage on the fertilization current of the <u>Mediaster</u> egg. An egg was fertilized at −70 mV and the fertilization current allowed to reach its peak inward level. The holding potential was then shifted to 0 mV where channel open probability is near zero. Hyperpolarizing voltage steps reopened fertilization channels producing an inward current which increased with time. Before fertilization no currents are elicited by hyperpolarizing voltage steps. The steady-state currents at the end of the voltage pulse are plotted in the I–V relations in the lower half of the figure.

The relationship between conductance and voltage (Fig. 9, n_{∞} = steady-state activation variable for the fertilization current) predicts that at the peak of the fertilization potential (+10 to +15 mV) only a fraction of the channels available at the resting potential are open.

Recently, the channel open probability of single
fertilization channels in an ascidian egg has been reported
to be approximately 0.1 near 0 mV (DeFelice, Dale 1983).
These results are consistent with a model in which
activation of fertilization channels occurs in two steps:
an initial binding of Ca ions followed by a voltage-
dependent opening.

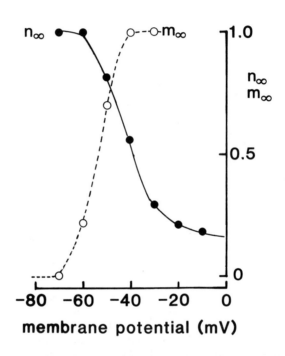

Figure 9. Voltage dependence of the steady-state
activation variables for the fertilization current (n) and
the calcium current (m).

What might the physiological role of the voltage-
dependent gating of the fertilization current be? There is
good evidence suggesting that one function of the
fertilization potential in the echinoderm egg is to prevent
polyspermic fertilization by blocking fusion of extra sperm
to the egg membrane (Jaffe 1976; Miyazaki, Hirai 1979).
The effectiveness of such a block would be determined by
how fast the egg membrane could depolarize to the blocking
voltage. During initiation of the fertilization potential,

the egg membrane begins to depolarize towards action potential threshold. Over this same voltage range, fertilization channels begin to close. Fig. 9 shows the voltage dependence of the steady-state activation variables for the fertilization current (n) and for the Ca current (m). At membrane potentials where I_{Ca} is maximally activated a substantial fraction of fertilization channels are closed. The voltage-dependent closure of fertilization channels insures a low total membrane conductance during generation of the action potential. This may serve to amplify the action potential generated by sperm-egg interaction.

In summary, two components of the starfish fertilization potential can be identified: action potential and sperm-dependent depolarization. The action potential has both a Ca component and a Na component which is activated by Ca entering the cell. The sodium current increases linearly with calcium current and its activation kinetics are sufficiently fast so that it follows I_{Ca}. Both currents increase after maturation but neither the sensitivity nor the number of channels carrying Na changes. The fertilization current is carried predominantly by Na. It flows through channels of low selectivity which indirect evidence suggests may be activated by intracellular Ca. Are sperm-dependent currents carried by the same channels that allow Na entry during Ca inward current? Certainly it would be more efficient to have a single channel type mediate these two similar functions. Further work should focus on the behavior of these currents in both immature and mature eggs, their selectivity and on their sensitivity to Ca ions and membrane potential. Detailed information on the mechanism of the electrical changes may then be related to the general program of egg activation and development.

REFERENCES

DeFelice LJ, Dale B (1983). Sperm activated channels in ascidian oocytes. Biophys J 41:48a.
Epel D (1978). Mechanisms of activation of sperm and egg during fertilization of sea urchin gametes. Current Topics in Developmental Biology 12:185.

Hagiwara S, Byerly L (1981). Calcium channel. Ann
  Rev Neurosci 4:69.
Hagiwara S, Jaffe LA (1979). Electrical properties of
  egg cell membranes. Ann Rev Biophys Bioeng 8:385.
Hagiwara S, Ozawa S, Sand O (1975). Voltage clamp analysis
  of two inward current mechanisms in the egg cell
  membrane of a starfish. J Gen Physiol 65:617.
Hagiwara S, Takahashi K (1967). Surface density of
  calcium ions and calcium spikes in the barnacle muscle fiber
  membrane. J Gen Physiol 50:583.
Jaffe LA (1976). Fast block to polyspermy in sea urchin
  eggs is electrically mediated. Nature 261:68.
Kanatani H (1973). Maturation inducing substances in
  starfishes. Int Rev Cytol 35:253.
Kozuka M, Takahashi K (1982). Changes in holding and
  ion channel currents during activation of an Ascidian egg
  under voltage clamp. J Physiol 323:267.
Lansman JB (1983). Voltage clamp study of the conductance
  activated at fertilization in the starfish egg. J Physiol
  (in press).
Lansman JB (1983). Calcium current and calcium-dependent
  inward current in the immature starfish oocyte. Biophys J
  41:61a.
Miyazaki S, Hirai S (1979). Fast polyspermy block and
  activation potential: Correlated changes during oocyte
  maturation of a starfish. Dev Biol 70:327.
Miyazaki S, Ohmori H, Sasaki S (1975a). Action
  potential and non-linear current-voltage relation in
  starfish oocytes. J Physiol 246:37.
Miyazaki S, Ohmori H, Sasaki S (1975b). Potassium
  rectifications of the starfish oocyte membrane and their
  changes during oocyte maturation. J Physiol 246:55.
Moody WJ, Lansman JB (1983). Developmental
  regulation of Ca and K currents during hormone induced
  meiotic maturation of starfish oocytes. Proc Natl Acad
  Sci USA (in press).
Shen S, Steinhardt RA (1976). An electrophysiological
  study of the membrane properties of the immature and
  mature oocyte of the Batstar, Patiria miniata. Dev
  Biol 48:148.
Steinhardt RA, Epel D (1974). Activation of sea urchin
  eggs by a calcium ionophore. Proc Natl Acad Sci
  USA 21:1915.
Whitaker MJ, Steinhardt RA (1982). Ionic regulation
  of egg activation. Quart Rev Biophys 15:593.

The Physiology of Excitable Cells, pages 247–256

SOME ELECTRICAL PROPERTIES OF THE XENOPUS LAEVIS OOCYTE

Raymond T. Kado

Laboratoire de Neurobiologie Cellulaire
Centre National de la Recherche Scientifique
91190 Gif sur Yvette, France

INTRODUCTION

Perhaps because they need to communicate with their surrounding cells as well as with more distant relatives, most cells are morphologically very complex. They tend to have highly convoluted membranes. On the other hand, all microelectrode electrophysiological techniques specifically measure only those electrical properties near the measuring sites which include averages for the more distant regions. Together, these properties combine to give such measurements a less than quantitative character. Complex cell geometries will necessarily lead to weighted averages for the membrane currents and voltages and the weighting factors will be determined by the cytoplasmic and membrane properties between the measuring site and more remote regions of the cell. The ideal cell is spherical and has uniform membrane properties.

A spherical cell which spends only a part of its time in close apposition with other cells is the oocyte. This cell probably has the distinction of being the first single cell to come under the scrutiny of physiologists but its use in membrane electrophysiology is relatively recent and the demonstration of excitability came even later (see Hagiwara, Jaffe 1979).

XENOPUS OOCYTES

One oocyte which has been extensively studied is that from the African clawed toad, Xenopus laevis. A drawing of the oocyte is shown in figure 1. It is large as cells go,

having a diameter as large as 1.4 mm when fully grown. It
is pigmented on the animal pole and the germinal visicle is

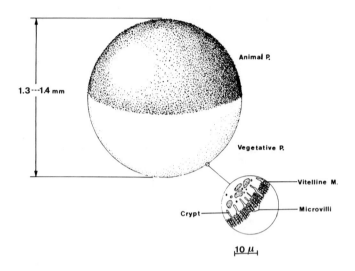

Figure 1.   Drawing of fully grown <u>Xenopus laevis</u> oocyte.

found in this pole. It is optically opaque due to the
nature of a thin layer just under the plasma membrane. The
plasma membrane consists of a great many microvilli and
crypts. Interwoven with the microvilli is a porous cover-
ing consisting of collagen-like fibers, the vitelline
membrane.

A great advantage of <u>Xenopus</u> is that at any time of the
year, fully grown oocytes may be obtained from pieces of the
ovary excised from mature females. The female gametes obtained in this way are not capable of being fertilized. The
oocyte must first pass through the meiotic maturation processes before it can be fertilized but the exact time during
this process when fertilization occurs appears to be
unclear (Hagiwara, Jaffe 1979; Masui, Clarke 1979).

OOCYTE MATURATION

It is known that oocytes can be made to undergo meiotic
maturation <u>in vitro</u> in the presence of appropriate hormones.

By this means it has been possible to establish that during the maturation process the membrane undergoes a morphological change which consists of the loss of a large part of its surface membrane in the form of microvilli and crypts. At the end of the process it has been shown in electron micrographs that the membrane presents an almost smooth surface (Dumont 1976). Cytochemically, maturation includes a wide variety of processes in this oocyte. Cell metabolism, protein synthesis, and RNA synthesis have been studied along with the changes in the membrane electrical properties (see review by Masui and Clarke, 1979).

The fact that a large amount of membrane could disappear without changing the size of the oocyte seemed to present the possibility of studying the effects of surface complexity in a geometrically simple cell.

We had already demonstrated the possibility of following the time course of changes in surface membrane area by the measurement of its capacitance (Jaffe et al. 1978). In that work we have shown that the addition of surface membrane by the attachment and opening of the cortical granules (subsurface vesicle-like organelles) in the eggs of the sea urchin, led to almost a doubling of the surface area. It therefore seemed reasonable to myself, René Ozon, head of the laboratory of reproductive physiology at the University of Paris VI and Krista Marcher from the laboratory, that we might be able to follow the decrease in membrane area which should accompany the disappearance of the microvilli and crypts in the Xenopus oocytes.

The dimensions of the microvilli are important. They are less than 0.1 micron in diameter and about 5 microns long. Depending on their membrane cable properties, they could be partially invisible to currents injected through a pipette into the oocyte. The crypts are much larger, about 0.6 microns in diameter and 5 microns deep. Since Xenopus oocytes are more than ten times greater in diameter than the sea urchin egg the surface area is more than 100 times larger. This difference made it necessary to use the more classical current pulse technique to measure the capacitance of the membrane.

A good agreement could be obtained between the measured capacitances and that which might be expected for the number of microvilli and crypts seen to exist in EM photographs of the oocyte surface (Dumont, Brummet 1978). A specific capacitance of 1 $\mu F/cm^2$ for about 5 microvilli and one crypt per square micron of apparent surface area will account for the measured capacitance. This study also revealed that this membrane has a very high specific resistance, on the order of megohms-cm^2, in spite of leakage produced by the electrode insertion technique.

Figure 2 shows the relative time courses for the changes in potential, resistance and capacitance for an oocyte undergoing meiotic progesterone-induced maturation. The bar indicates the period of establishment of a white spot in the animal pole which is the visible sign of germinal vesicle breakdown (GVBD) or early phase of meiotic division. The electrical parameters all began to change at the same time as GVBD, but this must not be taken to mean that the changes in the membrane properties are causally related to the maturation process. It has also been shown that cytoplasmic processes such as protein phosphorylation

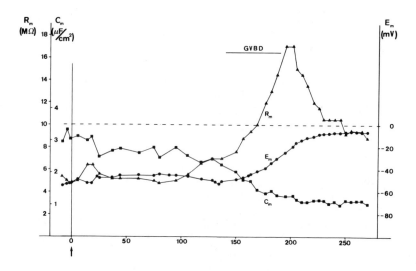

Figure 2. A 1.35 mm oocyte matured with 10^{-5} M progesterone in OR-2 medium starting at the arrow. Time is in minutes. The time of GVBD is indicated from the beginning of the white spot to its final form. (Courtesy of K. Marcher, unpublished figure).

show a very similar time course to the changes in resistance
(Belle et al. 1978). The appearance of a cytoplasmic maturation promoting factor is also known to occur at about one
half the time between the application of progesterone and the
appearance of GVBD (Wasserman, Masui 1975). The depolarization to about -10 mV from the initial resting potential of
around -60 mV is accompanied by an increase in the input
resistance (Kado et al. 1981). Both of these changes can be
explained by a decrease in the K permeability known to occur
in this period (Morrill, Ziegler 1980). The resistance is
seen to decrease again after GVBD without a repolarization
of the membrane. This may be explained by a shutting down
of the Na-K pump which has recently been reported to occur
during this period in the maturation of Rana oocytes
(Weinstein et al. 1982). It seems likely that the changes
seen in the membrane properties here are directed by the
processes in the cytoplasm during maturation.

OOCYTE DEVELOPMENT

The microvilli appear only after some early oocyte
growth and disappear again after meiotic maturation, indicating a role in oocyte growth. They have therefore been
studied in various ways. E.G. Dick and colleagues have
followed the development of the microvilli on Rana oocytes
by their water permeation and electron microscopic images
(Dick et al. 1970). For the Xenopus oocyte, Dumont has
followed the growth of surface structures with electron
micrographs in sufficient detail to be able to identify
stages in the development. Dumont numbered the succesive
stages from I to VI which include oocyte diameters from 50
to 1300 microns (Dumont 1972). We have also examined this
process by measuring the membrane capacitances of oocytes at
varying stages of growth (Marcher, Kado in preparation).
Most cells have specific capacitances of nearly one $\mu f/cm^2$,
larger values usually indicating membrane surfaces not
included in the area calculation. The table below illustrates the extra surface acquired by the growing oocyte as
indicated by the capacitance per cm^2 for areas calculated
from the mean diameter in the range.

The period of greatest increase in the number of
microvilli, as indicated by the increase in the apparent
specific capacitance, occurs when the oocyte has attained
500 to 700 microns in diameter. These diameters fall between stages III and IV which is the period of greatest

vitellogenic activity. Numerous endocytotic pits appear in the crypts and the base of the microvilli at this time (Brummett, Dumont 1976). Such pits may account for the peaking in the specific capacitance in the 500 to 600 micron diameter range. From this diameter range until about 1100 microns, the specific capacitance does not markedly change, suggesting that although the cell almost doubles in diameter it has probably not increased the number of microvilli or crypts. We are encouraged to have confidence in these capacitance measurements since during the two stages in the life of the oocyte where it is known to have small or no surface structures, our measurements give specific capacitances approaching 1 $\mu F/cm^2$. Oocytes in the 100 to 200 micron range give specific capacitances of 2.2 $\mu F/cm^2$ while at 1350 microns and after maturation, the value for the oocyte in figure 1 is 1.05 $\mu F/cm^2$. While the oocytes are still under about 200 microns in diameter, they have only

Dumont stage	diameter (micron)	capacity (nf)	uf/cm^2
I	100–200	6.5	2.3
	200–300	8.8	4.4
II	300–400	15.2	4.0
III	400–500	39.0	6.1
	500–600	71.3	7.5
IV	600–700	91.8	6.9
	700–800	113.3	6.4
	800–900	139.2	6.3
	900–1000	187.4	6.6
V	1000–1100	228.4	6.6
	1100–1200	240.7	5.8
VI	1200–1300	235.7	4.8
	1300–1400	229.2	4.0

Table I. Membrane capacitance in developing oocytes. The Dumont stages above have been assigned to the diameters we have measured in our experiments and correspond to within 50 microns of the diameters given by Dumont.

very stubby microvilli and after maturation they are very smooth again having only scattered surface irregularities (Dumont 1972).

The oocyte is therefore not a smooth, spherical cell ideally suited for membrane electrophysiology, and would have been abandoned except for a chance observation.

OOCYTE EXCITABILITY

The capacitance measurements had been made using small constant current pulses to obtain membrane potential shifts following the membrane time constant. In general, these current pulses had to be kept to one or two nanoamperes to avoid changes in potential due to the Na-K pump which was apparently activated or inactivated by the K ions carrying current across the membrane (Kado, Marcher and Ozon, unpublished observations). If sucessive large depolarizations were produced (to above +30 mV), the membrane no longer behaved in a purely passive fashion, but began to respond in a regenerative way and finally remained at positive potentials (Kado et al. 1979). The long lasting depolarization produced could reach +80 mV from an initial resting potential of -60 mV. This process is illustrated in Figure 3

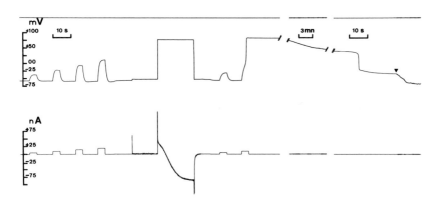

Figure 3. Channel induction by a voltage clamp depolarization imposed in the middle of the tracings. Upper trace membrane potential; lower, current. The breaks and downward arrowhead separate segments of recording at very slow speed. The switching from current to voltage clamp is evident in the two traces. (From C. Baud 1982 by permission).

where an oocyte initially in current clamp shows passive
potential responses to pulses of injected current. In the
second segment of this record, the membrane was voltage
clamped to +50 mV, resulting in a membrane current which
slowly changed from outward to inward. In the third segment the membrane was again under current clamp and a
current pulse which previously produced only a passive
response now elicited, in a regenerative way, a long
lasting depolarization.

This long lasting depolarization was shown to be due
to an electrically gated Na channel which was not present
in the membrane before the membrane was clamped to +50 mV
and which persisted for some time. Once "induced", the
channel remains electrically excitable at normal resting
potentials for as long as 15 minutes at temperatures lower
than about 15 °C. At higher temperatures this period
decreases.

This persistence of excitability explains the regenerative potential produced by a current pulse which before had
only evoked passive responses. That the channels close
again is evident at the end of the voltage clamp depolarization when repolarization to resting potential results in
a rapid inward transient followed by a return to zero
current. The results of many studies on this channel indicate that in spite of its rather unusual origins it is
indeed an ion selective channel with many similarities to
the other more well known channels. It also does not inactivate while the potential remains more positive than
about +50 mV (Kado, Baud 1981; Baud et al. 1982).

The origins of this channel are especially interesting
since the phenomenon of "inducing" a channel with membrane
potential change is not common. The activation of a "sleepy"
population of Na channels in the squid giant axon by depolarization has been reported (Matteson, Armstrong 1982).
These channels also show little or no inactivation and a
slow activation but, as with most such channels, they do not
seem capable of supporting membrane depolarizations which
last as long as 10 to 30 minutes. The induced oocyte
channels appear to be incapable of a normal inactivation so
that voltage clamp inward currents, such as that in the
middle of the current trace in figure 3, are maintained for
many minutes. The oocyte membrane has the advantage that
under conditions of temperature lower than about 15°C, this

is virtually the only channel that can be made to appear
(Baud, Kado in preparation). This property considerably
facilitates studying the induced channel since no pharmacological manipulations are necessary to eliminate other
channels that might produce interfering currents.

An induced excitability is clearly not the same as an
excitability which is in place and ready to respond to the
first depolarization. The sodium dependent, voltage gated
channel described here is most likely induced by the large
inversions of the membrane potential. If it is assumed
that this membrane already has an appropriate sodium selective channel through which it obtains sodium when required
by cytoplasmic processes, it is reasonable that such a
channel will be equipped with some kind of gating mechanism.
Such a gating mechanism is probably not electrically activated, at least not in the same way in which we have induced the channels, since there appears to be no way in
which the oocyte membrane will be able to produce the
necessary depolarizations. It probably has a gating
mechanism which is mediated cytochemically. Inversion of
the membrane potential could transform the internal gating
mechanism so that it becomes temporarily sensitive to
voltage. This is the biggest if, and the most intriguing.
The implications for the possible modification of channel
populations by sustained membrane potential shifts are very
broad indeed. Of more immediate interest is that the lack
of inactivation in this channel may make it a good model
for the study of selectivity mechanisms and, of course, if
the above assumptions are anywhere near valid, we may have
another way of studying electrical channel gating.

REFERENCES

Baud C (1982). Excitabilité membranaire dans l'ovocyte de
    Xenopus laevis, induite par depolarization prolongee de la
    membrane. Thèse Doc 3e cycle Univ Paris VI.
Baud C, Kado RT, Marcher K (1982). Sodium channels induced
    by depolarization of the *Xenopus laevis* oocyte. Proc
    Nat Acad Sci USA 79:3188.
Belle R, Boyer J, Ozon R (1978). Endogenous protein phos-
    phorylation in *Xenopus laevis* oocytes. Quantitative and
    qualitative changes during progesterone induced
    maturation. Biol Cellulaire 32:97.
Brummet AR, Dumont JN (1976). Oogenesis in *Xenopus laevis*

(Daudin). III. Localization of negative charges on the surface of the developing oocytes. J Ultrastruc Res 55:4.

Dick EG, Dick DAT, Bradbury S (1970). The effect of surface microvilli on the water permeability of single toad oocytes. J Cell Sci 6:451.

Dumont JN (1972). Oogenesis in Xenopus laevis (Daudin). I. stages of oocyte development in laboratory maintained animals. J Morphol 136:153.

Dumont JN, Brummet AR (1978). Oogenesis in Xenopus laevis (Daudin). V. Relationships between developing oocytes and their investing follicular tissues. J Morphol 155:73.

Hagiwara S, Jaffe LA (1979). Electrical properties of egg cell membranes. Ann Rev Biophys Bioeng 8:385.

Jaffe LA, Hagiwara S, Kado R (1978). The time course of cortical vesicle fusion in sea urchin eggs observed as membrane capacitance changes. Dev Biol 67:243.

Kado RT, Marcher K, Ozon R (1979). Mise en évidence d'une dépolarization de longue durée dans l'ovocyte de Xenopus laevis. C. R. Acad Sci Paris 288D:1187.

Kado RT, Marcher K, Ozon R (1981). Electrical membrane properties of the Xenopus laevis oocyte during progesterone-induced meiotic maturation. Dev Biol 84:471.

Kado RT, Baud C (1981). The rise and fall of electrical excitability in the oocyte of Xenopus laevis. J Physiol Paris 77:1113.

Masui Y, Clarke HJ (1979). Oocyte maturation. Int Rev Cytol 57:185.

Matteson DR, Armstrong CM (1982). Evidence for a population of sleepy sodium channels in squid axon at low temperature. J Gen Physiol 79:739.

Morrill GA, Ziegler D (1980). Na^+ and K^+ uptake and exchange by the amphibian oocyte during the first meiotic division. Dev Biol 74:216.

Wasserman WJ, Masui Y (1975). Effects of cycloheximide on a cytoplasmic factor initiating meiotic maturation in Xenopus oocytes. Exp Cell Res 91:381.

Weinstein SP, Kostellow AB, Ziegler DH, Morrill GA (1982). Progesterone-induced down-regulation of an electrogenic Na^+, K^+-ATPase during the first meiotic division in amphibian oocytes. J Memb Biol 69:41.

The Physiology of Excitable Cells, pages 257–265
© 1983 Alan R. Liss, Inc., 150 Fifth Avenue, New York, NY 10011

LITHIUM BLOCKS CURRENT THROUGH VOLTAGE-INDUCED, VOLTAGE-
GATED SODIUM CHANNELS, IN XENOPUS OOCYTES

Christiane Baud

Laboratoire de Neurobiologie Cellulaire
CNRS   91190 Gif-sur Yvette
France

The preceding report (R.T. Kado, this volume and Kado
et al. 1979) describes a membrane permeability change
occuring in the membrane of the Xenopus laevis oocyte,
during prolonged membrane depolarization. The membrane
which is normally selective for potassium ions becomes selective to sodium ions, provided it has been depolarized for
several seconds to a positive potential. Using voltage
clamp technique, it has been shown that this phenomenon can
be dissociated into two different processes taking place
successively (Kado, Baud 1981; Baud et al., 1982). The
first one is a slow process (on the time scale of seconds)
by which the channels become potentially gatable. It has
sigmoidal kinetics and occurs at positive membrane potentials. It has been referred to as "induction" of sodium
channels and has been studied in some detail (Baud and
Kado 1983). The second process is relatively more rapid
and corresponds to the opening of the sodium channels.
Comparison of the oocyte channels with more classical
sodium channels can be summarized as follows. They differ
from other sodium channels in that 1) they have to be
induced by a depolarization of several seconds to a positive
potential; 2) they do not inactivate; 3) they will slowly
disappear (time scale of several minutes) if the membrane
potential is maintained at a negative value. However, part
of the molecular structure is likely to be very similar to
the other sodium channels since 1) after induction the
channels will close if the membrane is stepped to a negative
potential, resulting in an exponential tail current; 2) they
will open with a time-course of tenths of seconds under
successive depolarizations; 3) they selectively allow sodium

ions to go through, as demonstrated by the 52 mV change in reversal potential for a ten-fold change in external sodium concentration.

In this report, the effect of substitution of lithium ions for sodium is described. Substitution of all the external sodium by lithium produces no shift in the reversal potential of the current through the channels. However, the current carried by lithium ions is much smaller than that carried by sodium ions. At relatively low concentration, lithium ions act as blockers of the sodium current, in a voltage dependent way. This behavior differs from that of other potential gated sodium channels where lithium is not discriminated from sodium, but it is reminiscent of the effect of lithium on the Na-K ATPase, where lithium ions are not transported.

The method of obtaining oocytes has been extensively described elsewhere (Baud and Kado 1983). They were surgically removed from the ovary and treated for two hours in Dispase (Protease, Neutrale, Grad II, Boehringer, 40 mg/100 ml) before manual removal of the follicular cells. They were then used during several days after dissection, and kept overnight at 5°C. The solution used for both maintenance and recording is the OR2 medium (Wallace et al. 1973) having the following ionic composition (mM): Na 84.5, K 2.5, Ca 1, Mg 1, PO$_4$ 1, HEPES 5, Cl 86.5; pH 7.4. The cells typically had resting potentials between -60 and -80 mV and input resistances ranging between 2 and 6 Megohms. After dissection of the cells out of the follicular envelope, a ouabain-sensitive sodium extrusion occurs leading to a decrease in internal sodium activity (Vitto, Wallace 1976; Wallace, Steinhardt 1977). The cells used displayed a reversal potential for sodium ions of +80 to +90 mV, corresponding to an internal sodium activity of 2 to 4 mM. A difficulty arose in the reproducibility of I-V curves in a given oocyte. It has been reported that induction time course as well as current amplitude obtained in a given cell during two successive trials were rarely exactly reproducible (Baud, Kado 1983). The difference between two I-V curves obtained under same conditions in the same cell with a 10 minute interval varies between 0 and 15% of the maximum inward current, and could be either larger or smaller during the second trial. We have no explanation for this. In the present report a given cell was used to obtain two sets of data, one in normal medium, the other in lithium solution 10 to 15 minutes later; as a consequence, the data are contaminated by this uncontrolled variability.

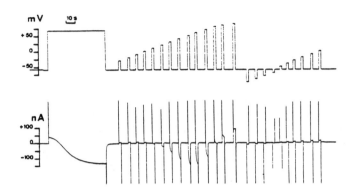

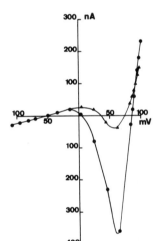

Fig. 1. A. Protocol to obtain I-V characteristics of the channels: upper trace membrane potential, lower trace current. A long lasting inducing depolarization to +60 mV is first applied for 50 seconds. Test pulses (5 sec) are then applied to different potentials. T°=12° in this experiment. B. I-V curve obtained after induction as described in a. Data from a different cell. T°=16.5°C. (●) in Na 84.5 mM (no Li), (▲) in Li 84.5 mM (no Na).

EFFECT OF TOTAL SUBSTITUTION OF SODIUM BY LITHIUM

Figure 1A shows the standard protocol by which current-voltage curves were obtained. The potential was first shifted to +60 mV for several seconds. The current displayed the sigmoidal shape that we have attributed to the induction plus opening processes. The membrane potential was then stepped back to resting level (-60mV), the current displayed an inward tail current due to the closing of the channels. The membrane was then again depolarized to a

positive potential and a rapid inward current was observed,
corresponding to the gating open of the channels. Potential
steps to different voltages were then applied, during 2-3
minutes, before the channels start to disappear. When the
current was plotted against membrane potential, an I-V curve
with a pronounced inward rectification was obtained. The
same protocol was then repeated on the same cell 10 minutes
later, after substitution of all the external sodium by
lithium. Figure 1B shows I-V curves obtained under such
conditions. The leakage current was not subtracted for any
of the data. The curves show that the reversal potential
for the permeant ions - +90 mV in this cell - was not
affected by the substitution. However, the peak current
intensity was reduced to about one fourth of its value in
sodium solution. Since the channels were induced and gated
open in lithium solution, the possibility existed that reduction of the current in lithium solution could be due
to an inhibitory effect of lithium on induction or on opening. These possibilities can be easily tested taking
advantage of the fact that the channels do not inactivate.
The external ionic composition can be changed while the
channels are in the open state. One such experiment is
shown in the inset of figure 2. Forty mM lithium has been
substituted for sodium. The beginning of the trace shows
induction in normal sodium solution. The lithium solution
was then introduced in the bath at the time indicated. The
inward current decreased upon arrival of the lithium solution
in the chamber. When the normal sodium solution was re-
admitted, the inward current returned to its original value.
(In some cases, the inward current reached a somewhat lower
value, due either to accumulation of sodium inside the cell
or to the activation of a very slow outward current which
has been described elsewhere (Baud and Kado 1983)).

DOSE-DEPENDENCE OF THE BLOCKING EFFECT

From experiments such as that described in the inset
of figure 2, a percentage of current reduction was calculated for different lithium and sodium concentrations. The
percentage of reduction was defined as

$$\frac{100 \times (I_{(Na)} - I_{(Na, Li)})}{I_{(Na)}}$$

Li/Na concentration ratio was varied systematically with the sum Li+Na being kept constant. This kept the reversal potential at a constant value and thus the assumption was made

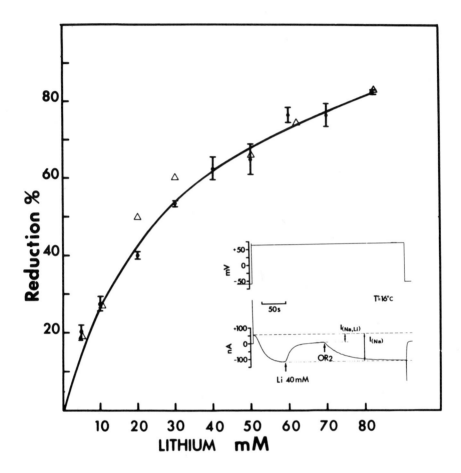

Figure 2. Inset. Reduction of inward current upon arrival of lithium in the chamber, while the channels are kept in the open state. Percentage of reduction of the inward current when sodium is replaced by equal amount of sodium. (●) data obtained from 2 cells following the protocol described in the inset; (Δ) data obtained by comparing I-V curves obtained following the protocol described in figure 1.

that the chord conductance in each case was proportional to
the current. All the experiments were done using a depolarization to +55 mV at 17°C. The percentage reduction vs
the lithium (and sodium) concentration is plotted in figure
2. The plot shows a clear dose dependence of the reduction
of current by lithium. The curve showed no saturation at
high substitution ratio; this was expected since lithium
acts as a blocker of the sodium current but does itself carry
current.

MEMBRANE POTENTIAL DEPENDENCE

The blocking effect at different membrane potentials
was analyzed in a systematic way by comparing I-V plots
such as the ones shown in figure 1B rather than using the
protocol described in the preceeding paragraph. A percentage of current reduction was calculated in the same way.
The two methods gave results in good agreement. In figure 2
data obtained using the two different methods are plotted
simultaneously and show a close correlation supporting the
idea that lithium ions exert their effects on the open
channel only and have no effect on induction or opening.
Inhibition at different potentials is shown in figure 3
for +30, +40, +50, +60, +70 mV. A different cell was used
for each lithium concentration. The curves in figure 3
show that the blocking effect is strongly potential dependent, being more pronounced at less positive membrane
potentials.

The data reported here show that in the sodium selective channels of the Xenopus oocyte membrane, lithium ions
are not distinguished from sodium ions on the basis of the
reversal potential. However, the current is much smaller
in lithium solution than in sodium solution and the presence of lithium ions reduces the sodium current amplitude.
Such effects have long been known in other channels, and
demonstrate the inadequacy of simple diffusion theory to
account for ion permeation in transmembrane channels
(Hille 1975). The interpretation of these data has to be
done in terms of interaction of the ions with channel
binding sites. The current can then be theoretically derived by calculating the rate of jump of the ions from site
to site, using chemical reaction rate theory and Eyring
rate theory. To account quantitatively for the curves presented in figures 2 and 3, two facts will complicate the

analysis. First, the current measured is carried by both
sodium and lithium, so equations have to be derived for both
ions; second, the sodium concentration is not constant along
the abscissa, which is itself expected to have an influence
on the sodium current. However, the analysis can be made
much simpler taking into account only the range of small
lithium concentrations; the current carried by lithium can
then be neglected and it can be assumed that all the current
is carried by sodium. The effect of varying sodium concentration on sodium current can be neglected too. Furthermore,
the measurements being done in steady state, the inhibition
effect arises directly from the fact that a given channel
is occupied by a lithium ion instead of a sodium ion. The
reduction of current is then directly proportional to the
fraction of channels being occupied by a lithium ion,
following the simple scheme:

Li + Channel $\rightleftharpoons$ Li-Channel

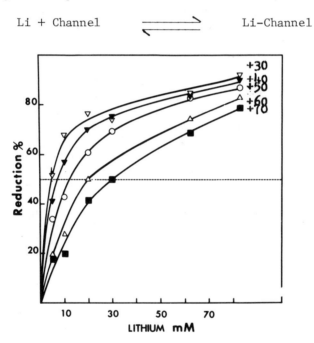

Figure 3. Percentage of inward current reduction at
different membrane potentials. All data obtained by comparing I-V curves.

The concentration of lithium at which half of the sodium current is blocked gives the apparent Km for the binding reaction between lithium and the channel. A Km of 5 mM can be estimated at +30 mV and of 7 mM at +40mV.

The fact that permeabilities estimated from reversal potential measurements are identical for lithium and sodium as is the case in other sodium channels (Hille 1972) might indicate that the molecular structure responsible for ion selectivity, the "selectivity filter", is similar in the <u>Xenopus</u> oocyte channel and in other sodium channels. On the other hand, the small amount of current carried by lithium indicates that once inside the channels, lithium ions travel at a much lower rate. This can be accounted for by a stronger interaction of lithium with an internal binding site. This feature is generally not observed in other sodium channels. Since the present channel also lacks an inactivation mechanism, it is possible that the two facts are linked and that the lack of an inactivation mechanism (which indicates a modification of the channel at the inner side of the membrane) reveals a site of high affinity for lithium ions.

Finally, lithium ions have been long known to have effects at low concentration in such different situations as manic-depressive illness and embryological development. The only well established difference between sodium and lithium is difference in binding affinity to the Na-K ATPase: sodium ions are actively transported by the ATPase out of the cell, while lithium ions are not and will accumulate inside the cell (Keynes, Swan 1959). Here I have shown that a sodium selective channel exists, with a very low conductance to lithium. This might help clarify the way by which lithium ions exert their drastic effects.

References

Baud C, Kado RT (1983). Induction and disappearance of excitability in the oocyte of *Xenopus laevis*: a voltage-clamp study. Submitted for publication.

Baud C, Kado RT, Marcher KS (1982). Sodium channels induced by depolarization of the *Xenopus laevis* oocyte. Proc Natl Acad Sci USA 79:3188.

Hille B (1972). The permeability of the sodium channel to metal cations in myelinated nerve. J Gen Physiol 59:637

Hille B (1975). Ionic selectivity, saturation and block

in Na channels, a four barrier model. J Gen Physiol 66: 535.

Kado RT, Baud C (1972). The rise and fall of an electrical excitability in the oocyte of *Xenopus laevis*. J Physiol (Paris) 77:1113.

Kado RT, Marcher KS, Ozon R (1979). Mise en evidence d'une dépolarisation de longue durée dans l'ovocyte de *Xenopus laevis*. C R Acad Sci Paris 288D:1187.

Keynes RD, Swan RC (1959). The permeability of frog muscle fibers to lithium ions. J Physiol 147:626.

Vitto A, Wallace RA (1976). Maturation in *Xenopus* oocytes. I. Facilitation by ouabaine. Exp Cell Res 97:56.

Wallace RA, Jared DW, Dumont JN, Sega MW (1973). Protein incorporation by isolated amphibian oocytes. III. Optium incubation conditions. J Exp Zool 184:321.

Wallace RA, Steinhart RA (1977). Maturation in *Xenopus* oocytes. II. Observation on membrane potential. Dev Biol 57:305.

Y. Fukushima; C. Baud; J. Lansman

The Physiology of Excitable Cells, pages 267–277
© **1983 Alan R. Liss, Inc., 150 Fifth Avenue, New York, NY 10011**

EXCITABILITY OF OVARIAN OOCYTES AND CLEAVING EMBRYOS OF THE
MOUSE

Shigeru Yoshida

Department of Physiology
Nagasaki University School of Medicine
Nagasaki 852, Japan

Abstract: Ovarian oocytes and cleaving embryos were taken
    from mice and intracellular recordings were performed
    using glass microelectrodes. Action potentials depend-
    ent on divalent (Ca^{2+}, Sr^{2+}, Ba^{2+} or Mn^{2+}) or monovalent
    (Na^+ or Li^+) cations were observed in these preparations.
    Monovalent-cation spikes were resistant to tetrodotoxin
    (TTX) and were blocked by polyvalent cations such as
    Co^{2+}, Cd^{2+}, Mn^{2+} and La^{3+}. A competition phenomenon
    was observed between Na^+ & Ca^{2+} and Na^+ & Mn^{2+}. It is
    concluded that the mouse oocyte has only Ca channels
    from ovarian oocyte through to 8-cell embryo and both
    divalent and monovalent cations can pass through the
    membrane to produce action potentials (Yoshida 1982;
    Yoshida 1983).

INTRODUCTION

    Electrical properties of egg cells or oocytes have be-
come a matter of increasing interest, and reviews have been
published concerning these properties from invertebrate and
vertebrate animals (Hagiwara, Miyazaki 1977; Hagiwara, Jaffe
1979; Hagiwara, Byerly 1981). However, there are only a few
papers dealing with mammalian oocytes (Powers, Tupper 1974;
Okamoto, Takahashi, Yamashita 1977; Powers 1982; Yamashita
1982).

    This report mainly describes the properties of mouse
ovarian oocytes. In addition, description of embryo properties from stage 1-cell through to 8-cell will also be
discussed.

METHODS

Mature female mice of ICR strain (older than 8 weeks) were used. The ovaries were dissected out, and the follicles were broken with sharp needles to obtain the cumulus clots containing oocytes. Oocytes were isolated by dispersing the follicular cells with 0.2 % hyaluronidase (Iwamatsu & Chang,1972). Ovulated oocytes were collected by opening the oviducts after injection, approximately 16 hrs previously, of 5-10 units HCG (human chorionic gonadotropin). The surrounding follicular cells were removed by the same treatment using 0.2 % hyaluronidase. Cleaving embryos were obtained at various intervals from mated female mice.

Composition of the standard solution was (mM): NaCl, 140; KCl,5; $CaCl_2$,2; $MgCl_2$,1 and glucose,5.6. The pH of the solution was buffered by 5 mM PIPES at 7.4. In Na^+-free solutions, NaCl was replaced by TMA·Cl (tetramethyl-ammonium chloride). In Ca^{2+}-free solutions, 2 mM EGTA was added to chelate residual Ca^{2+}. All solutions were prepared at pH 7.4.

Intracellular recordings were performed with single 3 M KCl microelectrodes (30-60 MΩ) using a bridge circuit to inject current into the oocyte or embryo for stimulation. All experiments were carried out at room temperature (20-23·C).

RESULTS

A. OVARIAN OOCYTES

a) Action potentials in standard solution

The resting potential of the isolated ovarian oocytes in standard solution was -7.0 ± 1.8 mV, and the membrane resistance ranged between 11 and 100 MΩ . The oocyte generated action potentials when the inactivation of ionic channels was removed by hyperpolarizing the membrane with a prepulse (Fig. 1A). The action potentials are very slow in time course.

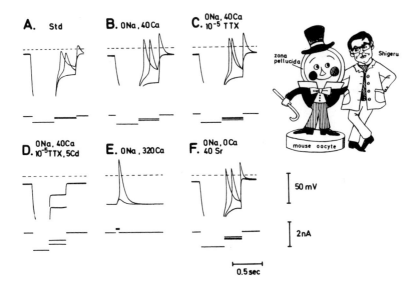

Fig. 1. Action potentials obtained from the mouse ovarian oocytes. In order to reveal action potentials, two-step current stimulation was performed by triggering the second pulse on the end of the first pulse. The first pulse was used to remove inactivation of ionic channels by hyperpolarizing the membrane, and the second pulse was used to elicit action potentials. Numbers indicate the concentration of ions in mM. Broken line: reference potential or zero potential level. A: spikes in standard solution. B-E: Ca spikes. F: Sr spikes. Voltage, current and time calibrations apply to all records.

b) Action potentials dependent on divalent cations

The isolated ovarian oocytes produced action potentials under Na^+-free conditions. Figs. 1B-E & 2 show properties of Ca spikes. The spikes illustrated in Fig. 1B were insensitive to high concentration (10^{-5}g/ml) of TTX (tetrodotoxin) (Fig. 1C) and were blocked by polyvalent cations such as Cd^{2+} (Fig. 1D), Co^{2+}, Mn^{2+} or La^{3+}. The resting potential varied in linear fashion with $\log[Ca^{2+}]_o$ (Fig. 2A, E_r), and the ovarian oocyte generated a spike with a short depolarizing pulse at 320 mM Ca^{2+} (Fig. 1E).

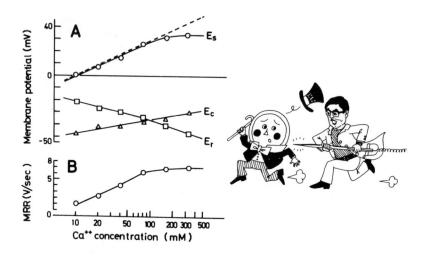

Fig. 2. Effects of $[Ca^{2+}]_o$ on parameters of Ca spikes.
A: spike overshoot(E_s), critical membrane potential(E_c)
and the resting potential(E_r). Broken line: Nernstian
slope of 29 mV for a tenfold increase in $[Ca^{2+}]_o$.
B: maximum rate of rise(MRR) of the spikes. $[Ca^{2+}]_o$ is
indicated in logarhythmic scale.

The threshold membrane potential or the critical level of
the spike (Fig. 2A, E_c) became more positive with the increase of $[Ca^{2+}]_o$. The spike overshoot showed a slope of
28 mV per tenfold change of $[Ca^{2+}]_o$, and it saturated when
$[Ca^{2+}]_o$ was highly elevated (Fig. 2A, E_s). The maximum
rate of rise(MRR) of the spikes changed directly with external Ca^{2+} concentration (Fig. 2B). Spikes persisted when
external Ca^{2+} was replaced by Sr^{2+} (Fig. 1F) or by Ba^{2+}.

It is therefore concluded that the ovarian oocyte membrane of the mouse has voltage-dependent Ca channels.

c) Action potentials dependent on monovalent cations

The isolated ovarian oocytes showed action potentials
dependent on monovalent cations under Ca^{2+}-free conditions.
The oocyte membrane became leaky in the absence of external

Ca^{2+}. The overshoot (E_s) and the maximum rate of rise (MRR) of the spikes increased and saturated with increase of external Na^+ concentration (Figs. 3A-C & 4). An overshoot sensitivity was 39 mV per tenfold change of $[Na^+]_o$ (Yamamoto & Washio,1979). The resting potential became smaller when $[Na^+]_o$ was increased (Fig. 4A, E_r) while the critical level did not substantially change (Fig. 4A, E_c).

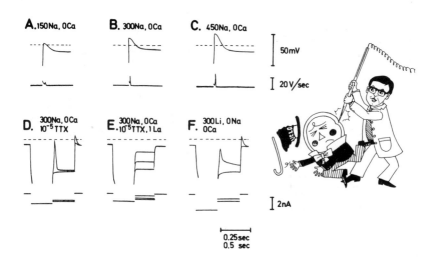

Fig. 3. Action potentials dependent on monovalent cations in the isolated ovarian oocytes of the mouse. Broken line: reference potential level. A-C: effects of $[Na^+]_o$ on the overshoot and the maximum rate of rise of the spikes under Ca^{2+}-free conditions. The Na spike was resistant to TTX (D) and was blocked by 1 mM La^{3+}(E). Li^+ substituted for Na^+ in producing an action potential (F). Voltage scale is the same for all records. Time calibration is 0.25 sec for A-C and 0.5 sec for D-F. Lower traces of A-C: differentiated records of the membrane potential.

The Na spikes were resistant to 10^{-5}g/ml tetrodotoxin (Fig. 3D) and were blocked by so-called Ca-blockers such as La^{3+}(Fig. 3E), Co^{2+}, Cd^{2+} or Mn^{2+}. Thus, it is considered that the Na^+ goes through the Ca channel instead of a separate Na channel in the ovarian oocyte membrane of the mouse. Li^+ substituted for Na^+ in producing action potentials (Fig. 3F). Such spike (Li spike) was also blocked by polyvalent cations. In contrast, Rb$^+$ did not maintain action potentials, suggesting that the permeability of Rb$^+$ is negligible in the mouse ovarian oocytes.

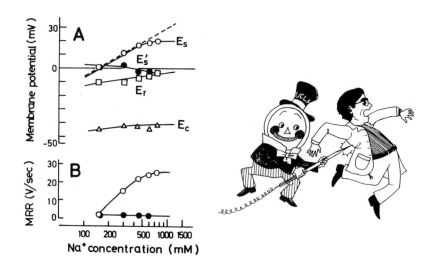

Fig. 4. Effects of [Na$^+$]$_o$ on parameters of Na spikes. A: spike overshoot(E$_s$), critical potential(E$_c$) and the resting potential(E$_r$). Broken line: slope of 39 mV for a tenfold increase in [Na$^+$]$_o$. Closed circles show the spike overshoot(E$_s$⁻) and the maximum rate of rise(MRR) of the spikes in the presence of 5 mM Ca^{2+}. External Na$^+$ concentration is indicated in logarhythmic scale.

Competition phenomenon was observed between divalent and monovalent cations. Fig. 4 shows that the spike overshoot(E$_s$⁻) and the MRR do not increase with [Na$^+$]$_o$ in the presence of 5 mM Ca^{2+} in the bathing solution (closed circles). The same phenomenon was also observed between Na$^+$ and Mn^{2+} which goes through the Ca channel to produce a Mn spike (not illustrated). The competition supports the idea that the Na$^+$ passes through the Ca channel. Divalent and monovalent cations would compete at the binding site of the Ca channel, and the affinity of monovalent cations to the binding site might be much smaller than that of divalent cations.

d) Actions of Mn^{2+}

Mn^{2+} showed two actions in the ovarian oocyte membrane of the mouse: (1) Mn^{2+} acted as a Ca blocker, and (2) Mn^{2+} acted as a charge carrier during excitation (Mn spike).

Spikes depending on either divalent (Ca^{2+}, Sr^{2+}, Ba^{2+}) or monovalent (Na^+, Li^+) cations were blocked by 10 mM Mn^{2+}.

Mn spikes were observed when the outward K current was suppressed by raising the external K^+ concentration from normal 5 to 155 mM or when external Mn^{2+} concentration was elevated to 80 mM.

e) Resting membrane of the mouse ovarian oocyte

It is described above that the resting potential of the ovarian oocyte varies with $[Ca^{2+}]_0$, the relationship having a negative slope (Fig. 2A, E_r), and with $[Na^+]_0$ having a positive slope (Fig. 4A, E_r). The resting potential changed with a small positive slope when external K^+ concentration was increased: -13.3 ± 5.9 mV at 5 mM K^+, -12.7 ± 6.4 mV at 55 mM K^+, and -11.3 ± 3.2 mV at 155 mM K^+ under Na^+-, Ca^{2+}-free conditions. It is suggested that the resting membrane is permeable to not only Na^+ but also to some extent to K^+, and that Ca^{2+} reduces the leakage due to a stabilizing effect (Okamoto et al.,1977).

f) Conclusion

The isolated ovarian oocyte membrane of the mouse has voltage-dependent Ca channels, and both divalent (Ca^{2+}, Sr^{2+}, Ba^{2+}, Mn^{2+}) and monovalent (Na^+, Li^+) cations can pass through the Ca channels in producing action potentials.

B. OVULATED AND UNFERTILIZED OOCYTES

Okamoto and his colleagues (Okamoto et al.,1977) studied mature and unfertilized oocytes collected from the mouse oviducts. The resting potential in standard solution was -23.1 ± 2.9 mV. Such ovulated oocytes showed regenerative responses which were dependent on divalent cations, Ca^{2+}, Sr^{2+} or Ba^{2+}. A small current carried by Mn^{2+} was revealed, but a Na current was concluded to be negligible.

The same material was studied in the present work, and it was found that the ovulated and unfertilized mouse oocytes can produce action potentials dependent on either divalent or monovalent (Na$^+$,Li$^+$) cations. The Na or Li spikes were insensitive to TTX and were blocked by Ca antagonists as in the case of ovarian oocytes. Thus, it is considered that the Ca channels remain in the oocytes after ovulation, and Na channels do not appear in this stage of development.

C. CLEAVING EMBRYOS

Excitability of cleaving mouse embryos was examined at stages of 2-, 4- and 8-cells. These fertilized oocytes showed action potentials dependent on divalent or mono-valent cations. The Na or Li spikes were resistant to TTX and were completely suppressed by Ca blockers at all stages.

Na channels could not be detected at all stages in the present study. It is expected that Na channels would appear later than the stage of 8-cell.

Fig. 5 shows Na and Ca action potentials observed during development.

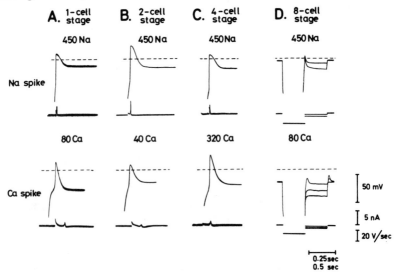

Fig. 5. Na and Ca spikes obtained at different stages of development. Time scale of 0.5 sec applies to records D.

DISCUSSION

It has been known that there is a change in the ionic
dependence of action potentials during development in neurons and muscle cells (see review by Spitzer, 1979). In a
number of cases, inward current is at first carried in
large part by Ca^{2+} , then a Na current appears later, and,
in some instances, the Ca current disappears at late stages
of development.

As for egg cells, Takahashi and his colleagues have
been studying the tunicate. The cells destined to become
muscle cells were followed during development, and it was
shown that the spikes depended on both Na^+ and Ca^{2+} in unfertilized egg cells and the spikes were exclusively dependent on Ca^{2+} in differentiated muscle cells (Miyazaki
et al, 1972, 1974). A study performed in the presumptive
muscle and ectodermal regions of the cleavage-arrested embryos of the tunicate revealed similar results (Takahashi
& Yoshii, 1981).

Tunicate eggs have the Na channels in addition to
the Ca channels (Okamoto, Takahashi & Yoshii, 1976). Recently, it was found that Na channels can be induced in
the membrane of the Xenopus oocyte by depolarization.
These Na channels have no inactivation mechanism (Baud,
Kado & Marcher, 1982).

The present study shows that Ca channels exist from
ovarian oocytes through to 8-cell embryos of the mouse,
and that no Na channels are present in these stages of development. The Ca channels of the mouse oocytes were found
to be permeable both to divalent and monovalent cations as
reported in insect muscle cells (Yamamoto & Washio, 1979)
and in cardiac preparations (see review by Reuter, 1979).
However, such Ca channels have not been described in other
eggs.

Further study is necessary to determine the stage of
Na channel appearance in the mouse embryos.

A part of the results presented here appears in the
following articles.
1) Yoshida S. Na and Ca spikes produced by ions passing
   through Ca channels in mouse ovarian oocytes.
      Pflügers Archiv, 395:84-86, 1982.

2) Yoshida,S. Permeation of divalent and monovalent cations through the ovarian oocyte membrane of the mouse. J. Physiol. (Lond.) (In press, 1983).

ACKNOWLEDGEMENTS

I wish to express gratitude to Dr. Yoshihiro Matsuda who participated in the research of cleaving mouse embryos, to Mr. Masaru Yogata and Mrs. Nui Momosaki for making figures, and to Mr. Robert M. Clay for reading the manuscript. This research was supported by a grant from the Japan Ministry of Education.

I remember my golden time in Dr. Susumu Hagiwara's laboratory spent between May 1979 and May 1981. He taught me many things, and one of them is that we should enjoy our experiments.

REFERENCES

Baud C., Kado R.T. & Marcher K. Sodium channels induced by depolarization of the Xenopus laevis oocyte. Proc. Natl. Acad. Sci. USA, 79:3188-3192, 1982.

Hagiwara S. & Byerly L. Calcium channel. Ann. Rev. Neurosci. 4:69-125, 1981.

Hagiwara S. & Jaffe L.A. Electrical properties of egg cell membranes. Ann. Rev. Biophys. Bioeng. 8:385-416, 1979.

Hagiwara S. & Miyazaki S. Ca and Na spikes in egg cell membrane. Prog. Clin. Biol. Res. 15:147-158, 1977.

Iwamatsu T. & Chang M.C. Sperm penetration in vitro of mouse oocytes at various times during maturation. J. Reprod. Fert. 31:237-247, 1972.

Miyazaki S., Takahashi K. & Tsuda K. Calcium and sodium contributions to regenerative responses in the embryonic excitable cell membrane. Science 176:1441-1443, 1972.

Miyazaki S., Takahashi K. & Tsuda K. Electrical excitability in the egg cell membrane of the tunicate. J. Physiol. (Lond.) 238:37-54, 1974.

Okamoto H., Takahashi K. & Yamashita N. Ionic currents through the membrane of the mammalian oocyte and their comparison with those in the tunicate and sea urchin. J. Physiol. (Lond.) 267:465-495, 1977.

Okamoto H., Takahashi K. & Yoshii M. Two components of the calcium current in the egg cell membrane of the tunicate. J. Physiol. (Lond.) 255:527-561, 1976.

Powers R.D. Changes in mouse oocyte membrane potential and permeability during meiotic maturation. J. Exp. Zool. 221:365-371, 1982.

Powers R.D. & Tupper J.T. Some electrophysiological and permeability properties of the mouse egg. Dev. Biol. 38:320-331, 1974.

Reuter H. Properties of two inward membrane currents in the heart. Ann. Rev. Physiol. 41:413-424, 1979.

Spitzer N.C. Ion channels in development. Ann. Rev. Neurosci. 2:363-397, 1979.

Takahashi K. & Yoshii M. Development of sodium, calcium and potassium channels in the cleavage-arrested embryo of an ascidian. J. Physiol. (Lond.) 315:515-529, 1981.

Yamamoto D. & Washio H. Permeation of sodium through calcium channels of an insect muscle membrane. Can. J. Physiol. Pharmacol. 57:330-222, 1979.

Yamashita N. Enhancement of ionic currents through voltage-gated channels in the mouse oocyte after fertilization. J. Physiol. (Lond.) 329:263-280, 1982.

Yoshida S. Na and Ca spikes produced by ions passing through Ca channels in mouse ovarian oocytes. Pflügers Archiv 395:84-86, 1982.

Yoshida S. Permeation of divalent and monovalent cations through the ovarian oocyte membrane of the mouse. J. Physiol. (Lond.) (In press, 1983).

Top: H. Sakata; S. Chichibu; K. Takahashi; K. Toyama; H.
Koike; P.L. Marchiafava
Bottom: L. Byerly; S. Yoshida; R. Meech; S. Hestrin
(back to camera)

The Physiology of Excitable Cells, pages 279–289
© 1983 Alan R. Liss, Inc., 150 Fifth Avenue, New York, NY 10011

DEVELOPMENT OF CALCIUM CHANNELS IN THE MEMBRANE OF THE
CLEAVAGE-ARRESTED EMBRYO OF ASCIDIANS

Tomoo Hirano and Kunitaro Takahashi

Laboratory of Neurobiology
Institute of Brain Research
University of Tokyo, Tokyo, Japan

Calcium channels, opened by membrane depolarization and
responsible for Ca influx, are an essential property of most
excitable cells (Hagiwara, Byerly 1981). Ca channels have
been shown to exist in invertebrate striated muscle, in
heart muscle, smooth muscle, neuroendocrine cells and special
regions of neuronal membrane, in axon terminals where transmitters are released, and in some parts of the neuronal soma
and dendrites. While striated muscle fibers in vertebrates
have been thought to generate only Na spikes and no Ca current, recent precise analysis of membrane currents after
suppressing the Na current with TTX and K current with TEA
has revealed a Ca-dependent inward current which is clearly
distinguished from Na current by its time course under
voltage-clamp (Sanchez, Stefani 1978; Almers, Fink, Palade
1981). Further, during ontogenesis the existence of Ca channels at early stages of development seems to be universal
among excitable cells from various preparations, such as
myotubes derived from a rat skeletal muscle cell line
(Kidokoro 1975), myotubes grown in vivo (Kano 1975), Rohon-
Beard cells in Xenopus tadpoles (Baccaglini, Spitzer 1977)
and dorsal root ganglion cells of vertebrates (Baccaglini
1978). For example, the presence of a Ca-dependent plateau
following a fast Na-dependent spike has been observed in the
myotube just after fusion of the mononucleated myoblasts
(Kidokoro 1975). Therefore, changes in the quantity and
quality of Ca channels can be expected during differentiation
of excitable cells.

The Ca channel was first described in crusteacean striated muscle fibers and then found in neurons of the molluscan

ganglion. Ca channels characteristically show high permeability to Ba ions and a long-lasting inward current with
much less inactivation than the Na channel (Hagiwara, Byerly
1981). Furthermore, in the Ca channel, the Ca current shows
more inactivation than the Sr or Ba current, indicating that
the inactivation is dependent on the ion species carrying the
current (Brehm, Eckert, Tillotson 1980; Ashcroft, Stanfield
1981; Eckert, Tillotson 1981). This is assumed to be due to
Ca-induced inhibition of Ca channels by the accumulation of
Ca ions inside the cell produced by the inward Ca current
itself (Brown, Morimoto, Tsuda, Wilson 1981). On the other
hand, recent studies have revealed another type of Ca channel
in oocytes of mammals, tunicates and echinoderms (Hagiwara,
Jaffe 1979; Okamoto, Takahashi, Yamashita 1977). This Ca
channel shows a potential-dependent inactivation which is
very similar to the Hodgkin and Huxley type found in the Na
channel. The time course and the degree of inactivation are
not dependent upon ion species carrying the current. Further,
the Ca channel is strongly permeable to Sr ions. Thus, the
egg Ca channel is clearly different from the Ca channel found
in other excitable membranes. Since the egg cell membrane is
actually situated at the starting point of differentiation,
the Ca channel may change its properties from the egg cell
type to another type during the differentiation of embryonic
excitable membranes.

It has been demonstrated that cytochalasin B, which is
known to inhibit cytokinesis, can arrest cell-cleavage in the
ascidian embryo at any stage of development while not interfering with the development of tissue specific enzymes, such
as AChesterase in muscle cells and tyrosinase in pigment cells
(Whittaker 1973). It has recently been demonstrated that the
membrane differentiation evidenced by production of action
potentials also occurs in the cleavage-arrested embryo
(Takahashi, Yoshii 1981). The blastomeres in cleavage-arrested
8- or 16-cell embryos can show differentiation to three types
of excitable cells: neural, muscular and epidermal types.
Ca channels are observed in each of these membrane types.
Even in large blastomeres of the cleavage-arrested 1-, 2- and
4-cell embryos, excitability appears. In those cases, a single
blastomere may have mosaic properties of various types, because
the fate of the blastomere is not uniquely determined. However,
in reality the differentiated unicellular embryo shows a characteristic long-lasting action potential, which is identified
as epidermal type. The inward current in the differentiated
one-cell embryo is less contaminated by outward current. Thus,

using the uncleaved but differentiated embryo in addition to neural or muscle type blastomeres in the 8- or 16-cell embryo, the analysis of changes in properties of Ca channels can be attempted during differentiation.

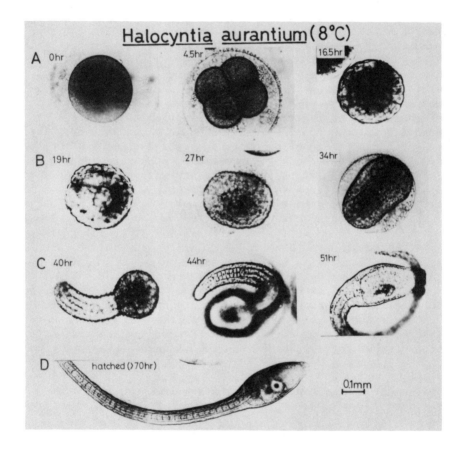

Fig. 1. Sequential photographs of the embryo or the tadpole larva, Halocynthia aurantium. The number in the upper left of each photograph is the developmental time after fertilization. The values of the time were estimated by averaging those of three series of cultures at a temperature of around 8°C.

Experimental Procedures

Adult ascidians, Halocynthia roretzi and Halocynthia aurantium, which had matured eggs and sperm were kept in an aquarium at 4°C. The eggs and sperm of H. roretzi were obtained from the spawning animals after 24 hours in a warm bath of 12°C; those of H. aurantium were dissected out. The eggs were fertilized by sperm from heteronymous animals. The developmental time was measured from this fertilization period in an incubation bath at a constant temperature of 8 to 9°C. In Fig. 1, the morphology of the control embryo of Halocynthia aurantium is illustrated in relation to the developmental times at 8°C. Before first cleavage or at the 8- or 16-cell stage, the embryos were transferred to another bath containing 1 µg/ml cytochalasin B in order to arrest cleavage and cultured further until the control intact tadpole larva hatched. After removing the chorion membrane by digestion with 10 mg/ml pronase, the cleavage-arrested embryos were kept in a high-proteinous external solution and were penetrated with two microelectrodes. The experiments were carried out with either constant current stimulation or voltage-clamp. The standard external solution was 400 mM NaCl. 100 mM $SrCl_2$, 10 mM KCl, 5 mM PIPES-Na (pH 7.0), 10 mg/ml bovine serum albumin (BSA) and 2 µg/ml cytochalasin B (Na-Sr solution). When it was necessary, sodium was replaced by equimolar tetramethyl ammonium (TMA) and equimolar calcium or barium was substituted for strontium. In order to analyse the inactivation process of the Ca channels, tail currents of the Ca current at various periods from the voltage-step were measured and the time course of the Ca conductance was estimated without contamination from the outward current.

Development of Ca channels in cleavage-arrested 16-cell embryos.

Until 50 hours after fertilization, intracellular communication between blastomeres in the cleavage-arrested embryo was so tight that the embryo behaved as a single cell and the sum of membrane currents on the embryonic surface was observed. The total Ca-channel current in an embryo was initially decreased until 20 hours of development and enhanced again after 35 hours. After 50 hours most of the intercellular communication disappeared and the membrane currents in individual blastomeres could be analysed (Takahashi, Yoshii 1981). The presumptive fate of each blastomere in the 16-cell embryo has been described classically and individual blastomeres have been named (Conklin 1905; Hirai 1941; Riverberi 1961; Satoh 1979).

In a cleavage-arrested 16-cell embryo eight blastomeres are included in the vegetal hemisphere. Two large blastomeres in the middle region, B_{5-1}'s, are known to include presumptive muscular region and to show strong positive staining of AChesterase by Karnovsky method. This is consistent with the fact that the muscle cells in the control larva are innervated by cholinergic axons. The two B_{5-1}'s always showed spike potentials of short duration. The spike was a Ca spike, identical to that found in the muscle cells of the control larva (muscular type differentiation; Miyazaki, Takahashi, Tsuda 1972). Under voltage-clamp the Ca current was always followed by a delayed K current. The posterior two small blastomeres, B_{5-2}'s, also include presumptive muscular regions, according to Conklin; however, they rarely showed Ca currents. The anterior four large blastomeres, $A_{5-1,2}$'s, include some neural region, but did not show any voltage-dependent inward current (non-excitable type differentiation).

In the animal hemisphere, anterior small blastomeres, $a_{5-3,4}$'s which include the presumptive brain region, showed an action potential which had a Na-dependent initial peak and a Sr-dependent plateau in Na-Sr solution. However, it was observed in only 59% of the examined population. We call this the neural type action potential. Under voltage-clamp, the inward current consisted of fast and slow components in Na-Sr solution. After replacing Sr with Mn plus Mg the slow component was abolished, and by removing Na ions the fast component was eliminated. Therefore, the fast component represents the current through Na channels, while the slow component was that through Ca channels. Thus, the neuron type membrane contained both Na and Ca channels. When the small blastomeres, $a_{5-3,4}$'s, did not show the neural type action potential, a long-lasting Sr-dependent action potential appeared in Na-Sr solution. Another group of small blastomeres, $b_{5-3,4}$'s, in the posterior of the animal hemisphere, always showed this Sr-dependent action potential, and under voltage-clamp a pure Ca channel current without delayed K current was recorded. Since the b group blastomeres are assigned to the presumptive epidermal region, the long-lasting action potential is called the epidermal type.

Development of Ca channels in the cleavage-arrested 8-cell embryo.

In the cleavage-arrested 8-cell embryos, the large blasto-mere in the posterior portion of the vegetal hemisphere, B_{4-1}, showed a Ca-channel current followed by a delayed K current in 30% of the observed embryos (muscular type differentiation). The same blastomere B_{4-1} was electrically coupled with the group of small blastomeres in the animal hemisphere and differentiated into the epidermal type in 40% of the observed embryos. Sometimes, the blastomere was isolated as in the case of the muscle type but did not show any excitability. In few cases, a small blastomere, a_{4-2}, in the anterior portion of the animal hemisphere was isolated and showed neural type differentiation. Most small blastomeres of the animal hemisphere were coupled together and differentiated to the epidermal type. AChesterase staining by the Karnovsky method was positive when the B_{4-1} differentiated to muscular type.

Development of Ca channels in the cleavage-arrested 1-, 2- and 4-cell embryos.

Although developmental fates of a single blastomere in 1- or 2- or 4-cell embryos of the control are not unique, the actual differentiation in the cleavage-arrested embryo was only the epidermal type. After culturing in seawater contain-ing cytochalasin B until the hatching time of the control larvae, the resting potential of the blastomeres in those differentiated "embryos" was about -70 mV, being at K equili-brium potential in Na-Sr solution. An action potential was evoked by depolarization above -30 mV, its duration being more than two seconds in Na-Sr solution. The action potential was abolished by eliminating Sr ions, but not changed by removal of Na ions. The overshoot was about 50 mV. No evidence of development of delayed K rectification was obtained. The action potential was exactly the epidermal type described above.

In summary, Ca channels of the cleavage-arrested embryos were found in all blastomeres that differentiated to neural, muscular or epidermal types. A relatively isolated Ca-channel current was observed in differentiated unicellular embryos.

Selectivity and inactivation of Ca channels in blastomeres dif-ferentiated to the epidermal type.

In Na-Ca solution in which Ca was substituted for Sr, or in Na-Sr solution, the membrane current in a one-cell embryo that had differentiated to epidermal type showed a marked outward component in addition to the Ca inward current. However, this outward current was abolished by injecting EGTA into the cell. Thus, the outward current was probably a Ca-dependent K current. After injecting EGTA, less inactivation of the current was observed with depolarization above -30 mV and the current showed a maximum at 15 mV in Na-Sr solution. In Na-Mn solution, in which Mn plus Mg were substituted for the Sr of Na-Sr solution, the inward current was eliminated but no outward current was observed, except for the non-time-dependent leakage current. Thus, the membrane current in the differentiated one-cell embryo, after EGTA injection, was identified as almost pure Ca-channel current (Meech, Standen 1975).

In Fig. 2, the Ca-channel currents in Na-free Ca, Sr and Ba solutions are illustrated for comparison with Ca-channel currents through the egg channel. The current records were obtained at membrane potentials where the inward currents became approximately maximal in the respective solutions. As shown in Fig. 2, the divalent cation selectivity of the Ca-channel in the differentiated one-cell embryo was Ca < Sr < Ba (Note the scale change for the Ba current). The selectivity ratio was 4.5 : 2.0 : 1.0 in average (n=6), being quite different from the selectivity ratio 1.1 : 1.9 : 1.0 in the case of egg Ca channels. In addition, the time course of the decay phase was much slower in the differentiated one-cell embryo in Ba or Sr solution, the inward currents having lasted for a second without significant reduction in any range of the membrane potential. Fast inactivation of the Ca channel current was only found in Ca solution, while the inactivation of the egg Ca channel current was potential-dependent and observed similarly in Ca, Sr and Ba solution (Fig. 2).

The Ca channel in the muscular type differentiated blastomeres

In the differentiated blastomeres of muscular type, such as B_{4-1} in the 8-cell embryo or B_{5-1} in the 16-cell embryo, the inward component of the membrane current was abolished by replacing Sr with Mn plus Mg, while the delayed outward current remained. In order to know whether the Ca channel in the muscular type was identical with that in the epidermal type of the differentiated one-cell embryo, the outward current was eliminated as much as possible by injecting both EGTA

and Cs and by keeping the embryo in TEA solution. Although
the inward current tended to decrease gradually during solution exchange, the selectivity sequence of the Ca channel in
the muscle type was Ca < Sr < Ba and only in Na-Ca solution,
was the fast inactivation of the current observed. Thus, the
Ca channel was considered to be similar to that of the epidermal type and we may call both of them the differentiated type.

Ca·Channel Current

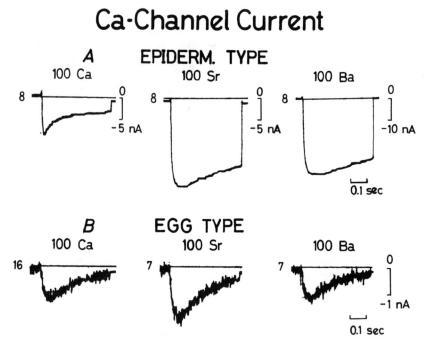

Fig. 2. Current through Ca channels in the egg (B) and the
cleavage-arrested 1-cell embryo (A) of Halocynthia roretzi.
Currents obtained around the potential level where they became maximal in respective solutions. A prepulse to -47 mV
(A) or -35 mV (B) of 400 msec duration was used to inactivate
the outward Na current. Ca induced K outward current was eliminated by using an EGTA/K_2 electrode for current injection in
case of A. 100Ca, 100Sr, 100Ba and 5Mn, 95Mg indicate Ca
TMA ASW, Sr TMA ASW and Mn Mg TMA ASW, respectively. The
figure on the left of each trace indicates the potential level
of the test pulse in mV. Note differences in current scale.

Discussion

After differentiation of excitable membranes in the cleavage-arrested embryos, Ca channels existed in all types of blastomeres except the non-excitable type, while Na channels remained only in the neural type differentiated blastomeres. It is noteworthy that the neural type blastomeres which develop Na-dependent action potentials also have significant amounts of Ca current, at least for some period before maturation. It is not yet known whether Ca channels disappear from neural cells after the maturation in this Halocynthia embryo. However, all the above results suggest that Ca channels in the membrane during development are an essential prerequisite for all blastomeres destined to become excitable cells.

As reported previously, from fertilization until about 10 hours at 15°C (gastrula stage), both Na and Ca currents were reduced on the entire surface of the embryo, while at later stages (after 17 hours at 15°C) both currents were enhanced again (Takahashi, Yoshii 1981). The enhancement corresponded to the differentiation of excitable cells. Thus, the segregation of Ca and Na channels on embryonic excitable membrane is not simple, but is associated with an initial decrease and later increase in their total numbers. In the present paper the Ca channels in the differentiated one-cell embryo were found to be very similar to those found in other differentiated excitable cells with respect to their divalent cation selectivity and inactivation mechanism (see Introduction), and definitely different from those found in various oocytes. During the differentiation of the excitable membranes in the Halocynthia embryo, the egg Ca channel must disappear during initial development while the differentiated Ca channel must appear in coincidence with the enhancement of the Ca current described above. Actually, in the case of the differentiation of the uncleaved 1-cell embryo of Halocynthia aurantium, these properties of the Ca channel, the selectivity and inactivation, were found to change suddenly from the egg type to the differentiated type at 40 hours developmental time after fertilization at 8°C. Thus, it is suggested that the alteration between two kinds of Ca channel or the modification of a kind of Ca channel can occur at a critical period in the differentiation of excitable cells.

The B_{5-1} blastomere in the cleavage-arrested 16-cell embryo or B_{4-1} in the 8-cell embryo showed muscle type differentiation with a Ca spike and AChesterase activity. In these blastomeres, preliminary observations by electron microscopy showed myofibrils underneath the plasma membrane and an accumulation of mitochondria, which is characteristic of muscle cells in the control tadpole larva. Thus, the entire large blastomere seemed to have only muscle characteristics and seemed not to show any mosaic properties mixed with other cellular types. Since B_{5-1} blastomere contains presumptive endodermal regions as well, it is suggested that the blastomere selects only one cellular type from more than one possibility. Further, in the case of the 8-cell embryo, the B_{4-1} blastomere actually chose any one of three cellular types equally, as far as the membrane excitability is concerned.

References

Almers W, Fink, R, Palade PT (1981). Calcium depletion in frog muscle tubules: The decline of calcium current under maintained depolarization. J Physiol 312:177.

Ashcroft FM, Stanfield PR (1981). Calcium dependence of the inactivation of calcium currents in skeletal muscle fibers of an insect. Science N Y 213:224.

Baccaglini PI (1978). Action potentials of embryonic dorsal root ganglion neurones in Xenopus tadpoles. J Physiol 283:585.

Baccaglini PI, Spitzer NC (1977). Development changes in the inward current of the action potential of Rohon-Beard neurones. J Physiol 271:93.

Brehm P, Eckert R, Tillotson D (1980). Calcium-mediated inactivation of calcium current in Paramecium. J Physiol 306:193.

Brown AM, Morimoto K, Tsuda Y, Wilson DL (1981). Calcium current-dependent and voltage-dependent inactivation of calcium channels in Helix aspersa. J Physiol 320:193.

Conklin EG (1905). The organization and cell-lineage of the ascidian egg. J Acad natn Sci Philad 13:1.

Eckert R, Tillotson DL (1981). Calcium-mediated inactivation of the calcium conductance in caesium-loaded giant neurones of Aplysia californica. J Physiol 314:265.

Hagiwara S, Byerly L (1981). Calcium channel. Ann Rev Neurosci 4:69.

Hirai E (1941). An outline of the development of Cynthia roretzi DRASCHE. Sci Rep Tohoku Imp Univ Biology 16:217.
Kidokoro Y (1975). Sodium and calcium components of the action potential in a developing skeletal muscle cell line. J Physiol 244:145.
Meech RW, Standen NB (1975). Potassium activation in Helix aspersa neurones under voltage clamp: a component mediated by calcium influx. J Physiol 249:211.
Miyazaki S, Takahashi K, Tsuda K (1972). Calcium and sodium contributions to regenerative responses in the embryonic excitable cell membrane. Science N.Y. 176:1441.
Okamoto H, Takahashi K, Yamashita N (1977). Ionic currents through the membrane of the mammalian oocyte and their comparison with those in the tunicate and sea urchin. J Physiol 267:465.
Reverberi G (1961). The embryology of ascidians. Advances in Morphogenesis 1:55.
Sanchez JA, Stefani E (1978). Inward calcium current in twitch muscle fibres of the frog. J Physiol 283:197.
Satoh N (1979). On the 'clock' mechanism determining the time of tissue-specific enzyme development during ascidian embryogenesis. I. Acetylcholinesterase development in cleavage-arrested embryos. J Embryol exp Morph 54:131.
Takahashi K, Yoshii M (1981). Development of sodium, calcium and potassium channels in the cleavage-arrested embryo of an ascidian. J Physiol 315:515.
Whittaker JR (1973). Segregation during ascidian embryogenesis of egg cytoplasmic information for tissue-specific enzyme development. Proc Nat Acad Sci USA 70:2096.

S. Hagiwara and K. Takahashi, 1966

SECTION III
PHYSIOLOGICAL ROLES OF ION CHANNELS AND INTRACELLULAR IONS

Top: J. Patlak; R. Horn; G. Wooden; G. Zampighi; G.
Eisenman; N. Standen
Bottom: K. Kusano; J. Fukuda; K. Ikeda; Y. Fukushima; Y.
Kidokoro; K. Negishi; A Watanabe

The Physiology of Excitable Cells, pages 293–294
© 1983 Alan R. Liss, Inc., 150 Fifth Avenue, New York, NY 10011

INTRODUCTION

A keen interest in the relation of ion channels to
overall cell physiology in excitable tissues has been
characteristic of research in Professor Hagiwara's laboratory. This is especially evident in experiments on the
Ca^{2+} channel, where the extraordinary lability of Ca^{2+}
currents has been studied in relation to such processes as
contraction in crustacean muscle fibers or hormone release
from pituitary cells. Those of us who have worked with
Hagi found that we were reminded of the cell physiology
underlying our experiments just as often as we were
corrected in our misconceptions of biophysical theory. The
papers in this section are concerned with a variety of
topics, ranging from optical activity measurements in nerve
fibers during excitation to ultrastructural studies of the
Na-K pump.

The first two papers concern the membrane molecules
mediating nerve excitation and their relationship to
other structural components of the cell. Akira WATANABE,
one of Hagi's earliest associates, reports changes in the
optical activity of stained nerve fibers during passage
of the nerve impulse. This method may yield information
as to the conformational changes in membrane molecules
which occur during the action potential. It will be
interesting in the future to compare data obtained in
this manner with kinetic schemes of channel opening obtained from single channel recordings. FUKUDA explores
not the ion channel molecules themselves, but their
possible relation with submembrane cytoskeletal elements.
It would be especially helpful to have more information
on this subject, given recent reports of clustering of
channel molecules at certain sites over the surface of
the cell membrane.

It is particularly true of cells or portions of
cells that have voltage-dependent Ca^{2+} channels that
calcium ions entering during excitation serve a physiological role in some cellular process such as secretion
or contraction. Therefore, it is important to understand the relationships between the ionic composition

of the cytoplasm, membrane ion currents, and events occurring inside the cell. The remainder of the papers in this section deal directly with this question. Although it is clear that the maintenance of the Na^+ and K^+ gradients across the membrane is required for nerve signaling and the active control of cytoplasmic pH and Ca^{2+} activity, little is known about the physical process by which the $(Na^+ + K^+)$-ATPase splits ATP and translocates ions across the membrane. ZAMPIGHI has investigated this question by using ultrastructural techniques to unravel the three-dimensional structure of the ATPase molecule. Using both ion-sensitive microelectrode and optical absorbance techniques, BROWN has measured the changes in cytoplasmic pH and Ca^{2+} activity which occur during illumination of barnacle photoreceptors. Since both ions have strong effects on the light-induced membrane conductance change, either or both could act to set the sensitivity of the photoreceptor. VERGARA et al. have used optical techniques to measure both the cytoplasmic Ca^{2+} transient and the T-system potential during excitation-contraction coupling. They have concentrated on active propagation of electrical signals in the T-system, and demonstrate here that when propagation of the T-system action potential is blocked by removal of external Na^+, the rise in cytoplasmic Ca^{2+} during excitation is significantly slowed. In the following paper, SAND et al. take up the question of the relation of Ca^{2+} action potentials and hormone release from the GH_3 cell, the same preparation in which single calcium channel currents have been studied (see OHMORI & HAGIWARA, this volume). They demonstrate that thyroliberin, which triggers hormone release from these cells, exerts a biphasic effect on the GH_3 cell, first hyperpolarizing then depolarizing the membrane, causing a late burst of Ca^{2+} action potentials. They further show that the Ca^{2+}-calmodulin complex inhibitor trifluoperazine probably exerts its inhibitory effect on hormone release by blocking Ca^{2+} action potentials.

In the final paper of this section, HENKART discusses the general problem of the fate of Ca^{2+} once it gets into the cell. She reviews morphological techniques for studying the location of Ca^{2+} in cells, and then presents compelling evidence that the endoplasmic reticulum of a variety of cells acts to sequester calcium. An interesting question remains, namely, does the ER release Ca^{2+} into the cytoplasm in these cells by a process analogous to excitation-contraction coupling?

The Physiology of Excitable Cells, pages 295-304
© 1983 Alan R. Liss, Inc., 150 Fifth Avenue, New York, NY 10011

CHANGE IN OPTICAL ACTIVITY OF NERVE FIBERS WITH EXCITATION

Akira Watanabe

National Institute for Physiological Sciences
Myodaiji-cho, Okazaki 444, Japan

INTRODUCTION

This is a preliminary communication which reports that
the optical activity of nerve, stained with certain dyes,
changes when the nerve is excited. It is known that on
excitation the nerve changes many of its optical properties,
such as turbidity, birefringence (Cohen et al. 1968), fluorescence (Tasaki et al. 1969) and absorbance (Ross et al.
1974). The present results suggest that optical activity
is to be added to the list. Since the optical activities of
many macromolecules are closely correlated with their conformation (Jirgensons, 1973), this finding may offer a new
tool for obtaining information on conformational changes of
the membrane macromolecules resulting from excitation.

THEORY

When a polarizer, a birefringent sample and an analyzer
are placed in series in a light beam of intensity I_0 (see
Fig. 1), the intensity, I, of the beam emerging from the
analyzer is given by

$$I = I_0 \{\cos^2(a-p) - \sin 2a \cdot \sin 2p \cdot \sin^2(\delta/2)\} \qquad (1)$$

where a and p are the azimuthal angles between the axis of
birefringence of the sample and the transmission axes of
analyzer and polarizer, respectively, and δ is the phase
angle of retardation of the birefringent sample (cf.
Jerrard, 1948). When the birefringence of the sample
changes by a small amount, $\Delta\delta$, the resultant change in light
intensity is obtained by differentiating Eq.(1) with respect

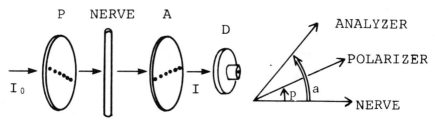

Fig. 1 Left: arrangement of polarizer, sample, analyzer and detector. P, polarizer; A, analyzer; D, photodiode; I_0, incident light; I, transmitted light. Right; diagram to show notations for the azimuthal angles $\underline{a}$ and $\underline{p}$.

to δ and multiplying by $\Delta\delta$; thus the change ΔI_d is

$$\Delta I_d = - \frac{I_0}{2} \sin 2p \cdot \sin 2a \cdot \sin \delta \cdot \Delta\delta \qquad (2)$$

When the optical activity of the sample changes by a small amount, $\Delta\alpha$, the resultant change in light intensity is obtained by differentiating Eq.(1) with respect to $\underline{p}$ and muliplying by $\Delta\alpha$; thus the change, ΔI_a, is

$$\Delta I_a = I_0 [\sin 2(a-p) - 2\cos 2p \cdot \sin 2a \cdot \sin^2 \frac{\delta}{2}]\Delta\alpha \qquad (3)$$

The settings of the polarizer and analyzer were chosen so that ΔI_d was at its minimum and ΔI_a was at its maximum. For this purpose one of the azimuthal angles in Eq.(2) was set to zero or 90°, whereas the difference between the azimuthal angles, $\underline{a}$ and $\underline{p}$, was set to 45°. When $\underline{a} > \underline{p}$, the combination of the azimuthal angles of the polarizer and analyzer will be called setting A. When the azimuthal angles of the polarizer and analyzer are exchanged, then $\underline{a} < \underline{p}$, and the combination of the azimuthal angles will be called setting B. One example of such pairs of settings is shown in Fig. 2. When, for example, $\underline{a} = 0°$ and $\underline{p} = 45°$, Eq.(3) reduces to

$$\Delta I_a = -I_0 \cdot \Delta\alpha \qquad (4)$$

When $\underline{a} = 45°$ and $\underline{p} = 0°$, Eq.(3) reduces to

$$\Delta I_a = I_0 (1 - 2 \sin^2 \frac{\delta}{2}) \Delta\alpha \qquad (5)$$

The second term in the parenthesis is usually smaller than one, and so the sign of ΔI_a reverses on exchanging the azimuthal angles of the analyzer and polarizer. Despite of the

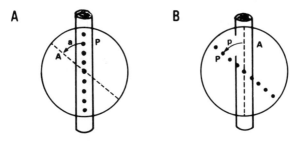

Fig. 2 Examples of settings of azimuthal angles used for the experiment. The nerve is held vertically. A, optical axis of analyzer. P, optical axis of polarizer. Left: Setting A. Right: Setting B.

reversal of the optical activity signal, the birefringence signal is unchanged by the exchange of the azimuths, as is expected from Eq.(2). This is true even when some of the nerve fibers are not strictly in parallel with the optical axis of the polarizer or the analyzer. Therefore, the difference between the signal obtained with setting B and that obtained with setting A gives the signal originating from the optical activity change.

METHODS

Experiments were done with nerves taken from walking legs of the spiny lobster, Panulirus japonicus. The nerves were cleaned and stained with dye solutions which required 10 - 30 minutes.

The dyes employed were fluorescein isothiocyanate (FITC; Sigma No. F-7250, Isomer 1) and mercuridibromofluorescein (Merbromin; Merck, Art. 5997). Acridine orange (Merck, Art. 1333) and one of the Merocyanine dyes (Nihon Kanko, NK2273) were also tried, but they did not give optical activity signals. FITC was dissolved in artificial sea water at a concentration of 0.5 mg/ml, and the pH was adjusted with HEPES-NaOH buffer to a value between 7 and 8 (usually at about 7.5). Merbromin was dissolved in artificial sea water at a concentration betwee 0.1 mg/ml and 0.025 mg/ml, and the pH was adjusted with Tris-HCl buffer to about 8.2.

Several chambers of different designs were employed, but only the one in current use will be described. The

chamber was made of a black lucite plate. A groove of 70 mm
length, 5 mm width and 4 mm depth was made in the plate. At
the central part the groove was widened to 40 mm, and a
rectangular hole was made at the bottom to introduce light.
The hole was closed by a cover slip, and sealed with petro-
leum jelly. There was a pair of platinum wire electrodes at
each end of the groove to stimulate the nerve and to record
extracellualr action potentials. The nerve was laid in the
groove, and a small plate of black lucite was placed at each
side of the nerve to block the light that did not go through
the nerve. A cover slip was placed at the top of the
central part of the nerve in order to flatten the water
surface and to prevent its movement by vibration of the
stage.

The light source was a 250 W quartz-halogen lamp driven
by a DC power supply at a regulated voltage of 24 volts. A
camera lens (focal length of 50 mm with an aperture of
f:1.2) was used to focus the image of the filament on the
nerve. An interference filter, an infrared-suppression
filter and a polarizer (HN 22 Polaroid film) were inserted
between the lens and the nerve. During the initial experiments, the light travelled horizontally and the nerve was
held vertically. During the later experiments the beam was
deflected by a surface mirror to travel vertically, and the
nerve was held horizontally. An analyzer of a similar
design to that of the polarizer was placed in the optical
path after the nerve. The light which emerged from the
analyzer was detected by a silicon photodiode (United Detector PIN 10). The output of the photodiode was fed to an I-V
converter of conventional design. The DC component of the
voltage was recorded at the output of the I-V converter, and
the changing component was amplified 640 times with an AC
amplifier. Its output was fed to a signal averager (Nicolet
1170). Normally 128 signal were averaged. The averaged
trace was then transferred to a desktop computer (HP 9845B),
and stored on a flexible disk.

The standard artificial sea water contained (in mM) 470
NaCl, 10 KCl, 10 $CaCl_2$ and 50 $MgCl_2$. Often $MgCl_2$ was replaced by $CaCl_2$; this 'high Ca artificial sea water' appeared sometimes to maintain excitability of the nerve longer
than the standard artificial sea water. Tris-HCl buffer at
a pH around 8.0 was added before the experiment. Experiments were done at 16 - 22°C.

RESULTS

a) Lobster nerves stained with Merbromin.

Optical records from a lobster nerve stained with 0.1 mg/ml Merbromin are shown in Fig. 3, B-J. The traces B, D, F, H and J were taken with setting A, and the traces C, E, G and I were taken with setting B. The figure shows that the optical response decayed quickly during the experiment; the action potential also decreased at about the same rate. Nevertheless the time courses of the optical records taken with setting A were clearly different from those taken with setting B; in the former, the transmitted light simply decreased after the stimulus artifact, but in the latter the transmitted light first increased transiently. The difference in time course suggests that on excitation the direc-

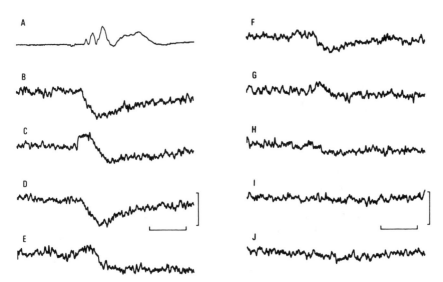

Fig. 3 Optical and electrical records taken from a nerve stained with 0.1 mg/ml Merbromin. A, externally recorded action potential taken at the beginning . B-J, Optical signals. B, D, F, H and J were taken with setting A ($\underline{a}$ = 45°, $\underline{p}$ = 0°); C, E, G and I were taken with setting B ($\underline{a}$ = 0°, $\underline{p}$ = 45°). Each trace is an average of 128 records. Vertical bar indicates 5 10^{-6} of the background light intensity with a single stimulus. Horizontal bar indicates 10 ms. Wavelength, 510 nm. (In this and following figures an upward deflection always indicates an increase in light intensity.)

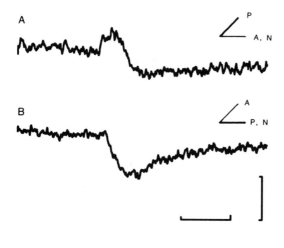

Fig. 4 Difference between optical signals obtained with setting A and setting B. Trace A is the average of C, E, G and I in Fig. 3. Trace B is the average of traces B, D, F, H and J in Fig. 3. Trace C is one half the difference of A and B.

tion of the E-vector of the transmitted light was rotated clockwise; the rotation decreased the light intensity when the azimuths of the polarizer and analyzer were in setting A, whereas the same rotation increased the light intensity when the azimuths were in setting B. The decrease or increase in light intensity was superimposed on the light intensity change caused by absorption and scattering changes, which were the same at these two settings. Fig. 4A shows the result of addition of all the data accumulated with setting A, and Fig. 4B shows the result of similar addition with setting B. The difference between Fig. 4A and Fig. 4B is proportional to the change in angle of rotation during excitation. One half the difference is shown in Fig. 4C; disregarding the second term in the parenthesis of Eq.(5), this gives an approximate value of the change in optical rotation, which was 3×10^{-6} in radians, or 1.7×10^{-4} in degrees. If the value was deduced from the initial part of the experiment the value of 3.2×10^{-4} degrees was obtained. The largest value of change so far obtained was 9.2×10^{-4} degrees (dextrorotation). At this dye concentration the response was usually an increased dextrorotation. However, when the nerve was stained with a 0.05 or 0.025 mg/ml dye solution, the response was an increased levorotation (Fig. 5). With these dilute dye solutions, the deterioration of nerve was reduced and birefringence signals could

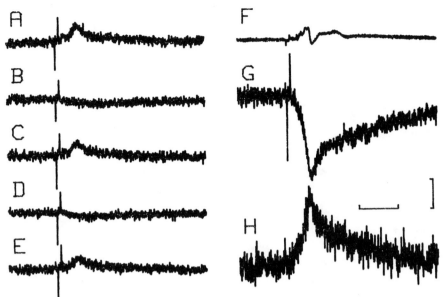

Fig. 5 Optical and electrical records taken from a nerve
stained with 0.025 mg/ml Merbromin. A-E, optical signals.
A, C and E were taken with setting A ($\underline{a}$ = 135°, $\underline{p}$ = 90°). B
and D were taken with setting B ($\underline{a}$ = $\overline{90}$°, $\underline{p}$ = 135°). F, an
externally recorded action potential. G, an optical signal
taken under the cross-polar condition ($\underline{a}$ = 135°, $\underline{p}$ = 45°).
G, the difference between the average of traces A, C and E
and the average of traces B and D. Time, 10 ms. Calib-
ration, 5×10^{-5} for traces A-E, 1.25×10^{-5} for G, and $6.25 \times$
10^{-6} for H, of the background light intensity, respectively.
Wavelength, 510 nm.

sometimes be recorded (Fig. 5, B). Traces C, E and G were
recorded with setting A, and traces D and F were recorded
with setting B. It is clear that the E vector was rotated
counterclockwise.

    b) Lobster nerves stained with FITC

    The results of an experiment with FITC-stained lobster
nerve are shown in Fig. 6. The signals obtained with setting A had a larger downward deflection than those obtained
with setting B, and the former often showed a small initial
upward deflection which was less frequent in the latter.
The accumulated signals (Fig. 6, F and G) showed a small

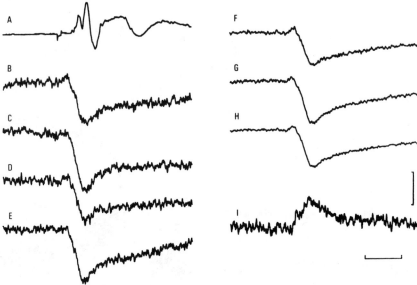

Fig. 6 Optical records taken from a nerve stained with 0.5 mg/ml FITC. A, externally recorded action potential. B and D, optical signals taken with Setting A (a = 45°, p = 0°). C and E, optical signals taken with Setting B (a = 0°, p = 45°). Each of the traces B–E was the result of averaging 128 signals. F and G are average of 896 responses taken with setting A and setting B, respectively. H is the average of F and G, and I is the difference between F and G multiplied by 5. Vertical bar indicates 5×10^{-6} of the background light per stimulus for traces B–H, and approximately 1×10^{-6} of the background light per stimulus for trace I. Horizontal bar shows 10 ms. Wavelength 540 nm.

difference (Fig. 6I), with an amplitude of about 10^{-6} of the background light intensity. The response obtained with FITC staining was an increased dextrorotation.

DISCUSSION

The experimental data shown above can easily be explained if one assumes that the stained nerve changes its optical activity during excitation. It is, however, necessary to consider other possibilities. For example, it may be argued that the retardation change can produce a signal which looks very much like the signal due to the optical activity change, for the following reasons.

The formula derived in the Theory section of this paper is based on a simplified optical model of the nerve, in which the specimen does not have the resting optical activity, and the change in optical activity is regarded as equivalent to a change in azimuthal angle of the incident E-vector. A more appropraite model would be to suppose that the specimen has a resting optical activity, which changes slightly when the nerve is excited. Using the method of the Jones matrix (cf. Shurcliff, 1962), a formula which describes the intensity of light emerging from the analyzer may be obtained for a specimen which is both optically active and birefringent (H. Watanabe, 1979). The formula predicts that a spurious optical change may be observed when a birefringence change takes place in the specimen if it is optically active in the resting state, even when the optical activity itself remains unchanged.

However, the size of the spurious optical activity signal is probably very small. With the intensity formula one can calculate the size of the spurious optical activity change from the magnitude of change in birefringence, if one knows the resting values of optical activity and retardation. At the time of recording of the data presented in this paper, these resting values had not been measured. Later measurements showed that stained nerves were indeed optically active in the resting state. When the E-vector of incident light was vertical to the nerve, the direction of rotation was clockwise with angles of rotation ranging between 0.1 and 1.5 degrees. Assuming that the nerve, from which the records in Fig. 5 were taken, had the angle of rotation of 1.5 degrees, the spurious optical activity signal is calculated to be less than 5.5×10^{-7} of the background light intensity. Since the observed optical activity signal in Fig. 5 is about 1.7×10^{-5}, the influence of the spurious signal on the observed signal is slight. A full account of measurement of the resting optical parameters of the nerve will have to be described elsewhere.

In fact, the time courses of the birefringence change and the optical activity change are often different. For example, the peak of the birefringence signal in Fig. 5 G came about 1 ms later than that of optical activity signal in Fig. 5 H. The latter also showed a more rapid rate of decay to the baseline than the former. Such differences in time course also suggest their different origins. Essentially similar arguments may be made to exclude influences from other sources, i.e., change in absorption, turbidity, or linear dichroism.

It is known that estimation of the molecular confor- mation from the optical activity measurement is a difficult task even when the sample is membrane fragments (cf. Wallach and Winzler, 1974; Tinoco, 1980). The living nerve is a more complicated system than the membrane fragments. Al- though it seems safe to conclude that the optical activity of the nerve changes during excitation, further experimental and theoretical works are needed to clarify the molecular basis of such a change.

ACKNOWLEDGMENTS

I thank Dr. C. Edwards for kindly correcting the English of the MS, and Dr. S. Terakawa for helpful discus- sion. This work is supported by a Grant-in-Aid for Scien- tific Research No. 00548096 and No. 57480106 given by the Ministry of Education, Science and Culture of Japan.

REFERENCES

Cohen LB, Keynes RD, Hille B (1968). Light scattering and birefringence changes during nerve activity. Nature 218: 438.

Jerrard HG (1948). Optical compensators for measurement of elliptical polarization. J Opt Soc Am 38: 35.

Jirgensons B (1973). Optical activity of proteins and other macromolecules. 2nd ed. New York: Springer.

Ross WN, Salzberg BM, Cohen LB, Davila HV (1974). A large change in dye absorption during the action potential. Biophys J 14: 983.

Shurcliff WA (1962). Polarized light. Cambridge Mass: Harvard University Press.

Tasaki I, Carnay L, Watanabe A (1969). Transient changes in extrinsic fluorescence of nerve produced by electric stimulation. Proc Nat Acad Sci 64: 1362.

Tinoco I Jr, Bustamante C (1980). The optical activity of nucleic acids and their aggregates. Ann Rev Biophys Bioeng 9: 107.

Wallach DFH, Winzler RJ (1974). Evolving strategies and tactics in membrane research. New York: Springer.

Watanabe H (1979). Electro-optical responses of chiral substances. In Jennings BR (ed): "Electro-optics and dielectrics of macromolecules", New York: Plenum, p.43.

The Physiology of Excitable Cells, pages 305–316
© 1983 Alan R. Liss, Inc., 150 Fifth Avenue, New York, NY 10011

A POSTULATED RELATIONSHIP BETWEEN SODIUM CHANNELS AND
MICROFILAMENTS IN CULTURED MAMMALIAN NERVE CELLS

Jun Fukuda

Department of Physiology, Faculty of Medicine,
University of Tokyo, Bunkyo-ku,
Tokyo 113, Japan

Tissue-cultured nerve cells derived from dorsal root
ganglia of adult mammals generate action potentials with
both Na and Ca spike components (Fukuda, Kameyama 1979,
1980a). When the action potential was compared among nerve
cells of various ages in tissue culture, it was noticed that the
contribution of the Na and Ca spike components was altered
in a nerve cell associating with neurite growth in tissue
culture (Fukuda, Kameyama 1979). For example, when the
amplitude of Na and Ca currents through the membrane was
determined by measuring the V_{max} of pure Na and Ca spikes
elicited in the nerve cells, the Ca currents were enhanced
during a particular period of tissue culture. The increment
in Ca currents appeared to be correlated with neurite growth.
The Na currents stayed at an almost steady level during this
time (Fukuda, Kameyama 1979, 1980a). These findings suggest
that ionic channel molecules, especially Ca channels, can
alter their number or features rather flexibly in the plasma
membrane of the nerve cells.

Indeed, the amplitude of Na and Ca currents in the nerve
membrane is changed within a few days when the nerve cells
are exposed to a chemical which depolymerizes particular
components of cytoskeletal filaments in a highly specific
fashion (Fukuda, Kameyama, Yamaguchi 1981). For example,
depolymerization of microfilaments after incubation of the
cultured nerve cells with cytochalasin B for 2-3 days caused
a marked reduction in the V_{max} of pure Na spikes but not in
that of the Ca spikes. By contrast, breakdown of microtubules
by vinka alkaloids (such as colchicine) reduced the V_{max} of
the Ca spikes but not that of the Na spikes. Thus it seems

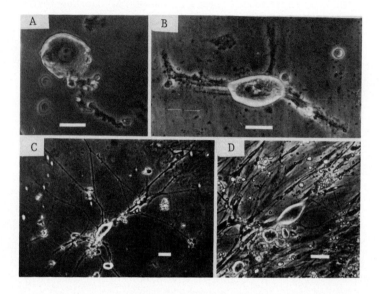

Fig. 1. <u>Nerve cells grown in culture</u>. Nerve cells were
isolated from dorsal root ganglia of adult guinea pigs and
were grown in plastic dishes. A) Immediately after isolation. B) 2 days, C) 5 days, and D) 15 days in culture.
Bars are 50 μm.

possible that the Na channels are intimately associated with
microfilaments located just underneath the plasma membrane
while they are rather independent of microtubules (Fukuda
et al. 1981). Ca channels, on the other hand, appear to be
independent of microfilaments but are related to the development of microtubules of the nerve cells. These observations indicate that changes in Na channels in nerve cells
are somehow associated with alteration of the microfilaments
in the nerve cells (Fukuda et al. 1981).

As we have reported previously, there are two types of
Na channels in the plasma membrane of the cultured nerve
cells, tetrodotoxin (TTX)-sensitive Na channels and TTX-
resistant Na channels (Fukuda, Kameyama 1978a, b, 1980a, b).
These two types of Na channels appear to remain unchanged
during several weeks of tissue culture (Fukuda, Kameyama
1980a), but little is known about how these two types of Na
channels are affected by breakdown of the cytoskeletal filaments. This paper summarizes results of our recent studies

of these relationships, and also reports a new finding, that
the TTX-resistant Na channels are also interrelated with the
microfilaments in the cultured nerve cells, as is the case
with the TTX-sensitive Na channels.

Figure 1A shows a nerve cell isolated from a dorsal root
ganglion of an adult guinea pig by means of trituration after
incubation of the ganglion with collagenase in L-15 medium
for 1 hour at 37°C. The nerve cells immediately after isolation are spherical, with an average diameter of 40 µm (Fukuda,
Kameyama 1980b). An axon-like fiber is often seen emerging
from the nerve cell soma. The nerve cells started to regenerate their axons on collagen-coated plastic dishes when they
were incubated in a growth medium (which contained 75% Eagle's
medium, 15% fetal calf serum, 9% chick embryo extract and 1%
antibiotics) with 37°C air containing 5% CO_2. The length of
the neurites was often several hundred µm from the soma after
48 hours in culture (Fig. 1B). The nerve cells extended their
neurites further (Fig. 1C) and survived in vitro for more than
2 months (Fukuda, Kameyama 1979, 1980a). Around 4-6 days in
culture (Fig. 1C), non-neuronal cells started to increase
their population on the dishes and occupied most of the available area of the culture dishes in 2 weeks (Fig. 1D). The
nerve cells, however, could easily be differentiated from
these non-neuronal cells on the basis of the large size of
the cell soma.

It was not difficult to penetrate the cultured nerve
cells with a glass microelectrode while viewing them with
an inverted, phase contrast microscope. The nerve cells
exhibited resting membrane potentials ranging between -45
and -65 mV in the standard tissue culture conditions. These
values stayed appoximately level during 4 weeks. Input
resistance and capacitance of the cultured nerve cells,
which were measured by intracellular passage of a small
current pulse while holding the intracellular potential
around -100 mV, were calculated to be 20-40 MΩ and 40-70
pF, respectively, assuming that the membrane behaved as a
simple RC circuit (Fukuda, Kameyama 1979, 1980a, b, 1981).
The values of the resistance and capacitance stayed within
these ranges during the 4 weeks in culture.

Action potentials elicited in the nerve cells by depolarizing pulses consisted of Na and Ca spike components (Fukuda,
Kameyama 1979, 1980b). In order to study the Na spike component, the Ca component was blocked by bathing the nerve

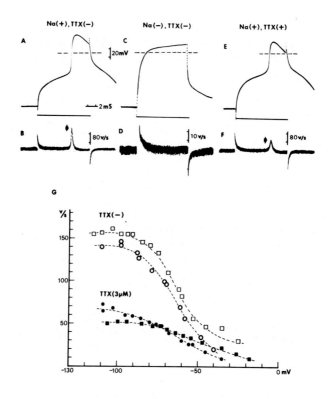

Fig. 2. TTX-resistant component of pure Na spikes in cultured nerve cells from adult mammals. Nerve cells of 5 days in culture were bathed in a solution which contained 143 mM NaCl, 10 mM TEA-Cl, 1.0 mM $CoCl_2$, 1.0 mM $CaCl_2$, 5.5 mM KCl, 5.5 mM glucose and 2.0 mM Na-Hepes, pH 7.4 at 37°C. Intracellular passage of a depolarizing current pulse elicited a pure Na spike (A); the membrane potential of the nerve cell was hyperpolarized to a level around -100 mV prior to the current pulse. Total replacement of Na^+ in the bathing solution caused blockage of spike generation (C), while a small Na spike was elicited in another nerve cell even when 3 μM TTX was added to the bathing solution (E), i.e., a TTX-resistant Na spike. The V_{max} of each spike is indicated by arrows in the lower, electronically differentiated traces (B and F). In G, the relationships between the holding membrane potential and the V_{max} of the Na spikes (open symbols) and that of the TTX-resistant Na spikes (filled symbols) are plotted.

cells in a cobalt-containing solution (1 mM). Tetraethylammonium (10 mM) was also added to the bathing solution to
suppress K currents. The spike elicited under these conditions is a pure Na spike (Fukuda, Kameyama 1980a). Indeed,
total replacement of Na ions in the bathing solution with
either $tris^+$ or $tetramethylammonium^+$, which are considered
to be impermeant through Na channels, caused the nerve cells
to fail to generate any spike (Fig. 2C). No sign of a regenerative inward current was observed in the lower, electronically differentiated trace of the intracellular record (Fig.
2D).

The nerve cells bathed in the Co^{++}-containing solution
were able to generate pure Na spikes even when 3 μM TTX was
also added to the solution (Fig. 2E), indicating that this
is a TTX-resistant Na spike (Fukuda, Kameyama 1980a). The
V_{max} of the Na spikes (arrows in the lower traces, Fig. 2B
and F), which represents the approximate value of the transient inward current associated with the spike, becomes
20-40% of that elicited without TTX, that is, about 20-40% of
the inward current was resistant to TTX in the cultured nerve
cells. Figure 2G illustrates the relationship between the
holding membrane potential and the V_{max} of the Na spikes in
the presence (filled symbols) and absence of 3 μm TTX (open
symbols). The inactivation curves of the two Na currents
(Fig. 2G) were essentially the same. The largest values of
the V_{max} thus obtained during hyperpolarization were compared
with nerve cells of various ages in tissue culture. Assuming
that the membrane capacitance stayed unchanged, the largest
value of the V_{max} was proportional to density of the Na
channels in the plasma membrane of the nerve cells.

Changes in the V_{max} during neurite outgrowth in tissue
culture were small for both the total Na spikes and the TTX-
resistant Na spikes (Fig. 3). This is in contrast to the
V_{max} of Ca spikes, which were also resistant to TTX; the Ca
currents were small in amplitude in an early period of tissue
culture and became larger during a week of incubation (Fukuda,
Kameyama 1979). The Ca channels appear to change in parallel
with neurite growth.

The cultured nerve cells were incubated for another 2
days with cytochalasin B (12 μg/ml), which depolymerizes
microfilaments in a highly specific manner (Fig. 4). The
incubation caused drastic changes in cellular morphology;
the nerve cell soma became spherical; neurite growth was

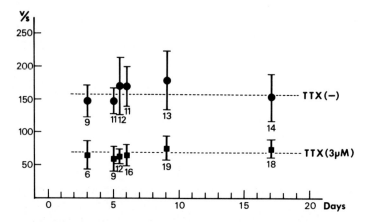

Fig. 3. Daily change in V_{max} of Na spikes (filled circles) and of TTX-resistant Na spikes (filled squares) during tissue culture. Each point represents mean value, bars indicate S.D., and numbers are sampled nerve cells.

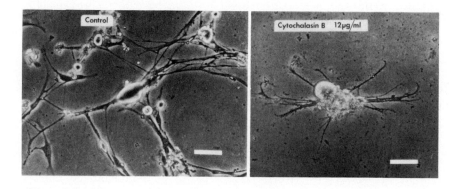

Fig. 4. Changes in cellular morphology after incubation with cytochalasin B (12 µg/ml). Nerve cells of 2 days in culture were exposed to the cytochalasin B for another 2 days. Bars are 50 µm. Control is a nerve cell of 4 days.

suppressed, and proliferation of non-neuronal cells was inhibited (Fukuda et al. 1981). These changes are considered to be due to breakdown of microfilaments (Spooner, Yamada, Wessells 1971; Laduena, Wessells 1973) and are different from

those induced by colchicine (Fukuda et al. 1981) which depolymerizes microtubules. Similar changes in morphology were also
induced by cytochalasin D and H$_2$-cytochalasin B (not illustrated), which depolymerize the microfilaments in slightly
different fashion (Yamada, Spooner, Wessells 1971; Laduena,
Wessells 1973; Ross et al. 1975).

Even though drastic changes in nerve cell morphology
were induced after incubation with cytochalasin B, the nerve
cells exhibited unchanged resting membrane potential, input
resistance and capacitance (Fig. 5). A marked change due to
cytochalasin B was that the nerve cells became less capable
of generating Na spikes (Fukuda et al. 1981). The V$_{max}$ of
the Na spikes was significantly reduced (Fig. 5) and the
duration was prolonged (Fukuda et al. 1981). However, no
change was induced in Ca spikes after incubation with the
cytochalasin B (Fig 5 and Fukuda et al. 1981). This was in
constrast to the changes in Ca spikes, but not in Na spikes,

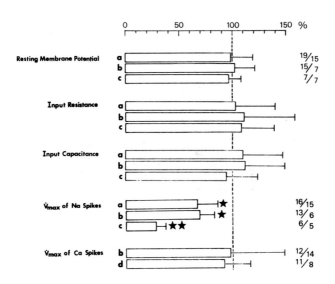

Fig. 5. Changes in nerve cell properties after incubation
with cytochalasin B. a) 4 µg/ml, 2 days; b) 6 µg/ml, 2
days; c) 12 µg/ml, 1 day, and d) 10 µg/ml, 3 days, respectively. Values of control, non-treated nerve cells are
taken as 100%. Numbers of sampled cells are indicated,
cytochalasin B treated cells/control cells. Bars are S.D.
*p<0.01 and **p<0.001, respectively in student t-test.

induced after incubation with colchicine (Fukuda et al. 1981).
We thus conclude that breakdown of microfilaments selectively
reduces Na currents through the plasma membrane of the nerve
cells.

The TTX-resistant component of Na spikes was also affected
by exposure of the cultured nerve cells to cytochalasin B
(Fig. 6). In this experiment, nerve cells of 3 days in culture
were exposed to cytochalasin B (10 µg/ml) for another 20 hours,
and V_{max} of the Na spikes was compared before and after exposure. The V_{max} of the Na spikes was reduced to 3/4 after
the cytochalasin B (open columns in Fig. 6), which was a
significant change (p<0.01 student t-test). The V_{max} of the
TTX-resistant Na spike component (filled columns) which was
about 23% of the whole V_{max} before the cytochalasin B, was
also reduced to 3/4 after the exposure, p<0.01. The TTX-
resistant component remained 21% of the total Na spike; this
is essentially the same value as that before cytochalasin B.
Similar observations were obtained after cytochalasin D

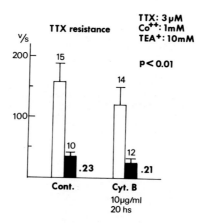

Fig. 6. Comparison of TTX sensitivity of Na spikes before
and after cytochalasin B exposure. Nerve cells of 3 days
in culture were exposed to cytochalasin B (10 µg/ml) for 20
hrs. Open columns: V_{max} of whole Na spikes; filled columns:
V_{max} of TTX-resistant Na spikes. Bars are S.D. Numbers of
sampled cells are indicated. Reduction of both whole Na
spikes and TTX-resistant Na spikes were significant in
student t-test (p<0.01), but the ratio remained unchanged,
from 0.23 to 0.21.

treatment (10 μg/ml, 2 days: not illustrated). This means
that breakdown of microfilaments reduces both TTX-sensitive
and TTX-resistant Na channels in an equal manner, suggesting
that interaction of these two types of Na channels with the
microfilaments may not differ significantly.

TTX is known as a specific inhibitor of voltage-sensitive Na channels and has been utilized as a tool for differentiating Na currents from other TTX-resistant membrane cur-
rents (Hagiwara, Nakajima 1966; Kao 1966; Evans 1972; and
Narahashi 1972, 1974). Na currents in some excitable mem-
branes are, however, known to be resistant to TTX (Redfern,
Thesleff 1971; Kidokoro, Grinnell, Eaton 1974; Miyazaki,
Takahashi, Tsuda 1974; Kostyuk, Krishtal, Pidoplichko 1975;
Lee, Akaike, Brown 1977). TTX-resistant Na currents are
also present in mammalian nerve cell membranes (Ranson,
Holtz 1977; Matsuda, Yoshida, Yonezawa 1978; Yoshida,
Matsuda, Samejima 1978; Fukuda, Kameyama 1978a, b).

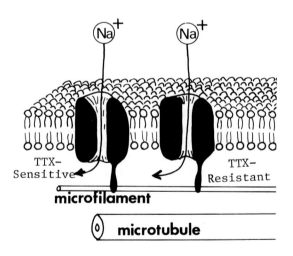

Fig. 7. Postulated relation between Na channels and cytoskeletal filamentous components. TTX-senstive and TTX-
resistant Na channels, microfilaments and microtubules are
indicated. Note that other proteins located underneath the
plasma membrane (such as ankyrin and spectrin) are abbrevi-
ated in this model.

Laduena MA, Wessells NK (1973). Cell locomotion, nerve elongation and microfilaments. Develop Biol 30:427.

Lee KS, Akaike N, Brown AM (1977). Trypsin inhibits the action of tetrodotoxin on neurones. Nature 265:751.

Matsuda Y, Yoshida S, Yonezawa T (1978). Tetrodotoxin sensitivity and Ca component of action potentials of mouse dorsal root ganglion cells cultured in vitro. Brain Res 154:69.

Miyazaki S, Takahashi K, Tsuda K (1974). Electric excitability in the egg cell membrane of the tunicate. J Physiol 238:37.

Narahashi T (1972). Mechanism of action of tetrodotoxin and saxitoxin on excitable membranes. Fed Proc 31: 1124.

Narahashi T (1974). Chemicals as tools in the study of excitable membranes. Physiol Rev 54:813.

Redfern P, Thesleff S (1971). Action potential generation in denervated rat skeletal muscle II: The action of tetrodotoxin. Acta Physiol Scand 82:7.

Ranson BR, Holz RW (1977). Ionic determinants of excitability in cultured mouse dorsal root ganglion and spinal cord cells. Brain Res 136:445.

Ross J, Olmsted JB, Rosenbaum JL (1975). The ultrastructure of mouse neuroblastoma cells in tissue culture. Tissue & Cell 7:107.

Schwartz JR, Ulbricht W, Wagner H-H (1973). The rate of action of tetrodotoxin on myelinated nerve fibres of Xenopus laevis and Rana esculenta. J Physiol 49:977.

Spooner BS, Yamada KM, Wessells NK (1971). Microfilaments and cell locomotion. J Cell Biol 49:595.

Tanaka M, Moore JW, Kao CY, Fuhrman FA (1966). Blockage of sodium conductance increase by tarichatoxin (tetrodotoxin). J Gen Physiol 49:977.

Yamada KY, Spooner BS, Wessells NK (1971). Ultrastructure and function of growth cones and axons of cultured nerve cells. J Cell Biol 49:614.

Yoshida S, Matsuda Y, Samejima A (1978). Tetrodotoxin-resistant sodium and calcium components of action potentials in dorsal root ganglion cells of the adult mouse. J. Neurophysiol 41:1096.

The Physiology of Excitable Cells, pages 317–326
© 1983 Alan R. Liss, Inc., 150 Fifth Avenue, New York, NY 10011

ULTRASTRUCTURAL ORGANIZATION OF THE MEMBRANE-BOUND (Na$^+$ + K$^+$)-ATPase ISOLATED FROM MAMMALIAN KIDNEY

Guido A. Zampighi

Department of Anatomy and Jerry Lewis Neuromuscular Research Center, University of California,
Los Angeles, California 90024

Sodium and potassium ion-activated triphosphatase (Na$^+$ + K$^+$)-ATPase is a membrane-bound enzyme responsible for the coupled active transport of sodium and potassium across the plasma membrane of animal cells. The resulting differences in the steady-state concentrations of these cations between the extracellular and cytoplasmic medium provide potential energy employed to maintain cell volume, to drive the uptake of nutrients, to move water, and to create the resting potential of most cells.

(Na$^+$ + K$^+$)-ATPase has been isolated in pure form from several tissues such as mammalian kidney (Kyte 1971; Jorgensen 1974), avian salt glands (Hopkins et al. 1976), elasmobranch rectal glands (Hokin et al. 1973), and electric organs (Dixon, Hokin 1974). All the enzymes purified so far are constructed from two different subunits: one is a large protein (α, MW = 110,000 ± 10,000) and the other is a smaller protein (β, MW = 50,000 ± 5,000) which together form an intimate complex. The α-subunit has been shown to span the membrane (Kyte 1975) and to contain the active site (Uesugi et al. 1971), the cardiac glycosides binding site (Ruoho, Kyte 1975), and the sulphydryl residue modification of which inactivates the enzyme (Winslow 1981). Moreover, the α-subunit is very similar to the protein which exclusively forms the Ca^{2+}-ATPase (MacLennan 1970) and the polypeptide which solely forms the (H$^+$, K$^+$)-ATPase responsible for the active transport of acid in the stomach Sachs et al. 1976).

The provocative questions that have yet to be answered

are how the enzyme translocates cations across the membrane and how it couples the movement of these cations to ATP hydrolysis. It is probable that the answer to these questions will require a detailed knowledge of the three-dimensional organization of the $(Na^+ + K^+)$-ATPase molecules.

We are investigating the structural organization of crystalline arrays constructed from $(Na^+ + K^+)$-ATPase molecules in the membrane-bound form, by electron microscopy and computer image processing. In this paper, information regarding the location of the enzyme molecules in the membrane as well as the size, shape and dimensions of the different domains of the enzyme will be presented.

Isolated, membrane-bound $(Na^+ + K^+)$-ATPase was isolated from microsomes of canine renal medulla (Kyte 1971) and purified by treatment with solutions containing $NaDodSO_4$ as described by Jorgensen (1974). Polyacrylamide gel electrophoresis of the fractions studied by electron microscopy showed that more than 90% of the total protein could be accounted for by the α and β polypeptides of the enzyme. The isolated $(Na^+ + K^+)$-ATPase fractions used in this study have a specific enzymatic activity (Kyte 1972) of 800 μmoles $mg^{-1} h^{-1}$.

It has been demonstrated that $(Na^+ + K^+)$-ATPase passes through a series of cyclic conformational states during active transport. It is now felt that the conformational changes that interconnect these states are responsible for the active movement of the respective cations across the membrane. The four conformations, E_1, E_1-P, E_2-P, and E_2 differ in their affinity for ATP (Moczydlowski, Fortes 1981), their suceptibility to trypsin (Jorgensen 1975), and their affinity for cardiac glycosides (Kyte 1972). It has been inferred that E_1 is the inward-facing form of the enzyme which normally receives sodium as a substrate and releases potassium as a product, while E_2 is the outward-facing form which releases sodium as a product and receives potassium as a substrate. It is believed that during the transition between the two forms, the cations translocate across the plasma membrane. The purified, isolated membrane-bound $(Na^+ + K^+)$-ATPase molecules were induced to form two-dimensional arrays by manipulating the ionic concentrations of the bathing media. Of the several different media that drive the enzyme into its different

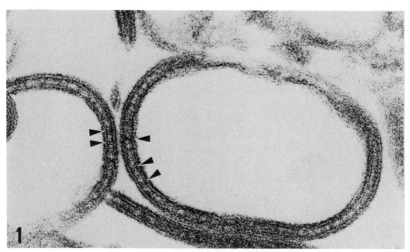

Fig. 1. A transverse section from a paired membrane complex formed by bathing the membrane-bound enzyme in solution favoring the E_2-P conformation. Note that the complex is comprised of two membranes separated by a gap of constant width. Arrows indicate electron-dense septa connecting both membranes. Magnification: x 200,000.

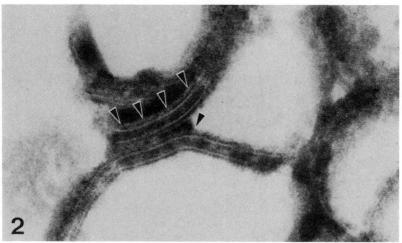

Fig. 2. A transverse view of a paired membrane complex decorated with the agglutinin from Ricinus communis. The heavily electron-dense material arises from the use of tannic acid in the fixation protocol. A new layer of material, associated with the free surface of the complex, appeared after treatment with the agglutinin . Magnification: x 200,000.

states (Winslow 1981), the one favoring the E_2-P conformation produced small crystalling arrays in a remarkably reproducible manner. Skriver et al. (1981) first reported the formation of arrays in those conditions after prolonged incubation (up to 4 weeks) with sodium vanadate in the presence of magnesium. We have improved upon those initial experiments and shortened the incubation time that induces crystallization to 1-2 hours.

The two-dimensional arrays produced by incubation of the membranes in ionic conditions favoring the E_2-P conformation were first studied by thin-section electron microscopy. An examination of numerous sections obtained from pellets of the crystalline membranes showed the appearance of a new type of membrane-membrane association having unique morphological characteristics. Figure 1 shows a transverse view of this new membrane complex. It is formed by two membranes separated by a narrow gap of constant width. The overall thickness of the membrane complex is 23-25 nm and the space between both partners is about 12 nm. Furthermore, electron-dense septa seem to connect the surface of both membranes in a repetitive fashion (arrows Fig. 1).

Further evidence consistent with the conclusion that the paired membrane complex described in Figure 1 is constructed from $(Na^+ + K^+)$-ATPase was obtained by decorating them with the bivalent agglutinin from Ricinus communis. This lectin is a large protein of molecular weight 150,000 which binds exclusively to the carbohydrate moiety of the β-subunit of the enzyme (Olsnes et al. 1974). Figure 2 shows paired membrane complexes decorated with agglutinin. The very electron-dense material present in the sections arose from treatment of the specimens with the glutaraldehyde-tannic acid procedure (Sealock 1982) which greatly improves the contrast of sectioned material. Figure 2 shows two paired membrane complexes which are joined together by layers of electron lucent material. The paired membrane at the top displays an additional layer of similar material decorating its free surface. Since these additional layers of densities were not observed in the controls, they have been interpreted as due to the binding of the agglutinin molecules to the polysaccharide of the β-subunit of the enzyme.

The location of these layers of agglutinin molecules

also defines the orientation of the two membranes in the pair. Because the layers from the agglutinin appeared on both free surfaces of the complex, it follows that the paired membranes are formed through the interaction of the cytoplasmic surfaces of the membranes.

From the measurement of many transverse views of the paired membranes decorated with the agglutinin from Ricinus communis and their respective controls, it was determined that the enzyme molecule is a rod-shaped molecule about 13 nm long and asymmetrically located in the bilayers. It protrudes more from the cytoplasmic than the external surface.

Samples of both crystalline and noncrystalline (Na$^+$ K$^+$)-ATPase were studied by negative staining. The method consists in directly attaching the isolated membranes to carbon-coated grids and staining them with uranyl acetate solutions. The noncrystalline specimens showed views corresponding to a membrane plane very similar to those presented previously by other investigators (Deguchi et al. 1976; Vogel et al. 1977). The fractions contained fragments of different size and shapes covered with randomly organized surface particles. These particles measured about 4-5 nm in diameter. The surface particles altered their organization in the plane of the membrane, after the (Na$^+$ + K$^+$)-ATPase was bathed in crystallizing solution. First, the surface particles aggregated in tightly packed clusters forming separated domains in the membranes. Later, the particles contained in these clusters arranged themselves in rows forming clear two-dimensional arrays. Figure 3 shows a membrane obtained from these experiments continuing several crystalline patches having unit cell parameters of a = 6.2 nm and b = 5.0 nm. The rows of particles intercept at angles of ∿ 116° which are characteristic of these arrays. The volume of the unit cell defined by this crystalline array of enzyme molecules can be defined as equal to a x b x c (6.2 nm x 5.0 nm x 13 nm). This volume can accomodate an (α-β) monomer plus its lipid complement.

Furthermore, an entirely different type of crystalline lattice was also observed in the membranes bathed in crystalizing solutions. Figure 4 shows several views of the type of molecular packing adopted by the membrane-bound (Na$^+$ + K$^+$)-ATPase. In this type of packing, the surface particles formed ribbons which were spaced 12-13 nm apart center-to-center. Each ribbon was composed of two adjacent rows of surface particles arranged in a staggered conformation with respect to each other. The spacing between the centers of the particles in the rows was 5.5 nm apart. The volume of the unit cell defined in this type of crystalline array (a = 12.5 nm, b = 5.5 nm, and c = 13 nm) can accomodate $(\alpha-\beta)_2$ asymmetric units.

The observations reported here on the structural organization of the membrane-bound $(Na^+ + K^+)$-ATPase are in close agreement with most of the evidence presented in the literature (Deguchi et al. 1977; Haase, Koepsell 1970; Van Winkle et al. 1976; Vogel et al. 1977). In particular, the observations presented here confirm the recent reports of Skriver et al. (1981) and Herbert et al. (1982) regarding the ionic conditions necessary for the formation of crystalline arrays of the enzyme, as well as the packing arrangement adopted by the enzyme molecules.

It has been proposed that the formation of two distinct types of crystalline arrays of the membrane-bound $(Na^+ + K^+)$-ATPase molecules is significant to the function mediated by the enzyme in vivo (Herbert et al. 1982). Although the formation of two dimensional arrays is an important step in the use of three-dimensional reconstruction methods by electron microscopy (Amos et al. 1982), it is unusual that the way molecules crystallize in the membrane bears relevance to their in vivo function. In fact, several membrane-bound proteins have been reported to adopt different packing conformations depending on the conditions used for crystallization. For example, the purple membrane protein of Halobacterium halobium forms hexagonal arrays in the native membrane (Unwin, Henderson 1975; Henderson, Unwin 1975) and orthodromic arrays in reconstituted systems (Michel et al. 1980). Therefore, while the formation of crystalline lattices provides a convenient specimen with which to determine the three-dimensional organization of the enzyme, the specific distribution assumed by the molecules in the array may well represent a secondary process with little, if any, physiological significance. The information necessary to explain the conformational changes which the enzyme undergoes during active transport must be contained in the structure of the asymmetric unit only. Such information can be obtained if a three-dimensional model of the $(Na^+ +$

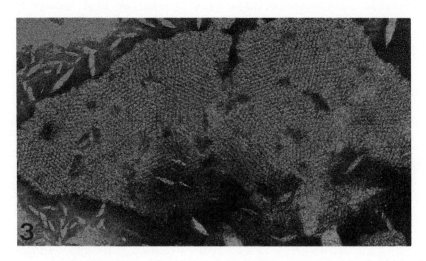

Fig. 3. A crystalline patch of the membrane-bound $(Na^+ + K^+)$- ATPase obtained by negative staining. Note that the surface particles form rows which intercept at oblique angeles. Magnification: x 200,000.

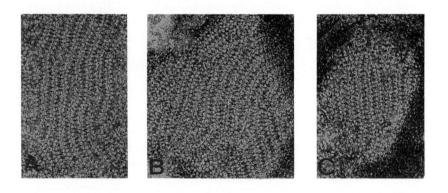

Fig. 4. A different type of packing adopted by the membrane-bound form of the $(Na^+ + K^+)$-ATPase when bathed in ionic conditions favoring the E_2-P conformation. Note that the surface particles associate into dimers which, in turn, form rows oriented parallel to each other. Magnification: x 200,000.

K$^+$)-ATPase can be calculated at a resolution high enough to separate the regions in the asymmetric unit arising from the α and β polypeptides and the rearrangements of the molecule during active transport.

REFERENCES

Amos LA, Henderson R, Unwin PNT (1982). Three dimensional structure determination by electron microscopy of two-dimensional crystals. Prog Biophys molec Biol 39:183.

Deguchi N, Jorgensen PL, Maunsbach AB (1977). Ultrastructure of the sodium pump. Comparison of thin sectioning, negative staining and freeze-fracture of purified, membrane-bound (Na$^+$ K$^+$)-ATPase. J Cell Biol 75:619.

Dixon JF, Hokin LE (1974). Studies on the characterization of the sodium-potassium transport adenonine triphosphatase. Arch Biochem Biophys 163:749.

Haase W, Koepsell H (1979). Substructure of membrane-bound (Na$^+$ + K$^+$)-ATPase protein. Pflugers Arch 381:127.

Henderson R, Unwin PNT (1975). Three-dimensional model of purple membrane obtained by electron microscopy. Nature 257:28.

Herbert H, Jorgensen PL, Skriver E, Maunsbach AB (1982). Crystallization patterns of membrane-bound (Na$^+$ + K$^+$)-ATPase. Biochem Biophysica Acta 689:571.

Hopkins BE, Wagner H(Jr), Smith TW (1976). Sodium- and potassium-activated adenonine triphosphatase of the nasal salt gland of the duck (Anas platyrhynchos). J Biol Chem 251:4365.

Hokin LE, Dahl JL, Deupree JD, Dixon JF, Hackney JF, Perdue JF (1973). Studies on the characterization of the sodium-potassium transport adenonine triphosphase. J Biol Chem 248:2593.

Jorgensen PL (1974). Purification and characterization of (Na$^+$ + K$^+$)-ATPase. III. Purification from the outer medulla of mammalian kidney after selective removal of membrane components by sodium dodecylsulphate. Biochim Biophysica Acta 356:36.

Jorgensen PL (1975). Purification and characterization of (Na$^+$ + K$^+$)-ATPase. V. Conformational changes in the enzyme. Transitions between the Na-form and the K-form studied with tryptic digestion as a tool. Biochem et Biophy Acta 401:399.

Kyte J (1971). Purification of the sodium- and potassium-dependent adenonine triphosphatase from canine renal medulla. J Biol Chem 246:4157.

Kyte J (1972). The titration of cardiac glycoside binding site of the (Na$^+$ + K$^+$)-adenonine triphosphatase. J Biol Chem 247:7634.

Kyte J (1975). Structural studies of sodium and potassium ion-activated adenonine triphosphatase. J Biol Chem 250:7443.

MacLennan DH (1970). Purification and properties of an adenonine triphosphatase from sarcoplasmic reticulum. J Biol Chem 245:4508.

Michel H, Oesterhelt D, Henderson R (1980). Orthorhomic two-dimensional crystal form of purple membrane. Proc Natl Acad Sci USA 77:338.

Moczydlowski EG, Fortes PAG (1981). Inhibition of sodium and potassium adenonine triphosphatase by 2', 3'-0-(2,4,6-Trinitrocyclohexadienyllidene) adenine nucleotide. J Biol Chem 256:2457.

Olsnes S, Saltnedt E, Pihl A (1974). Isolation and comparison of galactose-binding lectins from Abrus precatorius and Ricinus communis. J Biol Chem 249:823.

Ruoho A, Kyte J (1974) Photoaffinity labelling of the ovabain-binding site on (Na$^+$ + K$^+$)-Adenonine triphosphatase. Proc Natl Acad Sci USA 71:2352.

Sachs G, Chang HH, Rabon E, Schackman R, Saccomani G. (1976). A nonelectrogenic H$^+$ pump in plasma membranes of hog stomach. J Biol Chem 251:7690.

Sealock R (1982). Cytoplasmic surface structure in postsynaptic membranes from electric tissue visualized by tannic-acid-mediated negative contrasting. J Cell Biol 92:522.

Skriver E, Maunsbach AB, Jorgensen PL (1981). Formation of two-dimensional crystals in pure membrane-bound (Na$^+$, K$^+$-ATPase. FEBS Letters 131:219.

Uerugi S, Dulak NC, Dixon JF, Hexum TD, Dahl JL, Perdue JF, Hokin LE (1971). Studies on the characterization of the sodium-potassium transport adenonine triphosphatase. J Biol Chem 246:531.

Van Winkle WB, Lane LK, Schwartz A (1976). The subunit fine structure of isolated, purified Na$^+$, K$^+$-Adenonine triphosphatase. Freeze-fracture study. Exp Cell Res 100:291.

Vogel F, Meyer HW, Grocce R, Repke HRH (1974). Electron microscopic visualization of the arrangement of the two protein components of $(Na^+ + K^+)$-ATPase. Biochim et Biophys Acta 470:497.

Winslow J (1981). The reaction of sulphydryl groups of sodium and potassium opm-activated adenosine triphosphastase with N-ethylmaleimide. J Biol Chem 256:9522.

The Physiology of Excitable Cells, pages 327–341
© 1983 Alan R. Liss, Inc., 150 Fifth Avenue, New York, NY 10011

THE ROLE OF H^+ AND Ca^{2+} IN BALANUS PHOTORECEPTOR FUNCTION

H. Mack Brown, Ph.D.

Department of Physiology
University of Utah
Salt Lake City, Utah   84108

INTRODUCTION

Despite the fact that invertebrate and vertebrate
receptor potentials are the inverse of one another in the
sense that light initiates a membrane depolarization and
conductance increase in the former and a membrane hyperpolarization and conductance decrease in the latter, the
qualitative effects of H^+ and Ca^{2+} on both systems are remarkably similar. Raising Ca_o diminishes, and reducing Ca_o
increases the receptor potential of both and the effect is
exerted through the Na^+ conductance mechanism of the receptor membrane (see for e.g. Brown, Hagiwara, Koike and
Meech, 1970; Brown, Coles and Pinto, 1977). Similar effects
on both receptors have been observed by artificially changing the levels of Ca_i (Brown and Lisman, 1975; Yoshikami
and Hagins, 1973). Furthermore, light-elicited changes in
Ca_i have been documented directly in large invertebrate
photoreceptors with Ca^{2+} indicator dyes (Brown and Rydqvist,
1980; Brown, Brown and Pinto, 1977) and ion-sensitive
electrodes (Brown and Rydqvist, 1980, 1981). Extracellular
Ca^{2+} changes have been observed from illuminated vertebrate
photoreceptors that are thought to reflect an increase of
cytoplasmic Ca^{2+} (Gold and Korenbrot, 1980; Yoshikami,
George and Hagins, 1980). Small increases in intracellular
H^+ can reduce the receptor potential of an invertebrate
photoreceptor (Brown and Meech, 1975, 1976) and intracellular alkalinization causes an increase in receptor potential
amplitude of both types of receptors (Brown and Meech, 1979;
Pinto and Ostroy, 1978). In Balanus photoreceptor, an
intracellular acidification triggered by light has been

observed (Brown and Meech, 1979). The intracellular changes of H^+ and Ca^{2+} could be directly involved in the functional processes of visual transduction and light adaptation in photoreceptors or they could represent sequelae to certain biochemical events that are more directly linked to these processes (for summary, see O'Brien, 1982).

Several problems remain before the functional conse- quences of these changes are established. 1) Quantification of the changes are required. Intracellular Ca^{2+} changes in vertebrate photoreceptors have not been demonstrated directly and the estimates of the changes in invertebrate photore- ceptors have varied considerably. 2) The time course of the changes requires more exact measurement. Measurements with ion-sensitive electrodes (ISE) might be too slow to document the time course with accuracy. 3) The stoichiometry of the intracellular changes with light requires documentation. 4) The buffer capacity for H^+ and Ca^{2+} must be firmly estab- lished. 5) The locus of the intracellular changes must be determined.

This paper summarizes some of the work on Balanus photo- receptors that has been addressed to these questions.

METHODS

The Balanus eye is a simple ocellus consisting of three large photoreceptors about 100 μM in diameter (Fahrenbach, 1965). For study, the photoreceptors are isolated by remov- ing the back layer of pigment epithelium and tapetal cells to expose the photoreceptors. The preparation is placed corneal side down on a light source for illumination and continuously suffused with artificial saline. The photo- receptors are penetrated under visual control with micro- electrodes. Techniques used to investigate the properties of the cell have been described previously: 1) voltage- clamp analysis (see for example, Brown et al., 1970; Brown and Cornwall, 1975); 2) ion-sensitive electrodes (see Brown, 1976; Saunders and Brown, 1977; Brown, Pemberton and Owen, 1976); 3) indicator dyes (Brown and Rydqvist, 1980, 1981). The microspectrophotometer used in conjunction with dye indicators is of the same type originally described by Chance (1972) and modified by Brinley and Scarpa (1975) and Gorman and Thomas (1978).

RESULTS

Light-Induced Membrane Current

The behavior of the light-induced membrane current of a Balanus photoreceptor under voltage-clamp was first described in 1969 (Brown, Meech, Koike and Hagiwara, 1969) and was further documented in two other papers (Brown et al., 1970; Brown, Hagiwara, Koike and Meech, 1971). A set of these records at different light intensities is shown in Fig. 1, along with the current-voltage (I-V) relations that were obtained. Several conclusions can be drawn from these relations. 1) The light-induced membrane current, i.e. membrane current in the light (I_T) minus membrane current in the dark (I_D)is an inward current at both the peak (I_p) and the steady phase (I_s). 2) The steady-state I-V relations of both I_p and I_s are non-linear, i.e. the conductance change due to light is voltage-dependent. 3) At a given light intensity, I_p and I_s both reverse sign at the same E_m (approx. +25 mV). This occurs despite the fact that in an unclamped cell, V_p and V_s are at quite different voltage levels (inset), e.g. ~+10 mV and ~0 mV respectively during a bright light flash. 4) I_p and I_s increase with light intensity at $V_m = E_r$, but both reverse sign at the same membrane potential. This was an interesting finding since the membrane potential changes over a large range due to changes of light intensity. Additional studies including two-step voltage-clamp (Brown et al., 1969) were consistent with an electrical analog of the membrane consisting of a dark conductance, g_D (V,t), in parallel with a light modulated conductance, g_L (V,t,L). Ion substitution experiments indicated that the major current species involved during illumination was Na$^+$ since g_L varied in proportion to Na$_o$ (Brown et al., 1970).

Effects of Calcium

In the absence of Na$^+$, the Balanus photoreceptor membrane can function as a quasi-Ca^{2+} electrode, i.e. the zero-current potential varies about 20 mV for a ten-fold change in Ca$_o$. This is consistent with a small inward current (~5%) that persists when Na$^+$ is totally removed from the external medium.

The most salient effect of Ca^{2+} under normal conditions

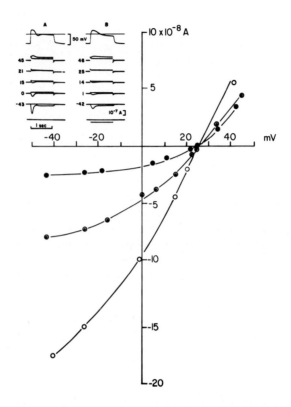

Fig. 1. Steady-state current-voltage relations of the membrane during illumination. The light-induced current $(I_T - I_D)$ is plotted on the ordinate; voltage-clamped membrane potential on the abscissa. Relation 100 msec (⊙) and 1,000 msec (●) after the beginning of illumination (intensity 1.65×10^3 lumen/m^2). Relation for light intensity 1.2×10^4 lumens/m^2 (o); 100 msec after the onset of illumination. Inset: membrane potential changes to two intensities of illumination. A, 1.2×10^4; B, 1.65×10^3 lumens/m^2. Beneath the voltage records are shown superimposed records of membrane current, with and without illumination, when the membrane potential was clamped from the resting level to the level indicated adjacent to each of the traces; inward current is displayed downward. Light was applied at a time corresponding approximately with the step change in membrane potential as indicated by the bottom traces. (Revised from Brown, Hagiwara, Koike, Meech 1971.)

is that it greatly modifies the light-induced current carried by Na$^+$. The voltage dependence of the LIC can be essentially abolished if Ca^{2+} is reduced by a significant amount as shown in the I-V relation of Fig. 2. In this case, a tenfold reduction of Ca^{2+} from 20 to 2 mM drastically increases the LIC at E_r and the I-V relation of the membrane becomes considerably more linear.

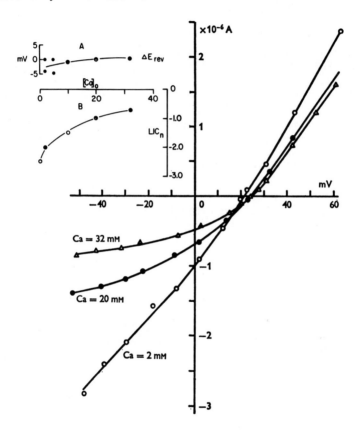

Fig. 2. Relations of light-initiated membrane current, measured 5 msec after the onset of light, and voltage-clamped membrane potential at the Ca^{2+} concentration indicated adjacent to each curve. Inset: A, change in reversal potential of 2 cells (o,•) as Ca^{2+} is varied in normal saline. B, normalized light-induced current (LIC$_n$) for 2 cells (o,•) as Ca^{2+} is varied in normal saline. Holding potential -50mV. (Revised from Brown, Hagiwara, Koike, Meech 1970.)

Under these conditions, i.e. Na^+ present in the external medium, the reversal potential is little affected by changes in Ca_o. These points are summarized in the inset which shows the change in reversal potential (A) and the normalized LIC (B) at different Ca^{2+} concentrations. It can be concluded that: 1) The major role of Ca^{2+} is to modify the Na^+ conductance. 2) Under conditions of low Ca^{2+}, the voltage-dependence of g_L is significantly reduced.

Intracellular Ca^{2+} changes may exert similar effects on the membrane. Lisman and Brown (1972) showed that intracellular injection of Ca^{2+} in Limulus ventral photoreceptors attenuated the receptor potential. In Balanus photoreceptors, there is an increase in Ca_i with light that we have documented with Ca-ISE (Brown and Rydqvist, 1980) of the type described by Brown, Pemberton and Owen (1976). Figure 3 shows a calibration curve of a Ca-ISE in a medium that approximates the known intracellular environment of Balanus photoreceptors determined with K^+, Cl^- and Na^+ ISE (Brown, 1976). The electrode response is linear to ~10^{-6}M Ca^{2+} and has a usable range below 10^{-7}M Ca^{2+} in a medium containing ionic interferents of the electrode. The most significant source of error for intracellular Ca^{2+} measurement is due to unknown amounts of Mg^{2+} in the photoreceptor. Preliminary studies indicate that free Mg^{2+} in Balanus photoreceptors is on the order of 0.5-1.0 mM (Brown and Rydqvist, 1982), which allows appropriate solutions to be designed for electrode calibration. Penetration of the cell with a Ca-ISE results in a potential change that corresponds to approximately 10^{-7}M Ca^{2+} in dark-adapted cells. Illumination increases Ca_i; in this cell a moderate flash of light increased Ca_i to pCa 6.5. The implications of these changes in Ca_i are not totally clear, but if the membrane behavior is symmetrical for Ca^{2+} changes, that is the inside behaves similarly to the outside, the increase in Ca_i could participate in a decrease in g_{Na} of the membrane which could contribute to the adaptive decline of the LIC from the peak to the steady phase. The time course of the true Ca_i changes are faster than Fig. 3 due to the relatively long time response of these electrodes, but changes of Ca-dye indicator absorbance changes indicate that the peak Ca^{2+} change does not occur until some time after the peak of the receptor potential (Brown and Rydqvist, 1981).

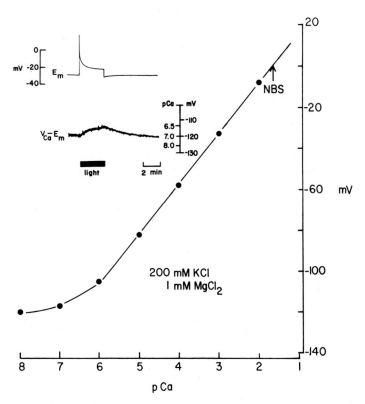

Fig. 3.   Inset: differential recording of Ca²⁺ electrode
and membrane potential changes elicited by light ($V_{Ca}-E_m$).
Light initiated a 7 mV positive change in $V_{Ca}-E_m$.

Calibration curve for t-HDOPP Ca-ISE. Calibrating
solutions contained background concentrations of the
following: 200 mM KCl, 1 mM MgCl₂ and 10 mM HEPES (pH 7.35).
$k_{Ca,K}$ = 2 x 10⁻⁶; $k_{Ca,Mg}$ = 2.5 x 10⁻⁴.   (From Brown and
Rydqvist, 1983).

Effects of pH

Unlike the effects of Ca²⁺, the effects of pH changes
are asymmetric in <u>Balanus</u> photoreceptor in the sense that
large extracellular pH changes have only small effects on
the resting potential and receptor potential of the cell.

However, both the receptor potential and resting potential
are very susceptible to subtle changes of intracellular pH.
For example, a pH_i change of only 0.2 pH units can attenuate
the peak of the receptor potential by 50%. Moreover, we
have shown with glass pH-ISE that light initiates a pH
change in Balanus photoreceptors (Brown and Meech, 1979).
The changes in membrane potential and intracellular pH are
shown in Fig. 4 for different light intensities in dark-
adapted photoreceptors.

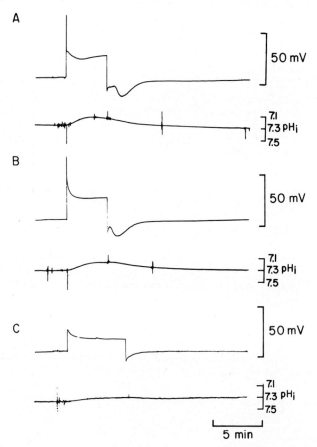

Fig. 4. The effect of different light intensities on intracellular pH (lower trace) and membrane potential (upper
trace). The photon flux (520 nm photons/photoreceptor/sec)
was: (A) 8.8×10^{10}; (B) 1.1×10^{10}; (C) 1.1×10^9. The
cell was dark-adapted for 15 min between light stimuli.
(From Brown and Meech, 1979)

The records show that the change in pH$_i$ is graded with light intensity. Saturating light can produce a pH change on the order of 0.2 to 0.3 of a pH unit. We measured the buffer capacity in this cell with pulses of CO_2 saline and found the cytoplasmic buffering capacity to be on the order of 15 slykes (15 mM/1 pH) which is of the same order for other cells measured with similar techniques.

Arsenazo III: pH$_i$ and Ca$_i$ Changes

The metallochromic indicator Arsenazo III (AIII) has been used fairly extensively as an intracellular Ca^{2+} indicator; AIII changes absorbance to pH changes as well because of protonic acceptor-donor sites on the molecules (Brown and Rydqvist, 1981). Thus, the dye is of potential use for measurement of pH changes as well as Ca^{2+} changes provided that the stoichiometry of the dye complex is known for the conditions of measurement and that the dissociation constant of the indicator-ion complex is established for the same conditions. These two parameters have received a great deal of attention, but there remains no clear concensus of the correct binding stoichiometry and K$_D$ under physiological conditions. We investigated this problem by measuring absorbance changes in conjunction with measurements of free Ca^{2+} (Ca$_f$) in the solutions with Ca-ISE. From the known total Ca^{2+} (Ca$_T$) added to the solutions, the Ca-dye complex could be directly calculated. Results from these experiments indicated a 1:1 binding stoichiometry and K$_D$ that varied in relation to [AIII]:ionic strength as summarized here.

Job plots are diagnostic for binding stoichiometry (1928). If the absorbance of equimolar mixtures of Ca^{2+} and AIII are measured, the relation between absorbance and Ca$_T$/Ca$_T$ + AIII$_T$ should have a peak at 0.5 for 1:1 binding, whereas 2 dye:1 Ca binding should peak at 0.33. Figure 5A shows absorbance data on the right ordinate (o) and Ca-dye formation calculated from Ca$_f$ measurements with Ca-ISE on the left ordinate (●) for AIII$_T$ + Ca$_T$ = 10$^{-3.5}$M (upper set) and 10^{-4}M (lower set). All solutions contained 200 mM KCl and were buffered to pH 7.25 with Tris (10 mM). It is evident at both concentration levels that the relations peak at Ca$_T$/Ca$_T$ +AIII$_T$ = 0.5. The results from the Ca-ISE were in good agreement with absorbance measurements.

Calculations of the K$_D$'s from Ca-ISE data indicated a

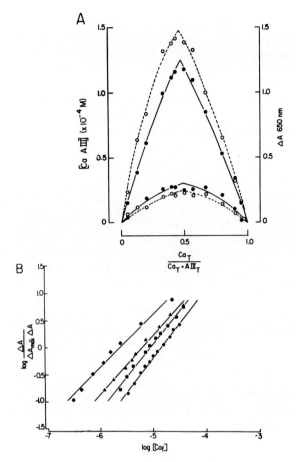

Fig. 5. A. Job plots. Abscissa: equimolar mixtures of AIII and CaCl$_2$. Ordinates: Ca-AIII concentration from Ca-ISE measurements (●) and absorbance measurements in Cary 11 (o); path length 0.5 cm. AIII$_T$ + Ca$_T$ = $10^{-3.5}$M for upper set of plots; AIII$_T$ + Ca$_T$ = 10^{-4}M for lower set. All solutions contained 200 mM KCl, pH 7.25.

B. AIII solutions with 1 mM HEPES + 200 mM KCl (pH 7.25). Linear regression analysis yielded lines for the different AIII concentrations with similar slopes but with different K$_D$ values: $10^{-4.3}$ AIII (●): 14.0 x 10^{-6}M; 10^{-4} AIII (■): 8.8 x 10^{-6}M; $10^{-3.5}$ AIII (▲): 6.5 x 10^{-6}M; 10^{-3} AIII (■): 2.5 x 10^{-6}M. (From Brown and Rydqvist, 1981)

large variation in K_D. To examine the inter-dependence of AIII concentration and ionic strength we titrated AIII with CaCl$_2$ and measured absorbance changes and Ca$_f$ with Ca-ISE. By rearrangement of the equilibrium equation according to Hill (1910), a relation can be obtained where the K_D is independent of the slope of the relation. This is important if K_D varies in any way with [AIII]:KCl.

$$\log \Delta A/\Delta A_{max} - \Delta A = nH \log Ca_f + \log \frac{1}{K_D} \qquad (1)$$

The slope of the relation yields the number of Ca^{2+} (n) that react with an AIII molecule.

Figure 5B shows data from 4 concentrations of dye in the presence of 200 mM background KCl. Four parallel relations are obtained that are displaced along the x-axis. From eq. 1, K_D = Ca$_f$ when $\log \Delta A/\Delta A_{max}$ - ΔA = 0. Thus, the largest K_D was obtained from the lowest AIII concentration and K_D increases systematically as AIII is increased in 200 mM KCl. Thus, KCl exerts a strong effect on the K_D of the dye and makes it pertinent for physiological measures to know both dye concentration and the KCl concentration in the cell (Brown, 1976). By measuring the absorbance of the dye at its isosbestic point (λ = 570 nm) after injection into the cell the dye concentration was obtained.

We attempted to exploit the pH sensitivity of the dye for making intracellular measurements by making simultaneous measurements of absorbance at 620 nm (pH sensitive wavelength). If the extinction of the dye is known for pH and Ca^{2+} at both of the wavelengths, the "pure" Ca^{2+} and pH changes can be calculated (Brown and Rydqvist, 1983). An experiment of this type is shown in Fig. 6.

This cell was injected with 0.75 mM Arsenazo III. A rather strong monitoring beam from the spectrophotometer was being used, therefore, the cell was relatively light adapted. A weak stimulating flash (log attenuation -2.0) elicited a receptor potential and the absorbance of AIII increased at 650 and 620 nm (referenced to the indifferent wavelength 720 nm). By increasing the intensity of the stimulating beam ten-fold, the peak transient phase of the receptor potential increased and absorbance changes at both wavelengths also increased. The intracellular changes in free Ca^{2+} and pH were calculated and are shown at the bottom of the illustration. The data indicate that both Ca$_i$ and H$_i$

increased with illumination. Increasing the light intensity
ten-fold approximately doubles the free Ca^{2+} change. The
pH change is also graded with intensity. The first flash
yielded an acidification of the cytosol of about 0.15 pH
units and when the light intensity was increased during
the second flash, the cytosol decreased in pH by about 0.2
pH units.

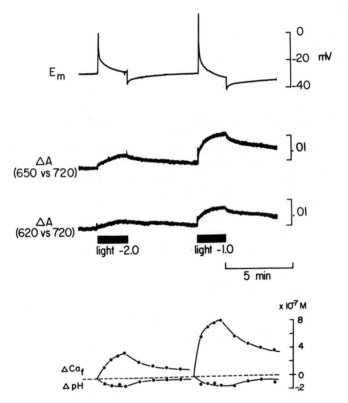

Fig. 6. Changes in intracellular Ca^{2+} and pH based on
simultaneous absorbance changes of AIII at two different
wavelength pairs. Top trace: membrane potential changes
to two steps of light (bars) of different intensity. Middle
trace: change in AIII absorbance (ΔA) at 650 vs. 720 nm.
Bottom trace: change in AIII absorbance at 620 vs. 720 nm.
At the bottom of the Figure are the calculated changes of
intracellular Ca^{2+} and pH based on the differences of
extinction of AIII at 650 and 620 nm for pH and Ca^{2+}.
(From Brown and Rydqvist, 1983)

DISCUSSION

Light elicits changes of intracellular Ca^{2+} and H$^+$ in
Balanus photoreceptors. These changes have been measured
with ion-sensitive electrodes (ISE) and ion-sensitive dyes.
Both indicators agree quite well concerning the magnitude
of the change. A saturating flash of light can raise Ca$_i$
from approximately 1 x 10^{-7}M Ca^{2+} to as much as 10^{-6}M Ca^{2+}
The same flash of light can decrease intracellular pH from
a resting level of about 7.35 (4.47 x 10^{-8}M H$^+$) to a pH
value of about 7.05 (9 x 10^{-8}M H$^+$). Although this change
in intracellular pH is small, it has been calculated from the
known buffer capacity of the cell that the "visible protons"
represent a change in H$^+$ in an unbuffered medium that would
reach the millimole range. This yields a hydrogen per
photon stoichiometry of about 100:1 (Brown and Meech, 1979).
This indicates that the H$^+$ evolved during light is not a
simple by-product of photopigment bleaching, but rather
must occur at some later phase in the photo-transduction
process. Our experiments to estimate the Ca^{2+} buffer capacity in *Balanus* photoreceptors (Brown and Rydqvist, 1983)
indicate that Ca^{2+} may not be as well buffered on the same
time scale as H$^+$. Furthermore, the Ca^{2+} stoichiometry with
relation to light appears to be less than H$^+$. This suggests
that the evolution of Ca^{2+} ions might occur at a different
locus in the transduction process than H$^+$ and that the two
ions are not simply co-related as in a simple ligand system.
The evolution of both products appear to have functional
significance for the photoreceptor in the sense that even
though the concentration changes are small, the inner membrane appears very sensitive to small changes in Ca$_i$ and H$_i^+$.
They could act by substantially altering the membrane conductance changes due to illumination and thereby reduce the
magnitude of the receptor potential to a given step of light.
In this sense one or both may be involved in setting the
sensitivity of the receptor.

ACKNOWLEDGEMENTS

The author wishes to thank and acknowledge the colla-
boration of the following investigators in this work:
Drs. S. Hagiwara, R.W. Meech, H. Koike, H. Sakata and
B. Rydqvist. The technical assistance by T. Gillett and
S. Marron is gratefully appreciated.

Brinley FJ, Scarpa A (1975). Ionized magnesium concentration in axoplasm of dialyzed squid axons. FEBS Letters 50:82.

Brown JE, Brown PK, Pinto LH (1977). Detection of light-induced changes of intracellular ionized calcium concentration in Limulus ventral photoreceptor using Arsenazo III. J Physiol 267:299.

Brown JE, Coles JA, Pinto LH (1977). Effects of injections of calcium and EGTA into the outer segments of retinal rods of Bufo marinus. J Physiol 269:707.

Brown JE, Lisman JE (1975). Intracellular Ca modulates sensitivity and time scale in Limulus ventral photoreceptors. Nature 258:252.

Brown HM (1976). Intracellular Na^+, K^+ and Cl^- activities in large barnacle photoreceptors. J Gen Physiol 68:281.

Brown HM, Cornwall MC (1975). Ionic mechanism of a quasi-stable depolarization in barnacle photoreceptor following red light. J Physiol 248:579.

Brown HM, Hagiwara S, Koike H, Meech R (1970). Membrane activities of a barnacle photoreceptor examined by the voltage-clamp technique. J Physiol 208:385.

Brown HM, Hagiwara S, Koike H, Meech R (1971). Electrical characteristics of a barnacle photoreceptor. Fed Proc 30:69.

Brown HM, Meech RW (1975). Effects of pH and CO_2 on large barnacle photoreceptors. Biophys J 15:276a.

Brown HM, Meech RW (1976). Intracellular pH and light adaptation in barnacle photoreceptor. J Physiol 263:128P.

Brown HM, Meech RW (1979). Light-induced changes of internal pH in a barnacle photoreceptor and effect of internal pH on the receptor potential. J Physiol 297:73.

Brown HM, Meech R, Koike H, Hagiwara S (1969). Current-voltage relations during illumination: photoreceptor membrane of a barnacle. Science 166:240.

Brown HM, Pemberton JP, Owen JD (1976). A calcium-sensitive microelectrode suitable for intracellular measurement of calcium(II) activity. Anal Chim Acta 85:261.

Brown HM, Rydqvist B (1980). Changes of intracellular Ca^{2+} in Balanus photoreceptors probed with Ca^{2+} microelectrodes and arsenazo III. Proc Internat'l Union Phys Sci 14:339.

Brown HM, Rydqvist B (1981). Arsenazo III-Ca^{2+}. Effect of pH; ionic strength and arsenazo III concentration on equilibrium binding evaluated with Ca^{2+}-ISE and absorbance measurements. Biophys J 36:117.

Brown HM, Rydqvist B (1982). Intracellular free Mg^{2+} in Balanus photoreceptors estimated with eriochrome blue.

Biophys J 37:196a.

Brown HM, Rydqvist B (1983). Simultaneous changes of pH and Ca^{2+} in Balanus photoreceptors assayed by Arsenazo III and Ca-ISE. (in preparation)

Chance B (1972). Principles of differential spectrophotometry with special reference to the dual-wavelength method. Methods Enzymol 3:169.

Fahrenbach WH (1965). The micromorphology of some simple photoreceptors. Z Zellforsch mikrosk Anat 66:233.

Gold GH, Korenbrot JI (1980). Light-induced calcium release by intact retinal rods. Proc Natl Acad Sci 77:5557.

Gorman ALF, Thomas MV (1978). Changes in the intracellular concentration of free calcium ions in a pacemaker neuron, measured with the metallochromic indicator dye Arsenazo III. J Physiol 275:357.

Hill AV (1910). A new mathematical treatment of changes of ionic concentration in muscle and nerve under the action of electric currents, with a theory as to their mode of excitation. J Physiol 40:190.

Job P (1928). Reserches sur la formation des complexes mineraux en solution et sur leur stabilite. Ann Chim 9:113.

Lisman JE, Brown JE (1972). The effects of intracellular iontophoretic injection of calcium and sodium ions on the light response of Limulus ventral photoreceptors. J Gen Physiol 59:701.

O'Brien DF (1982). The chemistry of vision. Science 218:961.

Pinto LH, Ostroy SE (1978). Ionizable groups and conductances of the rod photoreceptor membrane. J Gen Physiol 71:329.

Saunders JH, Brown HM (1977). Liquid and solid-state Cl$^-$ sensitive microelectrodes. Characteristics and application to intracellular Cl$^-$ activity in Balanus photoreceptor. J Gen Physiol 70:507.

Yoshikami S, George JS, Hagins WA (1980). Light-induced calcium fluxes from outer segment layer of vertebrate retinas. Nature 286:395.

Yoshikami S, Hagins WA (1973). Control of the dark current in vertebrate rods and cones. In Langer H (ed): "Biochemistry and Physiology of Visual Pigments", New York: Springer-Verlag, p. 245.

H.M. Brown; Mrs. Brown

The Physiology of Excitable Cells, pages 343–355
© 1983 Alan R. Liss, Inc., 150 Fifth Avenue, New York, NY 10011

OPTICAL STUDIES OF T-SYSTEM POTENTIAL AND CALCIUM RELEASE
IN SKELETAL MUSCLE FIBERS.

Julio Vergara, Michael Delay, Judith Heiny and
Bernard Ribalet

Department of Physiology, Jerry Lewis
Neuromuscular Research Center and Ahmanson
Laboratory of Neurobiology, University of
California at Los Angeles, Los Angeles, CA 90024

Excitation-contraction (E-C) coupling in skeletal
muscle, defined as the sequence of events that occurs
between the depolarization of the surface membrane and
tension generation, remains an incompletely understood
process. Within this process the following events can be
distinguished: action potential propagation along the
surface membrane of the fiber; radially inward spread of
the depolarization along the T-system membranes;
transmission at the level of communicating junctions (T-SR)
between the T-tubules and the terminal cisternae of the
sarcoplasmic reticulum (SR); and release of calcium (Ca) by
the SR as a consequence of the coupling.

In particular, it is now believed that the T-system
depolarization is a dynamic process, not occurring
simultaneously with the surface membrane depolarization,
and that conductances in the T-system, which are activated
by the tubular transmembrane potential, influence this
inward propagation. Unfortunately, the tubular membrane
potential, which is the relevant parameter for the
subsequent steps of the E-C coupling, cannot be measured
electrically in muscle fibers because the dimensions of the
T-tubules preclude the use of microelectrodes. In an
attempt to overcome this difficulty, optical methods have
been developed using potentiometric dyes (Cohen and
Salzberg, 1978) to investigate directly the characteristics
of the tubular depolarization in muscle fibers stimulated
to propagate action potentials and under voltage clamp
conditions (Nakajima and Gilai, 1980; Vergara and

Bezanilla, 1981; Heiny and Vergara, 1982).

The transmission at the T-SR junction, a structure which is generally thought to be responsible for the transduction between tubular potential and subsequent Ca release, constitutes one of the key events in the E-C coupling process. To better define this process we have, besides studying the tubular depolarization, recently directed our attention to the study of the Ca release process using optical methods with different Ca indicators (Blinks et al., 1978; Palade and Vergara, 1981).

In this paper we describe experiments made in our laboratory using optical methods to study the characteristics of some of these intermediate steps of the E-C coupling process.

Non-penetrating potentiometric dyes have been used for optical recording of tubular potential changes because they do not report signals from intracellular membrane compartments and because their response is linear with potential, as seen in calibration experiments in the squid giant axon (Ross et al., 1977) and voltage-clamped skeletal muscle fibers (Heiny and Vergara, 1982). Experiments have been performed in single skeletal muscle fibers from Rana catesbeiana using the Hille and Campbell technique (Hille and Campbell, 1976) modified for optical experiments (Vergara et al., 1978; Palade and Vergara, 1982; Heiny and Vergara, 1982). Briefly, as shown in Fig. 1, single fiber segments are mounted in a triple vaseline gap chamber, with their cut ends immersed in solutions which diffuse into the sarcoplasm. The muscle fiber is electrically stimulated under current or voltage clamp conditions using four electrodes, one in each pool of solution; the pools are separated by vaseline strings. The segment of the fiber in pool B is externally bathed with an isotonic K solution and is used as a ground reference. The segment of the fiber in the neighboring pool (pool A) is bathed with an external solution, the composition of which may be varied. The potential in pool A can be accurately measured and used as a feedback signal in voltage-clamp experiments. This segment of fiber is illuminated with monochromatic light focussed on a narrow spot of the muscle fiber (ca. 150 μm) using either a fiber optic (Vergara et al., 1978; and see Fig. 1) or a high power condensing lens. The image of the illuminated segment of fiber is formed on a photovoltaic photodiode in order to measure light intensity changes; both the electrical and the optical signals can be simultaneously acquired and stored in a laboratory computer for analysis and display.

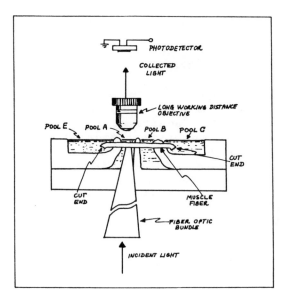

Fig. 1. Schematic diagram of experimental chamber and modifications to allow recording of optical signals. The drawing is not to scale; pool A is typically 150 μm long.

We will now describe experiments made using two types of optical measurements: a) Measurements of membrane potentials using the non-penetrating potentiometric dye NK2367. In these experiments the dye (at about 1 mM concentration in Ringer's solution) is externally applied for 15 minutes to the fiber segment in pool A to allow staining. This solution is later replaced with a lower dye concentration solution (about 1/8 of the staining concentration) used for the optical recording.

b) Optical measurements of the Ca release from the sarcoplasmic reticulum using the Ca indicators arsenazo III and antipyrylazo III. For these studies, these metallochromic indicators seem to be preferable to aequorin because they give larger signals, they bind faster to Ca, and they have a simpler and more definite dye-Ca stoichiometry (Blinks et al., 1978; Palade and Vergara, 1981). In addition, these dyes readily diffuse into the fiber from the cut ends. The dye concentration inside the fiber in each experiment can be estimated by measuring the resting dye absorbance.

Optical measurements of tubular potentials using the non-penetrating potentiometric dye NK2367.

Fig. 2 shows a typical result obtained from a fiber stained with this dye and stimulated under current clamp conditions to elicit an action potential. It can be observed that the membrane action potential recorded electrically shows a much faster time course than does the absorbance signal recorded at a wavelength of 670 nm. This

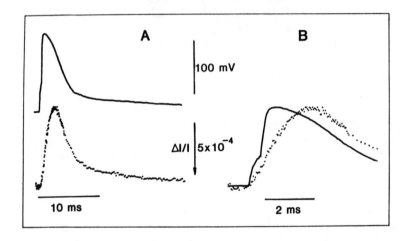

Fig. 2. (A): The upper trace shows the electrical signal, and the lower trace the T-system optical signal recorded with dye NK2367 at 670 nm, following action potential stimulation. (B): The two signals superimposed with a 4x magnified time base. The temperature was 9°C for these figures unless otherwise noted, and the resting potential was -100 mV.

result could be interpreted in two different ways: that the potentiometric dye is not able to respond fast enough to membrane potential changes or that the signal is reporting potential changes from a membrane compartment which is activated more slowly than the surface membrane (whose potential change is recorded electrically).   The first possibility seems unlikely since it has been shown for squid axon that the dye monitors potential changes with a response time at least better than 20 μsec. (Gupta <u>et al</u>., 1981).   In addition, Heiny and Vergara (1982) have demonstrated that in muscle fibers it is possible to record (preferentially at 720 nm) an early step in the optical signals which follows rapidly and linearly the voltage clamp step imposed across the surface membrane, even in cases in which the sodium conductance of the fiber is maximally activated by depolarizing pulses.   In contrast, Fig. 3A shows the signal recorded under similar voltage clamp conditions at 670 nm instead of 720 nm.   It can be observed that in the presence of Na, the time courses of the signals recorded for depolarizing and hyperpolarizing pulses are remarkably asymmetric.   During the 80 mV depolarization (left traces), the signal has a faster time course and reaches a peak at about 3 ms after the onset of

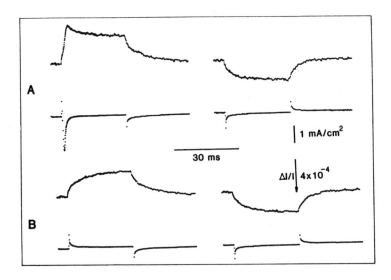

Fig. 3.   Electrical and optical signals recorded with NK2367 following stimulation with + 80 mV voltage-clamp pulses.   In (A) the external solution was 1/2 Na, 1/2 TMA Ringer's; in (B), 10^{-7} M TTX was added.

the pulse. During the hyperpolarizing pulse the signal
shows a monotonic, approximately exponential, time course.
This assymmetry in the signals recorded at 670 nm suggests
that the potential of the membrane compartment monitored at
this wavelength is not adequately controlled by the voltage
clamp circuit. This result can be explained in terms of
the second possibility described above. The T-system
membrane compartment in muscle fibers cannot be rapidly
controlled by the voltage clamp circuit; instead there is a
slow inward propagation of the depolarization. The
previous results can be summarized by suggesting that the
absorbance change recorded with NK2367 from skeletal muscle
fibers preferentially monitors the T-system membrane
potential at 670 nm.

It should be pointed out that in squid axon, a
preparation which has only one membrane compartment, the
optical signals have a time course independent of
wavelength (Gupta et al., 1981). However, Vergara and
Bezanilla (1981) and Heiny and Vergara (1982) have
demonstrated wavelength-dependent time courses compatible
with the proposal that both the surface and T-system
membranes contribute to the total optical change recorded
with dye NK2367 or WW375, but that the relative
contributions from these membranes is strongly wavelength-
dependent. The reasons for this spectral dependency are
currently under investigation in our laboratory.

As pointed out above, the absorbance signals recorded
at 670 nm in Fig. 3A show a marked assymmetry for
depolarizing and hyperpolarizing pulses. It is shown in
Fig. 3B that addition of TTX to the Ringer's solution
abolishes this assymmetry, and the resulting optical traces
show an exponential-like time course. TTX blockage of the
sodium conductance can be verified in the current traces
which now show the characteristics of a passive electrical
response without the conspicuous sodium current. A similar
result is obtained if the external sodium ions are replaced
by impermeant TMA or TRIS ions. It can be concluded that
there is a sodium conductance in the T-system, responsible
for the assymmetry in the optical records, that is
activated at depolarizing pulses and that makes the tubular
depolarization occur faster and more efficiently. These
results confirm and extend the suggestions of Costantin
(1970). It is important to point out that, for the first
time, we have been able to directly observe, using optical

techniques, the effect of the sodium conductance of the T-system on the electrical characteristics of this membrane compartment. These effects can be modelled mathematically as shown by Adrian and Peachey (1973), and this analysis can be extended to model the optically measured mean tubular potential by calculating a weighted average of the T-system potential at all radial positions. Fig. 4 shows an example of this modelling using the radial cable equations and the parameters of Campbell and Hille (1976) for the sodium conductance. The assymmetry in the optical signals is clearly predicted by including an activatable sodium conductance in the T-system. It should be pointed out, however, that no attempt has yet been made with this

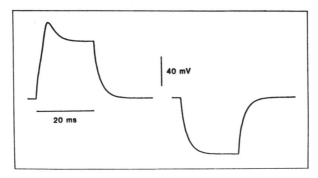

Fig. 4. Predicted average tubular potential in response to + 80 mV voltage-clamp steps.

model to fit the optical signals by modifying the parameters of the equations.

The results shown previously have been obtained in muscle fibers in which the delayed rectifier (potassium) conductance has been blocked by replacement of internal K by Cs. Fig. 5 shows results obtained from a fiber stained with NK2367 but perfused internally with solutions containing K. The current trace shows the typical early sodium current followed by a delayed outward potassium current. The optical trace shows a rising phase similar to that presented in Fig. 3 but now followed by a more pronounced decline in the signal. This falling phase in the presence of potassium suggests that the potassium conductance must play an important role in the repolarization of the T-system potential during normal activation. This is interesting because it was originally suggested by Adrian, Chandler and Hodgkin (1969) that the

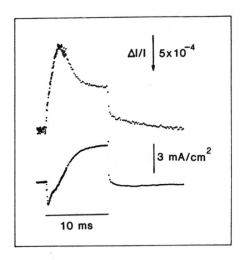

Fig. 5. T-system optical signal and net ionic current for a voltage-clamp step of +100 mV recorded from a fiber perfused internally with isotonic K and externally with 1/2 Na, 1/2 TMA Ringer's.

T-system might not have a delayed rectifier conductance. This assertion was later contested by Kirsch et al., (1977). Our results agree with those of Kirsch et al. and provide a direct method to quantitatively evaluate the magnitude and kinetics of this conductance.

Another T-system conductance that can be investigated with optical methods is the inward rectifier conductance (Hodgkin and Horowicz, 1959; Almers, 1972). Our results (Heiny et al., 1983) demonstrate that the conductance is responsible for a marked attenuation of the tubular potential when it is fully activated. This result is shown in Fig. 6. It can be observed that a depolarizing pulse elicits a smaller current and a larger optical signal than does the equivalent hyperpolarizing pulse. In this case, the mean tubular potential change recorded optically for an hyperpolarizing pulse reached only 70% of that for a depolarizing pulse. Further, the time course of the signal recorded for depolarizing pulses is slower than for hyperpolarizing pulses. Both results are compatible with the explanation that there is a faster time constant and a smaller space constant when the tubular conductance is

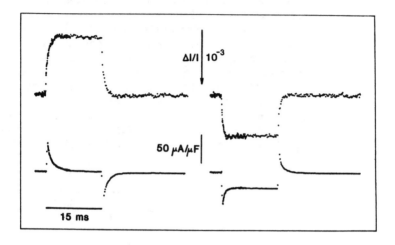

Fig. 6. T-system optical signals (top) and membrane current (bottom) recorded from a fiber perfused externally with a 100 mM KSO$_4$ Ringer's, for 15 ms voltage-clamp steps of ± 80 mV from the K equilibrium potential (-5 mV).

greatly increased by the activation of the inward rectifier.

In summary, our results strongly suggest that the non-penetrating potentiometric dye NK2367 largely monitors the tubular potential changes at 670 nm. These optical signals allow us to identify and eventually quantitate the role of tubular conductances in the inward spread of the depolarization in skeletal muscle fibers.

Optical studies of the release of Ca from the SR using Ca indicating dyes

Accurate measurements of the process of Ca release

from the SR are necessary in any attempt to understand the coupling at the level of the T-SR junctions. The metallochromic indicators arsenazo III and antipyrylazo III give extremely reproducible results and can therefore be used to this end.

Fig. 7A shows the time courses of an absorbance signal and membrane potential recorded from a fiber stimulated by an action potential and internally diffused with the Ca indicating dye antipyrylazo III. Not only does the optical signal have a significantly slower time course than does the action potential, but a lag of about 3 ms is present between the onset of the potential and that of the optical signal. The overall time course of this Ca-related absorbance signal is also significantly slower than that of the T-system potential signal shown in Fig. 2. Even though the results were obtained from two different fibers under

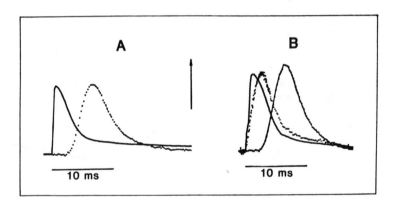

Fig. 7. (A): Optical recordings made at 660 nm, and membrane potential, recorded from a fiber internally perfused with antipyrylazo III following action potential stimulation. The vertical calibration bar corresponds to a membrane potential change of 110 mV, and a dye absorbance change of $\Delta A/A = 2.72 \times 10^{-3}$. (B): the calcium transient of (A) superimposed upon the traces of Fig. 2. The vertical calibration bar corresponds to a membrane potential change of 100 mV, a potentiometric dye absorbance change of $\Delta A = 2.17 \times 10^{-4}$, and a calcium-sensitive dye absorbance change of $\Delta A/A = 2.15 \times 10^{-3}$.

different conditions, it is interesting to compare both
signals in order to obtain a rough estimate of the coupling
lag. This is done in Fig. 7B where both the T-system and
Ca signals have been arbitrarily scaled to approximately
the same amplitude but retain the same time base. The lag,
and more generally the transfer characteristic between the
driving function and the observed Ca release, may in
principle depend upon the T-system voltage, the kinetic
behavior of dye-Ca interaction, and upon the coupling
across the T-SR junction. In order to obtain more
information on the properties of this coupling, we have
performed voltage-clamp experiments in which external Na is
replaced by TMA, thus preventing propagation of action
potentials in the T-system. Fig. 8 shows a comparison of
Ca release transients with and without Na. The Ca
transient is slowed in the presence of TMA, and the early
phase of the signal is considerably diminished in amplitude
for smaller applied depolarizations, presumably in part due
to the marked decrease in the early tubular potential when
compared to the situation with external Na. During the
later phase, the tubular potential will arrive at the same
value regardless of external Na, and this is reflected in

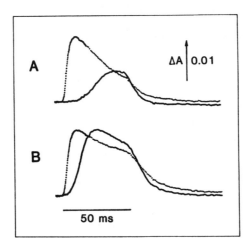

Fig. 8. Calcium transients recorded at 710 nm from a
voltage-clamped fiber internally perfused with antipyrylazo
III. The dotted traces were taken with 1/2 Na, 1/2 TMA
Ringer's; the solid, with TMA Ringer's externally. (A):
+70 mV step; (B): +100 mV step.

the similar amplitudes of the corresponding Ca transients. These results illustrate the great influence that the sodium conductance of the T-system has on the overall coupling process. It is clearly observed in Fig. 8 that in the absence of Na, the tubular depolarization can be the rate limiting process for the Ca release from the SR.

We would like to conclude from the experiments presented in this paper that the use of optical techniques in the study of the excitation-contraction coupling process in skeletal muscle can give valuable information not obtainable by other methods alone. A great number of experiments incorporating new and more refined optical methods will be necessary in the future to consolidate our understanding of this process.

REFERENCES

Adrian RH, Chandler WK, Hodgkin AL (1970). Voltage clamp experiments for striated muscle fibers. J Physiol 208:607-644.
Adrian RH, Chandler WK, Hodgkin AL (1969). The kinetics of mechanical activation in frog muscle. J Physiol 204:207-230.
Adrian RH, Peachey LD (1973). Reconstruction of the action potential of frog sartorius muscle. J Physiol 235:103-131.
Almers W (1972). Potassium conductance changes in skeletal muscle and the potassium concentration in the transverse tubules. J Physiol 225:33-56.
Blinks JR, Rudel R, Taylor SR (1978). Calcium transients in isolated amphibian skeletal muscle fibers: Detection with aequorin. J Physiol 277:291-323.
Campbell DT, Hille B (1976). Kinetic and pharmacological properties of the sodium channel of frog skeletal muscle. J Gen Physiol 67:309-323.
Cohen LB, Salzberg, BM (1978). Optical measurement of membrane potential. Rev Physiol Biochem Pharmacol 83:35-88.
Costantin LL, (1970). The Role of Sodium Current in the Radial Spread of Concentration on Frog Muscle Fibers. J Gen Physiol 55: 703-715.

Gupta RK, Salzberg BM, Grinvald A, Cohen LB, Kamino K, Lesher S, Boyle MB, Waggoner AS, Wang CH (1981). Improvements in optical methods for measuring rapid changes in membrane potential. J Membrane Biol 58:123-137.

Heiny JA, Vergara J (1982). Optical signals from surfce and T-system membranes in skeletal muscle fibers. Experiments with the potentiometric dye NK2367. J Gen Physiol 80:203-230.

Heiny JA, Ashcroft FM and Vergara J (1983). T-system optical signal associated with inward rectification in skeletal muscle. Nature 31:164-166.

Hille B, Campbell DT (1976). An improved vaseline gap voltage clamp for skeletal muscle fibers. J Gen Physiol 67:265-293.

Hodgkin AL, Horowicz P (1959). The influence of potassium and chloride ions on the membrane potential of single muscle fibers. J Physiol 148:127-160.

Kirsch GE, Nichols RA, Nakajima S (1977). Delayed rectification in the transverse tubules. Origin of the late after-potential in frog skeletal muscle. J Gen Physiol 70:1-21.

Nakajima A, Gilai A (1980). Action potentials of isolated single muscle fibers recorded by potential-sensitive dyes. J Gen Physiol 76:729-750.

Palade P, Vergara J (1981). Detection of Ca^{2+} with optical methods. Grinnell AD, Brazier MAB (eds.): "The Regulation of Muscle Contraction: Excitation-Contraction Coupling," New York: Academic Press, pp 67-77.

Palade P, Vergara J (1982). Arsenazo III and antipyrylazo III calcium transients in single skeletal muscle fibers. J Gen Physiol 79:697-707.

Ross WN, Salzberg BM, Cohen LB, Grinvald A, Davila HV, Waggoner AS, Wang CH (1977). Changes in absorption, fluorescence, dichroism and birefringence in stained giant axons: Optical measurement of membrane potential. J Membrane Biol 33:141-183.

Vergara J, Bezanilla F (1981). Optical studies of E-C coupling with potentiometric dyes. In Grinnell AD, Brazier M (eds): "The Regulation of Muscle Contraction: Excitation-Contraction Coupling," Academic Press, pp 67-77.

Vergara J, Bezanilla F, Salzberg BM (1978). Nile Blue fluorescence signals from cut single muscle fibers under voltage or current clamp conditions. J Gen Physiol 72:775-800.

Top: J. Heiny; F. Ashcroft
Bottom: J. Vergara; M. Bosma; B. Ribalet

The Physiology of Excitable Cells, pages 357–369
© 1983 Alan R. Liss, Inc., 150 Fifth Avenue, New York, NY 10011

EFFECTS OF THYROLIBERIN AND TRIFLUOPERAZINE ON HORMONE
RELEASE AND ELECTRICAL PROPERTIES OF CULTURED RAT
PITUITARY CELLS

Olav Sand[1], Seiji Ozawa[2], Kjersti Sletholt[3],
Egil Haug[4] and Kaare M. Gautvik[4]

[1]Department of Biology, University of Oslo,
P.O.Box 1066, Blindern, Oslo 3, Norway;
[2]Department of Physiology, Jichi Medical School,
Tochigi-ken, Japan; [3]Department of Physiology,
Norwegian College of Veterinary Medicine, Oslo
and [4]Institute of Physiology, University of Oslo

The anterior pituitary gland is composed of several
endocrine cell types secreting at least six different hormones. The hypothalamus exerts its influence on the gland by
secretion of stimulatory or inhibitory substances reaching
their target cells through the pituitary portal circulation.
The function of anterior pituitary cells is regulated by a
series of hormones in an integrated way. To simplify studies
of the regulation of anterior pituitary secretion, we have
therefore worked on permanent tumor cell lines which have
retained several of the functional characteristics of the
normal cell type. The present contribution describes recent
studies of the effects of thyroliberin (TRH) and trifluoperazine (TFP) on the GH_3 clonal cell line.

THE GH_3 CELL LINE

The initial establishment and culture of the GH_3 strain
of rat pituitary tumor cells were performed by Tashjian et
al. (1968). The cells produce both prolactin (PRL) and growth
hormone (GH), and have the ability to respond to several hypothalamic and steroid hormones, as well as to a range of other
secretagogues, in a manner analogous to normal anterior pituitary cells (see Tashjian 1979; see Haug et al. 1982).
Several laboratories have shown that the hypothalamic tripep-

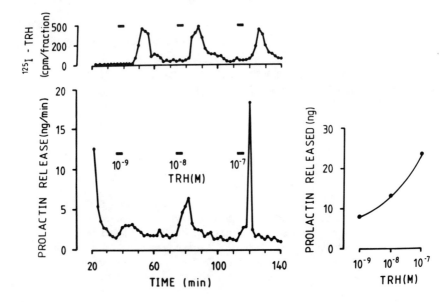

Fig. 1. PRL release from perfused Bio-Gel P-2 column containing about 70x10^6 GH$_3$ cells. Thick lines indicate 4 min infusion pulses of increasing concentrations of TRH (adjusted for the time-lag due to the dead volume of the system). ^{125}I-TRH was added to the infusions as an indicator for the presence of TRH in the eluate (upper part of the figure). The dose response curve on the right represents the total PRL release in excess of the background secretion.

tide TRH, which normally enhances the release of thyrotropin and PRL from the anterior pituitary gland (Boler et al. 1969; Burgus et al. 1970), stimulates the release of both GH and PRL from GH$_3$ cells.

The experiment presented in figure 1, based on the method of Lowry and McMartin (1974), shows the typical secretory response of GH$_3$ cells to TRH. A suspension of GH$_3$ cells was obtained from monolayer cultures, and the cells were stacked in a Bio-Gel P-2 column confined in a 5 ml syringe. The column was perfused at a rate of 0.5 ml/min and the fractions were collected at 2 min intervals. The experiment was conducted at 37°C. Infusions of increasing concentrations of TRH caused an immediate and dose-dependent increase in release of PRL as well as GH (data not shown). ^{125}I-TRH was added to the TRH infusions and used as a marker of TRH elution. Figure 1 shows that PRL elutes from the Bio-Gel P-2 column before TRH. The fractionation range for Bio-Gel P-2 is 100-1800 daltons, and PRL is therefore excluded from the column while TRH is retarded. Depolarizing concentrations of K^+ also cause an immediate increase of the hormone secretion from both normal anterior pituitary cells (Vale et al. 1967) and GH_3 cells (Ozawa, Miyazaki 1979).

ELECTROPHYSIOLOGICAL EFFECTS OF TRH

Several types of peptide producing endocrine cells, including both normal anterior pituitary cells and GH_3 cells, are able to generate partly Ca^{2+} dependent action potentials (Dean, Matthews 1970; Kidokoro 1975; Brandt et al. 1976; Tischler et al. 1976; Taraskevich, Douglas 1977; Ozawa, Sand 1978; Dufy et al. 1979; Ozawa, Miyazaki 1979; Ozawa, Saito 1980; Sand et al. 1981). The Ca^{2+} channels of the GH_3 cell membrane have recently been studied in detail using the patch clamp technique (Hagiwara, Ohmori 1982; Ohmori, Hagiwara this volume).

Exocytosis of secretory granules from endocrine cells is commonly triggered by Ca^{2+} influx (Douglas 1968). It is reasonable to suggest that this Ca^{2+} entry is associated with the Ca^{2+} dependent action potentials of the various endocrine cells. The situation would thus be analogous to the mechanism for the release of neurotransmitter. In agreement with this idea, TRH enhances the occurrence of action potentials in GH_3 cells (Kidokoro 1975; Dufy et al. 1979; Ozawa, Kimura 1979a; Sand et al. 1980). Ozawa and Kimura (1979a) showed that in the majority of GH_3 cells application of TRH produced a long-lasting enhancement of spike generation preceded by a transient membrane hyperpolarization lasting about 10 s. This hyperpolarization was due to an increase of the membrane permeability to K^+. Both short term (less than 10 s) and continuous TRH application produced the same response pattern.

The amplitude and duration of the transient hyperpolarizing response to TRH was augmented in Cl^--free solution (Ozawa 1981), and figure 2 gives examples of the TRH response under such conditions. During the experiments the culture medium was replaced with a Cl^--free saline obtained by replacing the chloride salts with the corresponding

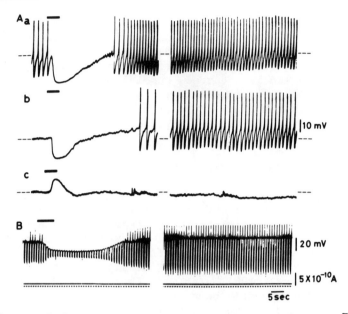

Fig. 2. TRH-induced membrane potential responses in Cl⁻-free saline. The TRH containing delivery pipette was located near the cell during the periods indicated by thick lines. (A) The 3 traces were obtained from the same cell at different membrane potential levels by passing DC current through the recording electrode. The dashed lines in a,b, and c indicate potential levels of -44 (a), -51 (b) and -83 (c) mV, respectively. (B) Changes of the membrane conductance. The input resistance was monitored by the amplitude of the hyperpolarizing potential deflections caused by constant current injections at 1 Hz (lower trace). The middle part of the traces are interrupted for 10 s. Modified from Ozawa (1981).

propionate salts. TRH was applied by diffusion from a delivery pipette (tip diameter 4-5 μm) positioned 40-60 μm from the cell. In an attempt to study the mechanism for the facilitating effect of TRH, this drug was applied to the same cell kept at different membrane potential levels by DC current injections. When the membrane potential was slightly more negative than the critical membrane potential for spike initiation (figure 2A_b), the transient TRH-induced hyperpolarization was followed by a depolarization eventually attaining a level causing spontaneous firing of action potentials. If the membrane was further hyperpolarized to reverse the transient hyperpolarization, the late depolarization was also reversed to a hyperpolarizing response (figure $2A_c$). The reversal potential of the transient hyperpolarization and the late depolarization was found to be the same, about -66 mV (Ozawa 1981). Since the transient hyperpolarization is due to an increased K^+ permeability of the membrane (Ozawa, Kimura 1979a), the late depolarization is probably caused by a decrease of the K^+ permeability.

This notion was confirmed by recordings of the TRH induced changes of the membrane conductance, as shown in figure 2B. The late depolarization was accompanied by a decreased membrane conductance, while the preceding hyperpolarization was associated with an increase. Both the membrane depolarization and the increased input resistance caused by the late decrease of the membrane K^+ permeability are favourable for spike generation, and it is thus likely that the sustained TRH-induced enhancement of spike activity is mainly due to the decreased K^+ permeability. Dufy et al. (1982) have recently reported a similar biphasic response pattern to TRH in cultured human pituitary cells.

EFFECTS OF TFP ON BASAL AND STIMULATED HORMONE RELEASE

A wide range of functions attributed to intracellular calcium is exerted via the calcium binding protein calmodulin, which regulates the activity of a number of cellular enzymes (see Cheung 1980; see Means, Dedman 1980). The calcium-calmodulin complex is inhibited by the phenothiazine drugs, the most potent being TFP (Levin, Weiss 1977). TFP has recently been reported to inhibit stimulated exocytosis in various secretory cells, including thyrotropin and insulin producing endocrine cells (Fleckman et al. 1981; Janjic et al. 1981; Valverde et al. 1981), synaptic terminals of the frog neuromuscular junction (Publicover 1982) and rat mast cells (Douglas, Nemeth 1982).

The biological action of the phenothiazines is not specific, and a number of membrane effects in addition to inhibition of the calcium-calmodulin complex have been reported (see Douglas, Nemeth 1982). Inhibitory effects of phenothiazines on Ca^{2+} dependent secretory processes should therefore not be interpreted as exclusive evidence for calmodulin involvement. In order to elucidate this problem, we have studied the effects of TFP on both the electrical membrane

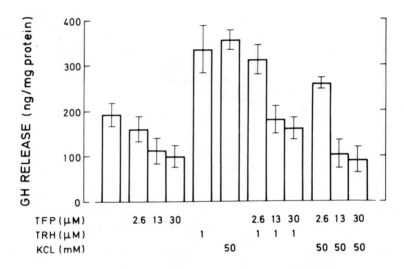

Fig. 3. Effects of TFP, TRH and KCl on GH release from cell suspensions of GH₃ cells prepared from monolayer cultures. An incubation period of 30 min was used to measure the release of stored hormone, and data are given as mean values ± S.D. of triplicates. Modified from Sand et al. (1983).

properties and the hormone release from GH$_3$ cells.

TFP affects the basal hormone release from GH$_3$ cells in a biphasic way dependent on the concentration employed. Less than 40 μM TFP inhibits the basal release, whereas higher concentrations cause a facilitation. This enhancement of the basal release is independent of extracellular Ca^{2+} (Sletholt et al. unpublished). Possible mechanisms for this effect of high TFP concentrations will not be discussed further in the present paper.

Figure 3 shows that TFP at a concentration as low as 13 μM clearly inhibited both the K$^+$ and TRH induced stimulation of the GH release, and similar results were obtained for the release of PRL (Sand et al. 1983). The stimulated hormone release caused by TRH or depolarizing concentrations of K$^+$ is completely blocked by the addition of EGTA or the Ca^{2+} antagonists Co^{2+}, D600 or verapamil (Gautvik, Tashjian 1973; Ostlund et al. 1978; Tashjian et al. 1978; Ozawa, Kimura 1979a,b; Gautvik et al. 1980). This stimulated hormone rel-

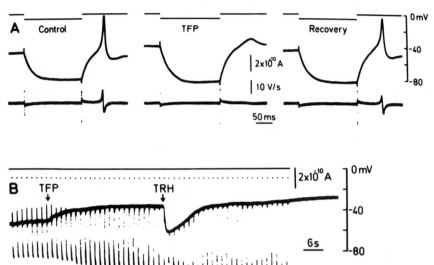

Fig. 4. (A) Intracellular recordings from a GH_3 cell in Na^+-free solution, showing Ca^{2+}-dependent action potentials evoked at the termination of a hyperpolarizing current pulse before and 10 s after termination of TFP application. The middle recording was obtained 5 s after start of TFP application by diffusion from a micropipette filled with 400 μM TFP and positioned 40 μm from the cell. The lower traces represent the first order derivative of the potential traces. The maximum rate of rise of the action potential, and hence the magnitude of the maximum inward Ca^{2+} current, was dramatically reduced by TFP. (B) Membrane response in Cl^--free solution to continuous TFP and TRH applications by diffusion from micropipettes positioned 40 μm from the cell and filled with 400 μM TFP and 100 μM TRH, respectively. Arrows indicate the timing for the positioning of the two different delivery pipettes. Hyperpolarizing current pulses (upper trace) were injected through the recording electrode in order to monitor the membrane resistance. Modified from Sand et al. (1983).

ease was also completely blocked at the higher TFP concentrations which induced a Ca^{2+} independent elevation of the basal hormone release (Sletholt et al. unpublished).

EFFECTS OF TFP ON ELECTRICAL MEMBRANE PROPERTIES

TFP at a concentration of 40 μM inhibits the K^+-induced $^{45}Ca^{2+}$ uptake by rat pituitary tissue (Fleckman et al. 1981), and it is reasonable to suggest that TFP may inhibit the stimulated hormone release primarily by blocking Ca^{2+} uptake rather than inhibiting the intracellular effects of Ca^{2+}. This conclusion was strengthened by studying the effect of TFP on electrically evoked Ca-spikes in Na^+-free solution. Figure 4A demonstrates how TFP application reversibly inhibits the Ca-spike in GH_3 cells. In addition to blocking the voltage sensitive Ca^{2+} channels in the membrane, TFP also caused a pronounced membrane depolarization and increased input resistance (Sand et al. 1983). These latter effects are most likely due to a decreased K^+ conductance.

We have also studied the effect of TFP on the normal membrane response to TRH, as seen in figure 4B. In these experiments two delivery pipettes, filled with TFP and TRH, respectively, were aimed at the penetrated cells. Within a few seconds after lowering the TFP pipette in position, the cells displayed a clear membrane depolarization and increased membrane resistance. The TRH pipette was positioned after a delay of about 30 s, and the normal biphasic TRH response was evident in spite of the continuous influence of TFP. However, the late depolarization caused by TRH did not initiate action potentials in TFP treated cells. In the absence of TFP, TRH induced spontaneous firing of action potentials in a similar way as demonstrated in figure $2A_b$. This experiment shows that the interaction between TRH and its receptor is intact. Furthermore, TFP at a concentration of 30 μM did not interfere with the binding of TRH to GH_3 cells (Sletholt et al. unpublished).

DISCUSSION

TRH-stimulated hormone release from GH_3 cells is preceded by an elevation of cellular cyclic AMP levels (Gautvik et al. 1980). Both the TRH-induced hormone release and the Ca^{2+} dependent action potentials are abolished by blockers of voltage sensitive Ca^{2+} channels (Ozawa, Miyazaki 1979; Ozawa, Kimura 1979a,b). The stimulated hormone release is also blocked in Ca^{2+}-free medium (Gautvik, Tashjian 1973; Gautvik et al. 1980), whereas the TRH evoked increase of cyclic AMP persists under such conditions (Gautvik et al. 1980). These data suggest that the biphasic membrane response to TRH, which eventually leads to increased firing rate of Ca^{2+}-dependent action potentials, may be connected to the elevation of cellular cyclic AMP. The increased Ca^{2+} influx normally associated with the action potentials seems, on the other hand, to be imperative for the TRH induced enhancement of hormone secretion.

The existence of a functional relationship between the TRH-stimulated action potentials and the TRH-mediated hormone release is further supported by studies of the effects of 4-aminopyridine (4AP) on the hormone release and electrical membrane properties of GH_3 cells (Sand et al. 1980). This drug inhibits the late K^+ current in a variety of excitable cells (see Lechat et al. 1982), and was found to mimick the facilitating effects of TRH on the action potentials without affecting the resting membrane properties. Furthermore, 4AP stimulated hormone release from GH_3 cells to the same extent as TRH, which strongly indicates that TRH enhances the hormone release via facilitation of the Ca^{2+}-dependent action potentials.

However, the membrane effects of TRH seem to be more complex. Gershengorn et al. (1981) have recently shown that TRH is able to enhance hormone release from GH_3 cells in medium containing only 0.02 µM free Ca^{2+}, whereas the K^+ induced hormone release was completely abolished. This result was interpreted as evidence for TRH dependent recruitment of free Ca^{2+} from intracellular sequestered stores, although such a model is unable to explain the effective inhibition of TRH-induced hormone release by blockers of voltage sensitive Ca^{2+} channels. Tan and Tashjian (1981), on the other hand, have presented evidence in favour of a TRH-induced Ca^{2+} release from a superficial compartment of GH cells. At extremely low extracellular Ca^{2+} levels, this effect of TRH could cause a local increase in the extracellular Ca^{2+} concentration sufficient to support an inward Ca^{2+} current through the gated, voltage dependent Ca^{2+} channels.

Clearly the mechanisms by which TRH enhances the level of intracellular free Ca^{2+} in GH_3 cells are at present not completely understood. A similar conclusion is also appropriate regarding the steps between the increased intracellular Ca^{2+} level and the final exocytosis of hormone. Experiments employing TFP application to intact cells, in order to detect a possible involvement of calmodulin in this process, should be interpreted with care. TFP inhibits the TRH and K^+ stimulated hormone release from GH_3 cells, but this effect may primarily be due to blocking of the voltage sensitive Ca^{2+} channels rather than to inhibition of an intracellular calcium-calmodulin complex. It is unlikely that the observed blocking of the Ca^{2+} channels by TFP could be due to inhibition of calmodulin-dependent enzymes necessary for the regulation of the functional state of these channels. Increased intracellular Ca^{2+} levels inactivate the voltage dependent Ca^{2+} channels in a variety of cell types (see Hagiwara, Byerly 1981). If this effect of intracellular Ca^{2+} is mediated via calmodulin-dependent enzymes, TFP would be expected to suppress the development of this inactivation process, thus increasing the inward Ca^{2+} current. The observed inhibition of the voltage-dependent Ca^{2+} channels is therefore probably due to a direct effect on the cell membrane, although the existence of calmodulin dependent processes regulating the Ca^{2+} conductance has not been excluded.

However, these results must of course not be interpreted as evidence against a possible role for calmodulin in the mediator function of Ca^{2+} in stimulated hormone secretion. Douglas and Nemeth (1982) have recently shown that TFP inhibits rat mast cell secretion elicited by drugs which increase the intracellular Ca^{2+} level independently of Ca^{2+} influx through voltage-dependent Ca^{2+} channels.

Acknowledgement. Financial support has been received from the Norwegian Research Council for Science and the Humanities and from the Norwegian Society for Fighting Cancer.

REFERENCES

Boler J, Enzman F, Folkers K, Bowers CY, Schally AV (1969). The identity of chemical and hormonal properties of the thyrotropin releasing hormone and pyroglutamyl-histidyl-proline amide. Biochem Biophys Res Commun 37:705.

Brandt BL, Hagiwara S, Kidokoro Y, Miyazaki S (1976). Action potentials in the rat chromaffin cell and effects of acetylcholine. J Physiol 263:417.

Burgus R, Dunn TF, Desiderio D, Ward DN, Vale W, Guillemin R (1970). Characterization of ovine hypothalamic hypophysiotropic TSH-releasing factor. Nature 226:321.

Cheung WY (1980). Calmodulin plays a pivotal role in cellular regulation. Science 207:19.

Dean PM, Matthews EK (1970). Electrical activity in pancreatic islet cells: Effects of ions. J Physiol 210:265.

Douglas WW (1968). Stimulus-secretion coupling: The concept and clues from chromaffin and other cells. Br J Pharmacol 34:451.

Douglas WW, Nemeth EF (1982). On the calcium receptor activating exocytosis: Inhibitory effects of calmodulin-interacting drugs on rat mast cells. J Physiol 323:229.

Dufy B, Vincent JD, Fleury H, Du Pasquier P, Gourdji D, Tixier-Vidal A (1979). Membrane effects of thyrotropin-releasing hormone and estrogen shown by intracellular recordings from pituitary cells. Science 204:509.

Dufy B, Israel JM, Zyzek E, Dufy-Barbe L, Guerin J, Fleury H, Vincent JD (1982). An electrophysiological study of cultured human pituitary cells. Mol Cell Endocrinol 27:179.

Fleckman A, Erlichman J, Schubart UK, Fleischer N (1981). Effect of trifluoperazine, D600, and phenytoin on depolarization- and thyrotropin-releasing hormone-induced thyrotropin release from rat pituitary tissue. Endocrinology 108:2072.

Gautvik KM, Tashjian Jr AH (1973). Effects of cations and colchicine on the release of prolactin and growth hormone by functional pituitary tumor cells in culture. Endocrinology 93:793.

Gautvik KM, Iversen JG, Sand O (1980).On the role of extracellular Ca^{2+} for prolactin release and adenosine 3':5'-monophosphate formation induced by thyroliberin in cultured rat pituitary cells. Life Sciences 26:995.

Gershengorn MC, Hoffstein ST, Rebecchi MJ, Geras E, Rubin BG (1981). Thyrotropin-releasing hormone stimulation of prolactin release from clonal rat pituitary cells. J Clin Invest 67:1769.

Hagiwara S, Byerly L (1981). Calcium channel. Ann Rev Neurosci 4:69.

Hagiwara S, Ohmori H (1982). Studies of calcium channels in rat clonal pituitary cells with patch electrode voltage clamp. J Physiol 331:231.

Haug E, Gautvik KM, Sand O, Iversen JG, Kriz M (1982). Interaction between thyrotropin-releasing hormone and prolactin producing cells. In McKerns KW, Pantic V (eds): "Hormonally Active Brain Peptides," New York: Plenum Publ Corp, p 537.

Janjic D, Wollheim CB, Siegel EG, Krausz Y, Sharp GWG (1981). Sites of action of trifluoperazine in the inhibition of glucose-stimulated insulin release. Diabetes 30:960.

Kidokoro Y (1975). Spontaneous Ca action potentials in a clonal pituitary cell line and their relation to prolactin secretion. Nature 258:741.

Lechat P, Thesleff S, Bowman WC (1982). "Aminopyridines and Similarly Acting Drugs: Effects on Nerves, Muscles and Synapses". Oxford: Pergamon Press.

Levin RM, Weiss B (1977). Binding of trifluoperazine to the calmodulin-dependent activator of cyclic nucleotide phosphodiesterase. Mol Pharmacol 13:690.

Lowry PJ, McMartin C (1974). Measurements of the dynamics of stimulation and inhibition of steroidogenesis in isolated rat adrenal cells by using column perfusion. Biochem J 142:287.

Means AR, Dedman JR (1980). Calmodulin - an intracellular calcium-receptor. Nature 285:73.

Ostlund Jr RE, Leung JT, Hajek SV, Winokur T, Melman M (1978). Acute stimulated hormone release from cultured GH_3 pituitary cells. Endocrinology 103:1245.

Ozawa S (1981). Biphasic effect of thyrotropin-releasing hormone on membrane K^+ permeability in rat clonal pituitary cells. Brain Res 209:240.

Ozawa S, Kimura N (1979a). Membrane potential changes caused by thyrotropin-releasing hormone in the clonal GH_3 cell and their relationship to secretion of pituitary hormone. Proc Natl Acad Sci USA 76:6017.

Ozawa S, Kimura N (1979b). Ca antagonist (verapamil) inhibits the stimulating effect of thyrotropin-releasing hormone on the release of prolactin from clonal rat pituitary cells. In Ito M (ed): "Integrative Control Functions of the Brain," Tokyo: Kodansha, p 296.

Ozawa S, Miyazaki S (1979). Electrical excitability in the rat clonal pituitary cell and its relation to hormone secretion. Jap J Physiol 29:411.

Ozawa S, Saito T (1980). Sodium and calcium action potentials in human anterior pituitary cells. Experientia 36:1235.

Ozawa S, Sand O (1978). Electrical activity of rat anterior pituitary cells in vitro. Acta Physiol Scand 102:330.

Publicover SJ (1982). Inhibitory effect of the "calmodulin inhibitor" trifluoperazine on stimulated transmitter release at the frog neuromuscular junction. J Physiol 330:4P.

Sand O, Haug E, Gautvik KM (1980). Effects of thyroliberin and 4-aminopyridine on action potentials and prolactin re-release and synthesis in rat pituitary cells in culture. Acta Physiol Scand 108:247.

Sand O, Ozawa S, Gautvik KM (1981). Sodium and calcium action potentials in cells derived from a rat medullary carcinoma.

Acta Physiol Scand 112:287.

Sand O, Sletholt K, Gautvik KM, Haug E (1983). Trifluoperazine blocks calcium-dependent action potentials and inhibits hormone release from rat pituitary tumour cells.
Eur J Pharmacol 86:177.

Tan KN, Tashjian Jr AH (1981). Receptor-mediated release of
plasma membrane-associated calcium and stimulation of calcium uptake by thyrotropin-releasing hormone in pituitary
cells in culture. J Biol Chem 256:8994.

Taraskevich PS, Douglas WW (1977). Action potentials occur
in cells of the normal anterior pituitary gland and are
stimulated by the hypophysiotropic peptide thyrotropin-
releasing hormone. Proc Natl Acad Sci USA 74:4064.

Tashjian Jr AH (1979). Clonal strains of hormone-producing
pituitary cells. Methods Enzymol 58:527.

Tashjian Jr AH, Yasumura Y, Levine L, Sato GH, Parker ML
(1968). Establishment of clonal strains of rat pituitary
tumor cells that secrete growth hormone. Endocrinology 82:
342.

Tashjian Jr AH, Lomedico ME, Maina D (1978). Role of calcium
in the thyrotropin-releasing hormone-stimulated release of
prolactin from pituitary cells in culture. Biochem Biophys
Res Commun 81:798.

Tischler AS, Dichter MA, Biales B, DeLellis RA, Wolfe H
(1976). Neural properties of cultured human endocrine
tumor cells of proposed neural crest origin. Science 192:
902.

Vale W, Burgus R, Guillemin R (1967). Potassium-induced stimulation of thyrotropin release in vitro. Requirement for
presence of calcium and inhibition by thyroxine. Experientia 23:855.

Valverde I, Sener A, Lebrun P, Herchulez A, Malaisse WJ
(1981). The stimulus-secretion coupling of glucose-induced
insulin release. XLVII. The possible role of calmodulin.
Endocrinology 108:1305.

Top: M. Bosma; F. Ashcroft; O. Sand; Y. Fukushima;
S. Hagiwara
Bottom: K. Nicolaysen; S. Krasne; C. Baud; H. Ohmori;
O. Sand

The Physiology of Excitable Cells, pages 371–383

DOWNSTREAM FROM THE CALCIUM CHANNEL: IDENTIFICATION OF
INTRACELLULAR CALCIUM STORES

Maryanna Henkart

Immunology Branch
National Cancer Institute
Bethesda, MD.   20205

It has repeatedly been demonstrated in this symposium
that Dr. Hagiwara's intellectual influence has had far-rea-
ching effects on the development of thinking about the many
roles of calcium in cellular physiology. My work is one
more example. Since I left his laboratory I have focussed
on the identification and function of intracellular calcium stores in a variety of cell types. I would like to
present here an overview of what has been learned so far
about the distribution of calcium in cellular organelles
and to discuss some of the implications of that calcium
distribution for the physiology of the cells.

The list of cellular functions known to be controlled
by calcium is steadily increasing. In addition to such
well known functions as the control of actin-myosin interactions and secretion, the control of microtubule polymerization, and the permeability of cell membranes to other
ions, there is an ever growing list of enzymes found to be
regulated by calcium-calmodulin complexes (Means, Dedman
1980). Such a list, however, can hardly convey the complexity of the interactions of calcium with cell components and
the handling of calcium by the cell. Another attempt is
shown in Fig. 1 which is a complex (yet highly simplified)
diagram of a cell showing some of the interactions of
calcium-controlled functions and cellular mechanisms of
calcium control.

Because intracellular functions regulated by calcium are
so numerous and are mediated by unbiquitous calcium binding
proteins such as calmodulin, an understanding of these

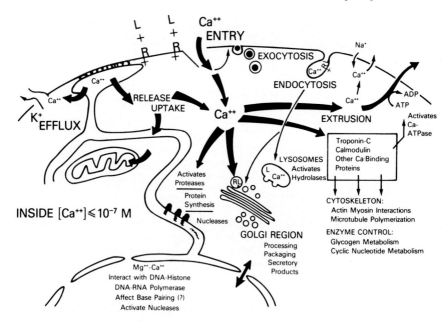

Fig. 1. Pathways of calcium movements in cells and some sites of control by calcium of cellular functions. R=receptor, L=ligand

varied processes may require more than a knowledge simply of the average cytoplasmic free calcium concentration. That is, the regulation of local calcium concentrations on the ultrastructural scale may be important. It would, therefore, be important to study the ultrastructural localization of calcium and how it it changes with physiologically interesting stimuli. Since the physiological roles of calcium in cells often involve specifically triggered changes in membrane permeability and shifts of calcium from one compartment to another, the localization of in situ calcium at the ultrastructural level presents a formidable, if not insurmountable, technical problem. A consideration of the time scales of some of these these processes may make this clearer. In aqueous solution a calcium ion can

diffuse a distance equal to a cell diameter in a few 10's
of msec. In a striated muscle an entire contraction-relaxa-
tion cycle can be complete in 100 msec. That is, calcium
can be released from internal stores, diffuse throughout
the cell cytoplasm, mediate its physiological function,
and be recaptured in about one tenth of a second.

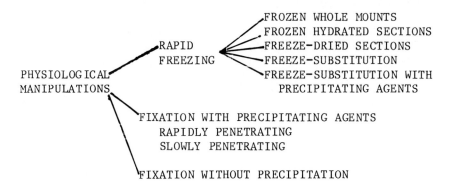

Fig. 2   Experimental strategies for studying calcium in
   its morphologic context.

      Fig. 2 illustrates the experimental techniques that
have been used to identify intracellular calcium stores.
They are arranged from top to bottom in order of decreasing
probability that they will preserve the physiological calcium distribution.   Since physiologically interesting
calcium stores can be so labile, it is obviously necessary
to use extremely rapid methods to immobilize them and to retain them in an immobile state during preparation for electron microscopy.   Rapid freezing provides the only feasible
way of doing this.   On the other hand, rapid freezing, itself, has disadvantages.   Rapid freezing techniques succeed
in preserving the ultrastructural integrity of only a
fairly thin outer layer (10-15 μm) of tissue or cells.

      Unfortunately the techniques that are most likely to
retain calcium in its physiological locations (frozen
hydrated whole mounts or frozen hydrated sections) provide
the least interpretable images in the electron microscope
because of their poor inherent contrast.   At the other
extreme are the classical techniques of tissue preparation

for morphology which involve fixation in aqueous solutions.
Long soaks in fixatives which slowly break down membrane
permeability barriers will allow wash out or redistribution
of most of the soluble cellular components including calcium. Long soaks in fixatives containing slowly penetrating
precipitating agents may retain calcium in the tissues,
but the distribution of the final precipitate may not bear
any relation to the distribution of calcium in the cells
before fixation began. The techniques between these extremes are all levels of compromise. Not shown in Fig. 2
is the necessary final step: identification of calcium
in these preparations by some analytic technique. Energy
dispersive electron probe x-ray microanalysis has been
most widely used for this purpose (e.g. Hall et al. 1974),
but electron energy loss spectroscopy is being developed
in a number of laboratories (e.g. Somlyo et al. 1982).
Although there is some disagreement on this point, at
present it appears that the practical limits of detection
of calcium is of the order of 1 to about 10 mM (Somlyo et
al. 1982; Leapman et al. 1982), and this imposes another
limitation on our ability to describe the changes in the
distribution of calcium associated with interesting physiological processes.

There is, however, a distinction between localization
and quantitation of physiological levels of calcium and
identification of calcium-containing compartments. The
former may, perhaps, prove impossible in some of the most
interesting systems. For example, quantitation of calcium
influx across presynaptic membranes at sites of exocytosis
during normal synaptic transmission may well be beyond the
time resolution of preparative techniques and below the
limits of spatial resolution and detection sensitivity of
analytic systems now available or likely to be developed.
If, however, we limit ourselves to the latter objective,
the experimental approaches available may be equal to the
task. The results I will discuss here will be in answer to
the rather carefully posed questions: (1) when cells are
loaded with calcium by one means or another, which organelles take it up, and (2) which organelles in cells contain
detectable levels of calcium under the preparative and
analytic conditions that we have available?

THE ENDOPLASMIC RETICULUM SEQUESTERS CALCIUM

In the Squid Giant Axon

Although it has been recognized for some time that the sarcoplasmic reticulum is the site of intracellular calcium sequestration in striated muscle (Ebashi, Endo 1968) it was thought that in other cell types the mitochondria were the primary calcium sequestering organelles (Lehninger, 1970). The possiblilty that the endoplasmic reticulum (ER) might be involved in calcium uptake in other cells was first suggested to me by some observations of the morphology of squid giant axons. The ER in freshly dissected squid axons occupies about 5% of the axon volume while the mitochondria occupy about 1%. In axons fixed under various conditions known to produce calcium loading (reviewed by Brinley 1978) the endoplasmic reticulum appeared swollen (Henkart 1972, 1975). In order to approach more directly the question whether this was due to calcium uptake into the ER, I took advantage of the longstanding observation that calcium uptake by isolated vesicles of the sarcoplasmic reticulum of striated muscle was enhanced by oxalate (Makinose, Hasselback 1965). Calcium which is actively accumulated by the SR vesicles is precipitated inside by the oxalate, and the concentration gradient against which the Ca pump must work is, therefore reduced. Apparently oxalate penetrates the SR membranes. It seemed likely that if an analogous calcium uptake mechanism existed in the squid giant axon, oxalate injection into a calcium loaded axon might produce calcium oxalate precipitates in the ER that could be identified in the electron microscope.

The experiments that were done (Henkart et al. 1978) may be summarized as follows: Squid axons were loaded by stimulation for various periods of time in artificial sea water containing 112 mM calcium or by soaking in sodium free artificial sea water containing 10 mM calcium. Some axons were injected with an isosmotic solution containing potassium oxalate. Some axons were prepared without calcium loading but injected with oxalate, and some were neither loaded nor injected. All axons were frozen using the device described by Heuser et al. (1976), freeze-substituted in acetone containing osmium, and embedded in Araldite. Sections were cut floating on glycerol (Ornberg, Reese 1980) and examined in transmission and scanning transmis-

sion modes in a Philips 400 or Hitachi 500 electron microscope equipped with an energy dispersive x-ray spectrometer.

In calcium loaded axons the endoplasmic reticulum as
well as the mitochondria contained electron opaque deposits
that contained calcium as indicated by x-ray analysis.
Calcium loaded axons without oxalate injection also contained calcium in the endoplasmic reticulum and mitochondria,
but the images suggested that some deposits had been lost
during preparation of thin sections. In axons loaded to
intermediate degrees there were intermediate amounts of deposits in the ER and mitochondria. In axons that were not
calcium loaded but were injected with oxalate there were
small amounts of calcium-containing deposits in the ER but
not in the mitochondria. In axons neither calcium loaded
nor oxalate injected, calcium was below the level of detec-
tability of the analytic system.

Our interpretation of these results was that in calcium loaded axons the ER as well as mitochondria take up calcium. Oxalate was not required for calcium uptake, but it
may have enhanced uptake and certainly seemed to enhance
the preservation of calcium-containing deposits during
preparation for microscopy. In axons not specifically
calcium loaded the ER still contained detectable levels of
calcium when oxalate was injected, so it seems likely that
the ER is capable of calcium uptake under more physiological conditions.

There is now evidence from many cell types suggesting
that at physiological concentrations of calcium and magnesium the ER is an important calcium sequestering organelle.
Microsomal fractions of kidney, liver, and fibroblasts are
capable of accumulating calcium by ATP-dependent mechanisms
(Moore et al. 1974; Moore et al. 1975; Moore, Pastan 1977),
and a similar function has been studied extensively in a
microsomal fraction from lysed synaptosomes (Blaustein et
al. 1978). In the latter system morphologic evidence has
been presented that the vesicles are derived from smooth
endoplasmic reticulum (McGraw et al. 1980).

In the squid axon (Henkart 1980) as in many types of
nerve cell bodies (e.g. Rosenbluth 1962; Henkart et al.
1976) cisterns of ER sometimes approached and formed morphologically specialized appositions with the surface membrane (subsurface cisterns). In muscle, similar appositions

between the surface membrane and sarcoplasmic reticulum
are thought to mediate excitation-contraction coupling
(Franzini-Armstrong 1964; Peachey 1964). The presence of
such morphologically specialized appositions suggests that
they may function in the coupling of surface membrane
stimulation to release of calcium from the ER by a mechanism analogous to excitation-contraction coupling in muscle.

In Other Cell Types

        I have studied a number of other cell types and will
show an example of results obtained in macrophages. The
macrophage is another cell in which calcium regulation of a
number of functions has been studied. Externally applied
chemotactic factors activate a calcium-dependent potassium
conductance (Gallin, Gallin 1977); the direction of cell
movement may be controlled by calcium gradients (Stendahl,
Stossel 1980) and plasma membrane calcium transport has
been demonstrated (Lew, Stossel 1980).

        In order to approach the question of which organelles
contain calcium in other cell types I have, on occasion, re-
sorted to a technique of tissue preparation that is farther
down on the scale of believability (Fig. 2). The second
approach is shown as the lower pathway in Fig. 3

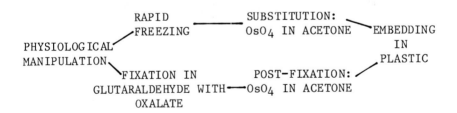

Fig. 3 Comparison of procedures for rapid freezing and
        fixation in the presence of oxalate: Two approaches for
        identification of intracellular calcium compartments.

When results with this technique were compared with
those obtained with the rapid freezing, freeze-substitution
method on squid axons and mouse striated muscle, qualitatively similar distributions of deposits were obtained. The
deposits often were found by x-ray microanalysis to contain
detectable levels of calcium. There were, however, many
electron-opaque deposits in which only osmium could be
demonstrated. One cannot tell whether those compartments
contained calcium in amounts below the limits of detection
in the presence of the high background due to effects of
osmium. These results clearly do indicate, however, that
the appearance of a deposit cannot be equated with the
presence of calcium.

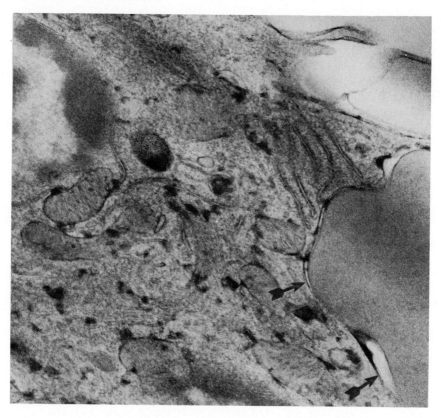

Fig. 4. Electron micrograph of a macrophage in the spleen
of a mouse that was perfused with fixative containing oxalate. The arrows indicate the border of a phagosome.

When macrophages in culture or in tissues of mice or guinea pigs were prepared for electron microscopy as in the lower part of Fig. 3, electron-opaque deposits were found in a number of intracellular compartments as illustrated in Fig. 4. As was the case in the squid axon, the ER contained deposits in which calcium was detectable by x-ray analysis. Also, as was the case in the squid axon as well as in other cell types including fibroblasts (Henkart, Nelson 1979), subsurface cisterns of the ER were common in macrophages. Evidence from a number of sources is now accumulating to suggest that mobilization of intracellular calcium stores may be involved in such apparently disparate systems as responses of secretory cells to secretagogues (reviewed by Putney 1979; Wollheim, Sharp 1982) and responses of B lymphocytes to cross-linking of their surface immunoglobulin by anti-immmunoglobulin antibodies (Tsien et al. 1982). It would be interesting to determine whether the receptors that trigger mobilization of these calcium stores might be concentrated over or near subsurface cisterns. The observations that subsurface cisterns are localized specifically beneath certain synapses (Hand 1970; Bodian 1972; reviewed by Pappas, Waxman 1972) suggest that this might be the case.

OTHER MEMBRANE-BOUND ORGANELLES ALSO CONTAIN CALCIUM

Another class of cellular compartment which contains calcium includes organelles whose membranes circulate directly through the surface membrane. These include pinosomes and phagosomes. Fig. 4 shows part of a phagosome in a mouse spleen macrophage which had internalized a red blood cell. Membrane bound dense cored granules (which were also found to contain calcium) sometimes appeared to be emptying their contents into such phagosomes. X-ray microanalysis often demonstrated calcium at the borders of internalized red cells. Some cisterns and vesicles in the Golgi region contained electron opaque deposits, but these are generally too small to identify individually with the electron probe. Lysosomal residual bodies also contain calcium as do secretory granules in many cell types (e.g. Borowitz et al. 1965).

The calcium content of this class of organelle may be
derived from the extracellular fluid contained in endocytic vesicles. It may also be due, however, to the presence in the internalized surface membrane of calcium extrusion mechanisms which would be oriented so as to pump
calcium into the lumens of endocytic compartments (Lew,
Stossel 1980). The contents of the Golgi apparatus could
receive calcium by transfer from endocytic vesicles which
is known to occur (reviewed by Farquhar 1981), by transfer
from the lumen of endoplasmic reticulum along with secretory products for packaging (reviewed by Palade 1975), or
it may also have calcium uptake mechanisms (Baumrucker,
Keenen 1975). Calcium in secretory granules may be used
as a counter ion in packing secretory product. The calcium
would then be effectively extruded from the cell in the
process of secretion. This would have the further effect
of raising the local calcium concentration immediately outside the plasma membrane during active exocytosis, which
might serve to amplify the secretory process by increasing
the amount of calcium available to enter in response to
the secretagogue.

The observations that I have summarized have led to
two general conclusions: (1) The endoplasmic reticulum
sequesters calcium in many types of cells, not only in
muscle. A possible corollary is that in cells other than
muscle the endoplasmic reticulum may release calcium by a
mechanism analogous to excitation-contraction coupling in
striated muscle. There is evidence from both morphology
and physiological studies to suggest that the coupling of
surface membrane events to cytoplasmic processses via
release of sequestered Ca may occur in many types of cells.
(2) A second system of membrane-bound organelles contains
calcium. These are the organelles whose membranes circulate
directly through the surface membrane. These organelles
include endocytic vesicles, some vesicles and cisterns
associated with the Golgi apparatus, vesicles and granules
destined for exocytosis, and lysosomes. Membranes which
cycle through the surface membrane would carry with them
small amounts of extracellular fluid, and possibly also
surface membrane calcium extrusion mechanisms. In structures formed by endocytosis these would become membrane
bound calcium-sequestering compartments. Calcium present
in organelles destined for exocytosis would be extruded
from the cell in the process of secretion.

References

Baumrucker CA, Keenan TW (1975). Membrane of mammary gland X. Adenosine triphosphate dependent calcium accumulation by Golgi apparatus rich fractions from bovine mammary gland. Exp Cell Res 90:253.

Blaustein MP, Ratzlaff RW, Kendrick NC, Schweitzer ES (1978). Calcium buffering in presynaptic nerve terminals I. Evidence for involvement of a nonmitochondrial ATP-dependent sequestration mechanism. J Gen Physiol 72:15.

Bodian D (1972). Synaptic diversity and characterization by electron microscopy. In Pappas GD, Purpura DP (eds): "Structure and Function of Synapses," New York: Raven Press, p 45.

Borowitz JL, Fuwa K, Weiner N (1965). Distribution of metals and catecholamines in bovine adrenal medulla subcellular fractions. Nature 205:42.

Brinley FJ, Jr (1978). Calcium buffering in squid axons. Ann Rev Biophys Bioeng 7:363.

Ebashi S, Endo M (1968). Calcium ion and muscle contraction. Prog Biophys Mol Biol 18:125.

Farquhar, MG (1981). Membrane recycling in secretory cells: implications for traffic of products and specialized membranes within the Golgi complex. In Hand AH, Oliver C (eds): "Basic Mechanisms of Cellular Secretion," Methods in Cell Biology, New York: Academic Press, 23:399.

Franzini-Armstrong C (1964). Fine structure of sarcoplasmic reticulum and transverse tubular system in muscle fibers. Fed Proc 23:887.

Gallin EK, Gallin JI (1977). Interaction of chemotactic factors with human macrophages. Induction of transmembrane potential changes. J Cell Biol 75:227.

Hall T, Echlin P, Kaufmann R (eds) (1974). "Microprobe analysis as applied to cells and tissues." New York: Academic Press.

Hand AR (1970). Nerve-acinar cell relationships in the rat parotid gland. J Cell Biol 47:540.

Henkart M (1972). Structure and function of the endoplasmic reticulum in the squid giant axon. Proc Second Ann Meet Soc for Neurosci p 108 (abs).

Henkart M (1975). The endoplasmic reticulum of neurons as a calcium sequestering and releasing system: morphological evidence. Biophys J 15:267 (abs).

Henkart MP (1980). Identification and function of intracellular calcium stores in axons and cell bodies of neurons. Fed Proc 39:2783.

Henkart M, Landis DMD, Reese TS (1976). Similarity of junctions between plasma membranes and endoplasmic reticulum in muscle and neurons. J Cell Biol 70:338.

Henkart MP, Nelson PG (1979). Evidence for an intracellular calcium store releasable by surface stimuli in fibroblasts (L cells). J Gen Physiol 73:655.

Henkart MP, Reese TS, Brinley FJ, Jr (1978). Endoplasmic reticulum sequesters calcium in the squid giant axon. Science 202:1300.

Heuser JE, Reese TS, Landis DMD (1976). Preservation of synaptic structure by rapid freezing. Cold Spring Harbor Symp Quant Biol 40:17.

Leapman RD (1972). Applications of electron energy loss spectroscopy in biology: Detection of calcium and fluorine. In Bailey GW (ed): Proc 40th Ann Mtg Elec Micr Soc Am, Baton Rouge: Claitor's, p 412.

Lehninger AL (1970). Mitochondria and calcium ion transport. Biochem J 119:129.

Lew PD, Stossel TP (1980). Calcium transport by macrophage plasma membranes. J Biol Chem 255:5841.

Makinose M, Hasselbach W (1965). Der einfluss von oxalat auf den calcium-transport isolierter vesikel des sarkoplasmatischen reticulum. Biochem Z 343:360.

McGraw CF, Somlyo AV, Blaustein MP (1980). Localization of calcium in presynaptic terminals. An ultrastructural and electron microprobe analysis. J Cell Biol 85:228.

Means AR, Dedman JR (1980). Calmodulin--an intracellular calcium receptor. Nature 285:73.

Moore L, Chen T, Knapp HR, Jr, Landon EJ (1975). Energy-dependent calcium sequestration activity in rat liver microsomes. J Biol Chem 250:4562.

Moore L, Fitzpatrick DF, Chen TS, Landon EJ (1974). Calcium pump activity of the renal plasma membrane and renal microsomes. Biochim Biophys Acta 345:405.

Moore L, Pastan I (1977). Regulation of intracellular calcium in chick embryo fibroblast: calcium uptake by the microsomal fraction. J Cell Physiol 91:289.

Ornberg RL, Reese TS (1980). A freeze-substitution method for localizing divalent cations: examples from secretory systems. Fed Proc 39:2802.

Palade G (1975). Intracellular aspects of the process of protein synthesis. Science 189:347.

Pappas GD, Waxman SG (1972). Synaptic fine structure-morphological correlates of chemical and electrotonic transmission. In Pappas GD, Purpura DP (eds): "Structure and Function of Synapses," New York: Raven Press, p 1.

Peachey LD (1964). Transverse tubules in excitation-
   contraction coupling. Fed Proc 24:1124.
Pozzan T, Arslan P, Tsien RY, Rink TJ (1982). Anti-immuno-
   globulin, cytoplasmic free calcium, and capping in B
   lymphocytes. J Cell Biol 94:335.
Putney JW, Jr (1979). Stimulus-permeability coupling: role
   of calcium in the receptor regulation of membrane permeability. Pharmacol Rev 30:209.
Rosenbluth J (1962). Subsurface cisterns and their relationship to the neuronal plasma membrane. J Cell Biol 13:405.
Somlyo AP, Shuman H (1982). Electron probe and electron
   energy loss analysis in biology. Ultramicroscopy 8:219.
Stendahl OI, Stossel TP (1980). Actin-binding protein
   amplifies actomyosin contraction, and gelsolin confers
   calcium control on the direction of contraction. Biochem
   Biophys Res Commun 92:675.
Wollheim CB, Sharp GWG (1981). Regulation of insulin
   release by calcium. Physiol Rev 61:914.

Top: S. Hagiwara
Bottom: O. Sand; M. Henkart; P.L. Marchiafava

SECTION IV
SYNAPTIC MECHANISMS

Top: T. Szabo; H. Sakata; Y. Fukushima
Bottom: R. Meech; C. Edwards; Y. Saito

The Physiology of Excitable Cells, pages 387–391
© 1983 Alan R. Liss, Inc., 150 Fifth Avenue, New York, NY 10011

INTRODUCTION

Although the research in Hagiwara's laboratory has
been focused largely on the ion permeation channels involved in action potentials, he and his collaborators have
also done important, ground-breaking research on synaptic
mechanisms. Again, the approach has been both analytical
and comparative. Hagi's first research, in fact, was on
neuromuscular connections in the sound producing muscles of
cicadas and the wing muscles of locusts, showing that some
muscle fiber membranes have focal innervation and generate
action potentials, while others do not produce action potentials but have distributed nerve endings and show contraction proportional to the magnitude of a non-selective
postsynaptic depolarization (Wakabayashi, Hagiwara 1953;
Hagiwara 1953; Hagiwara, Watanabe 1954). Later work in
Hagi's lab was directed at understanding the basis of the
membrane potential and postsynaptic responses in moth muscle,
where the high haemolymph K^+ concentration and relatively
low Na^+ concentration pose interesting problems (Rheuben
1972). The paper by RHEUBEN and KAMMER in this volume is a
continuation of this line of investigation, directed now at
an explanation for the intriguing differences in amplitude,
time course, and facilitation properties of synaptic potentials evoked by different excitatory nerves even on the
same postsynaptic muscle fibers.

In another important early contribution, Hagi, in
collaboration with Bullock (Bullock, Hagiwara 1957) and
Tasaki (Hagiwara, Tasaki 1958) developed the squid preparation
that, more than any other, has permitted a rigorous analysis of the coupling between presynaptic ion fluxes and
transmitter release. Hagi was also among the first to
identify and study electrotonic synapses. He introduced
the crayfish giant motor axon preparation (Hagiwara 1958)
subsequently shown by Furshpan & Potter (1959) to be a
rectifying electrical junction. He studied electrical
interaction and synaptic integration in the lobster cardiac
ganglion (Hagiwara, Bullock 1957; Hagiwara, Watanabe,
Saito 1959; Hagiwara 1961); and he demonstrated electrical
transmission between giant neurons of the leech ganglion
(Hagiwara, Morita 1962). In addition he and his

collaborators analyzed the membrane conductance changes
responsible for postsynaptic inhibition in the crayfish
stretch receptor (Edwards, Hagiwara 1959; Hagiwara, Kusano,
Saito 1960) and in molluscan neurons (Kusano, Hagiwara 1961;
Hagiwara, Kusano, Saito 1961; Hagiwara, Kusano 1961); studied
photoreceptor and hair cell transmitter release (Ozawa,
Hagiwara, Nicolaysen, Stuart 1975; Sand, Ozawa, Hagiwara
1975), used rat chromaffin cells to investigate neurosecretory mechanisms (Brandt, Hagiwara, Kidokoro, Miyazaki 1977),
and collaborated with K. Ikeda and S. Ozawa (Ikeda et al.
1976) in an early study of a single gene *Drosophila*
"paralytic" mutant, showing that the temperature-dependent
failure was in the synaptic release mechanism. In the present volume, IKEDA and KEONIG describe a further investigation of this mutant, suggesting that the mutation affects
transmitter release by slowing down or stopping endocytosis.
The potential for further dissection of release mechanisms
in this way is exciting.

In recent years, there has been growing recognition
that excitable cells interact with one another in ways that
go beyond the simple transmission of coded information and
evoked electrical (and/or contractile) activity in the postsynaptic cell. For example, how are the appropriate connections made in the first place? How are the size and
strength of synaptic inputs regulated and matched to the
postsynaptic cell's properties? And what other trophic and
inductive influences are exerted by a nerve on its target
cell and *vice versa* (cf. reviews by Purves 1976; Vrbová,
Gordon, Jones 1978; Lømo, Jansen 1980; Grinnell, Herrera
1981; Van Essen 1982).

Several of the papers in this volume are directed at
these problems. Using the familiar frog neuromuscular
junction as experimental material, LINDEN and LETINSKY
describe several aspects of development: the timing of
appearance of nerve and muscle fibers, the sequence of
events during synapse formation, and the timing of synapse
elimination. GRINNELL and his collaborators describe an
extensive series of experiments aimed at understanding the
regulation of size and amount of transmitter release from
frog motor nerve terminals.

It is well known that innervation is necessary for the
induction and maintenance of many sensory receptors and for
the final differentiation and trophic maintenance of muscle.

SZABO and KIRSCHBAUM present interesting evidence that the specialized muscle that forms the electric organ of a weakly electric fish can differentiate in the absence of innervation, at least when denervation takes place after contact has initially been made. It will be of interest to see whether the initial contact is necessary and a differentiating signal can be retained for long periods after this, or whether differentiation of this organ is really independent of innervation.

Development and differentiation clearly involve complex interaction between cells, both in terms of selective recognition and adhesion to certain surrounding cell surfaces of basal lamina, and, in many cases, at least, in response to diffusible trophic factors. The best known example of the latter is nerve growth factor (NGF). Exciting progress has been made in developing cell culture techniques to analyze the mechanism of action of NGF and other environmental factors on neuronal differentiation (cf. Patterson 1978). A significant new step in this direction has been taken by O'Lague and Huttner and their colleagues (HUTTNER ET AL.) with the development of huge fused PC-12 cells that show many neuronal properties and sympathetic neuron-like responses to NGF.

Finally, we include in this section a paper on steady transcellular ion currents: the newly devised techniques for measuring them, where they occur, and potential significance (NUCCITELLI). These steady currents reflect asymmetric distribution of ion channels, and may be of major importance in a wide variety of phenomena, including the establishment of polarity and gradients in development, and localization of membrane proteins.

The papers in this section obviously touch only certain aspects of current research into synaptic function and neuronal differentiation, but are representative of the comparative approach and innovative techniques characteristic of, and clearly influenced by, Susumu Hagiwara.

References

Brandt BL, Hagiwara S, Kidokoro Y, Miyazaki S (1977). Action potentials in the rat chromaffin cell and effects of acetylcholine. J Physiol 263:417.

Bullock TH, Hagiwara S (1957). Intracellular recording from the giant synapse of the squid. J Gen Physiol 40:565.

Edwards C, Hagiwara S (1959). Potassium ions and the inhibitory process in the crayfish stretch receptor. J Gen Physiol 43:315.

Furshpan EJ, Potter DD (1959). Transmission at the giant motor synapses of the crayfish. J Physiol 145:289.

Grinnell AD, Herrera AA (1981). Specificity and plasticity of neuromuscular connections: long-term regulation of motoneuron function. Prog Neurobiol 17:203.

Hagiwara S (1953) Neuromuscular transmission in insects. Jap J Physiol 3:284.

Hagiwara S (1958). Synaptic potential in the motor giant axon of the crayfish. J Gen Physiol 41:1119.

Hagiwara S (1961). Nervous activities of the heart in crustacea. Ergebnisse der Biologie 24:287.

Hagiwara S, Bullock TH (1957). Intracellular potentials in pacemaker and integrative neurons of the lobster cardiac ganglion. J Cell and Comp Physiol 50:25.

Hagiwara S, Kusano K (1961). Synaptic inhibition in giant nerve cell of Onchidium verruculatum. J Neurophysiol 24:167.

Hagiwara S, Kusano K, Saito S (1960). Membrane changes in crayfish stretch receptor neuron during synaptic inhibition and under action of gammaaminobutyric acid. J Neurophysiol. 23:505.

Hagiwara S, Kusano K, Saito S (1961). Membrane changes on Onchidium nerve cell in potassium-rich media. J Physiol 155:470.

Hagiwara S, Morita H (1962). Electrotonic transmission between two nerve cells in leech ganglion. J Neurophysiol 25:721.

Hagiwara S, Tasaki I (1958). A study on the mechanism of impulse transmission across the giant synapse of the squid. J Physiol 143:114.

Hagiwara S, Watanabe A (1954). Action potential of insect muscle examined with extracellular electrode. Jap J Physiol 4:65.

Hagiwara S, Watanabe A, Saito S (1959). Potential changes in synctial neurons of lobster cardiac ganglion. J Neurophysiol 22:554.

Ikeda K, Ozawa S, Hagiwara S (1976). Synaptic transmission reversibly conditioned by a single-gene mutation in Drosophila Melanogaster. Nature 259:489.

Kusano K, Hagiwara S (1961). On the integrative synaptic potentials of Onchidium nerve cell. Jap J Physiol 11:96.

Lømo T, Jansen J (1980). Requirements for the formation and maintenance of neuromusuclar connections. Current Topics in Developmental Biology 16:253.

Ozawa S, Hagiwara S, Nicolaysen K, Stuart AE (1975). Signal transmission from photoreceptors to ganglion cells in the visual system of the giant barnacle. Cold Spring Harbor Symposia Vol XL:563.

Patterson PH (1978). Environmental determination of autonomic neurotransmitter functions. Ann Rev Neurosci 1:1.

Purves D (1976). Long-term regulation in the vetebrate peripheral nervous system. Int Rev of Physiol 10:125.

Rheuben MB (1972). The resting potential of moth muscle fibre. J Physiol 225:529.

Sand O, Ozawa S, Hagiwara S (1975). Electrical and mechanical stimulation of hair cells in the mudpuppy. J Comp Physiol A102:13.

Van Essen D (1982). Neuromuscular synapse elimination: structure, functional, and mechanistic aspects. In Spitzer NC (ed.): "Neuronal Development," New York: Plenum Press, p 333.

Vrbová G, Gordon T, Jones R (1978). "Nerve-Muscle Interaction," Chapman and Hall, London.

Wakabayashi T, Hagiwara S (1953). Mechanical and electrical events in the main sound muscle of cicada. Jap J Physiol 3:249.

Top: Scientific Session
Bottom: (Back Table): Pat Ulrich; G. Zampighi; C. Gallion
         (Front Table, facing center): Mrs. M. Hirose,
         Satoko Hagiwara; S. Hagiwara; T. Wiesel

The Physiology of Excitable Cells, pages 393–409
© **1983 Alan R. Liss, Inc., 150 Fifth Avenue, New York, NY 10011**

MECHANISMS INFLUENCING THE AMPLITUDE AND TIME COURSE OF
THE EXCITATORY JUNCTION POTENTIAL

Mary B. Rheuben and Ann E. Kammer*

Dept. of Anatomy, Michigan State University,
East Lansing, MI 48824; *Division of Biology,
Kansas State University, Manhattan, KS 66506

INTRODUCTION

Susumu Hagiwara was one of the first investigators
to describe the shape of the junction potential and action
potential of insect muscle from intracellular recordings.
Using the sound muscle of cicadas and the wing muscle of
locusts (Hagiwara 1953; Hagiwara, Watanabe 1954), he
obtained results that verified the presence of multiple,
distributed nerve endings and that indicated that the role
of the electrical properties of the muscle membrane varied
in different insect muscles, with some recorded responses
to nerve stimulation being the result of nonselective
permeability increases, akin to the vertebrate endplate
potential, and others exhibiting the additional properties
of an action potential. This kind of variability in
properties among the different muscles has provided insect
neurobiologists with ample opportunity to examine compara-
tively the different synaptic mechanisms that could be
used to control contraction--ranging from those needed to
control the very high frequencies of the sound muscles
down to the slow crawling muscles of caterpillars.

In subsequent investigations of arthropod muscle,
three fundamental types of synaptic potential have been
recorded - slow excitatory, fast excitatory, and inhibitory. The time course of a typical slow excitatory junction potential (e.j.p.) can be 4 times longer than that of
a fast e.j.p., and the slow e.j.p. is much smaller than
the normally suprathreshold fast e.j.p. The contractile
mechanism of muscle is sensitive to the time course and

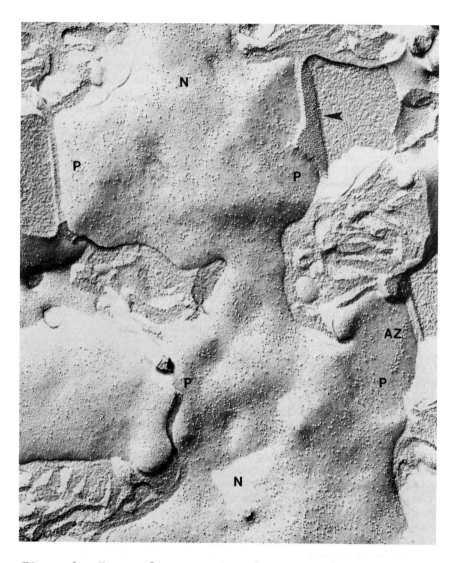

Figure 2. Freeze-fracture view of the cytoplasmic leaflet of a fast nerve terminal (N). Several regions of contact between nerve and muscle (plaques, P) are shown. In one, the presynaptic specialization, the active zone, is shown (A Z). In another plaque (upper right) a portion of the external leaflet of the muscle membrane remains, showing the densely packed particles characteristic of the region's postsynaptic receptors (Arrow). 55,100X.

of contact. This particle-covered patch presumably represents the area occupied by receptors. Freeze-fracture
studies of fast and slow terminals show them to be qualitatively similar, but there are some quantitative differences. The length of the narrow band of particles in the
presynaptic active zones is essentially the same in both
types of nerves, but the area of the individual postsynaptic
patches is greater in fast terminals than in the slow, 0.46
μm^2 vs. 0.32 μm^2 (Rheuben 1982 and in preparation).

In low magnification freeze-fracture views of the two
types of terminals, another difference is seen (Figs. 3 and
4). The slow terminals are more irregular and varicose in
shape and appear to be shorter. (This shape may be the
source of the differences in terminal diameter observed in
thin section by Titmus (1981) and Hill and Govind (1981).)
The varicose shape may be determined by the muscle fiber
since fast nerve terminals innervating tonic fibers are also varicose (Schaner, Rheuben in preparation). (The examples we have figured above, as well as those from lobster
and locust, are fast nerve on phasic muscle and slow nerve
on tonic muscle.) Measurements of terminal lengths made
using the scanning electron microscope confirm that slow
terminals are significantly shorter than fast, 20 μm vs. 57
μm, and are straight rather than Y-shaped (Rheuben 1982).

Determination of the number of plaques and active zones
per micrometer allows us to calculate the total area of
postsynaptic receptors and total length of active zone per
junction for the two types. The total length of active
zone in fast terminals is almost 4 times that in slow, and
the summed area occupied by receptors is almost 5 times
larger. The short total length of active zone in slow
junctions may be a factor limiting synchronous release,
leading to the smaller peak amplitude of the slow e.j.p.
However, no obvious anatomical correlations with the slow
time course suggest themselves. The length of the junction is short relative to an estimated length constant, so
conduction velocity within the terminal is not likely to
be important in determining the time course of release
(Rheuben, in preparation).

IMMATURE JUNCTIONS

In immature fast junctions from the pupal stage of
Manduca, the e.j.p. is small, 5-15 mV, and quite prolonged

Figure 3. Low magnification freeze-fracture view of the
surface of the subalar muscle and of a fast junction. The
nerve terminal (N) lies in the center of the junction. The
muscle-nerve contacts, or plaques, occur at regular intervals and are marked with arrows. Between the plaques are
the identations made by processes of the glial cells (G)
that cover the junction. Other regularly spaced indenta-
tions in the muscle membrane are T-tubules. 7,800X.

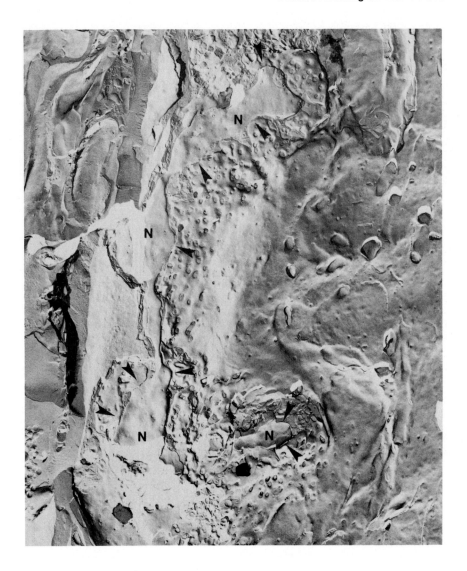

Figure 4. Low magnification view of a junction formed by
a slow nerve terminal on a tonic muscle fiber (middle third
axillary). While numerous nerve-muscle contacts (arrows)
are present, there is no regular alternation between muscle
plaque and glial processes. Nerve, N. Note the irregular
shape of the nerve terminal itself. 8,900X.

(Rheuben, Kammer 1981), not unlike a typical slow e.j.p., but the duration may approach 500 msec. In the developmental stage described, the length of the junction approaches that of the adult (only 13% shorter), and the length of each active zone is about the same as in the fully developed fast terminals. However, the plaques, the contact regions between nerve and muscle, are substantially smaller than those of the adult, 0.38 μm in diameter compared to 0.65 μm of the adult (Fig. 5). In many of the plaques the particles do not fill out the entire contact area. In addition, the diameter of the nerve terminal itself is smaller and the degree of complexity of the glial processes that lie between the plaques is much less. Another factor, aside from the nerve terminal structure, that is likely to influence the shape of the perceived electrical signal is present in immature muscle. The muscle fibers are electrically coupled via periodically spaced projections that are linked to each other with gap junctions (Rheuben, Kammer 1982). If the arrival of the nerve impulse at the immature terminals is asynchronous from fiber to fiber, then the summed electrical activity of distributed nerve endings of coupled fibers will be of prolonged time course. Similarly, one could hypothesize that the time course of the adult slow e.j.p. could be influenced by the asynchronous arrival of the nerve impulse to junctions on the same fiber or on adjacent fibers if they also are coupled.

In both cases of long e.j.p.'s we are unable to assess morphologically the possibility that the differences in time course are due to differences in the properties and availability of the receptor-ion channel complexes. Extracellular recordings of miniature e.j.p.'s that would shed light on the time course of the quantal event are not yet available from these muscles. In the case of the immature terminals the numbers of postsynaptic particles per plaque are so few as to suggest that they are a limiting factor in the peak amplitude of the response. We are also unable to assess by these morphological methods the possibility that the smaller postsynaptic patches and the absence of an inactivating enzyme promote repetitive binding and thereby produce longer junction potentials. It is entirely fitting to the theme of this symposium that the next experiments that suggest themselves would examine properties of the postsynaptic ion channels and of the receptor patches relative to the geometry of the plaques.

Figure 5. Fracture through an immature nerve terminal. The cytoplasmic leaflet of the nerve (N) and the external leaflet of the muscle (M) are shown. The contact regions (P) are much smaller than those of the adult and there are fewer postsynaptic particles in a plaque. 63,000X.

MODULATORS

In addition to the processes that mediate synaptic transmission, interactions between presynaptic and post-synaptic cells may be modulated by compounds supplied by cells outside the synapse. This phenomenon, which has become appreciated only recently, is exemplified in insects by the action of octopamine on neuromuscular junctions.

The e.j.p. elicited by stimulating the slow motor neuron to the extensor tibiae of a grasshopper, Schistocerca americana gregaria, is enhanced by bath-applied octopamine (O'Shea, Evans 1979). (The transmitter at this and most other excitatory neuromuscular junctions in insects is thought to be glutamate, although O'Shea and Bishop (1982) have implicated proctolin as a transmitter in a slow coxal motor neuron.) DL-octopamine (10^{-6}M) increases the amplitude of the slow e.j.p. by small and variable amounts, in some cases approximately 10%; under the same conditions tension is increased approximately 30%. Depending on stimulus frequency, octopamine increases the rate of relaxation of a twitch contraction or reduces the amount of maintained tension (Evans, Siegler 1982). Similar effects are obtained by intracellular stimulation of DUMETi (O'Shea, Evans 1979; Evans, Siegler 1982), a neuron that contains octopamine (Evans, O'Shea 1978) and that has an axon in the motor nerve supplying the extensor tibiae (Hoyle et al. 1974).

In adult Manduca sexta, the normal electrical response elicited by stimulation of the fast motor neuron to the dorsal longitudinal muscle is suprathreshold and would not be noticeably increased by octopamine. The small response of immature junctions, however, is markedly enhanced (Klaassen, Kammer, in preparation). Application of 10^{-6}M octopamine to day 16 pharate moths (moths 3 days from eclosion) increases the amplitude of the e.j.p. of immature fast motor neurons from approximately 15 mV to more than 25 mV (Fig. 6). Immature junctions fatigue more readily than those of adults; octopamine, at least in part by increasing the amplitude of the e.j.p., also increases the frequency at which repeated stimulation of the motor neuron elicits a suprathreshold e.j.p., active membrane responses, and muscle contraction. With day 18 moths (one day before eclosion) a dose of only 10^{-10}M octopamine increases the maximum frequency followed from 0.5 Hz to 1.1 Hz.

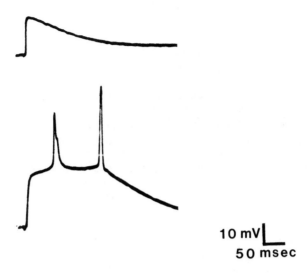

10 mV

50 msec

Figure 6. The effect of 10^{-6}M octopamine on the amplitude
of the immature e.j.p. Upper trace shows the response
recorded from the dorsal longtudinal muscle of a day 16
pharate moth. The lower trace is the e.j.p. after application of 10^{-6}M octopamine.

The modulatory role of other biogenic amines in neuromuscular transmission has been demonstrated in a variety of
preparations. Serotonin facilitates release of excitatory
transmitter at crustacean neuromuscular junctions (Dudel
1965; Florey, Rathmayer 1978; Battelle, Kravitz 1978;
Livingstone et al. 1980; Enyeart 1981; Jacobs, Atwood 1981;
Glusman, Kravitz 1982). Dopamine increases the amplitude
of e.j.p.'s in _Aplysia_ gill musculature (Swann et al.
1978), and it enhances nerve-evoked contractions of crustacean foregut muscles (Lingle 1981). In some vertebrate
skeletal muscles catecholamines increase twitch tension
(Bowman, Nott 1969); in frog muscle noradrenaline enhances
transmitter release (Jenkinson et al. 1968).

Biogenic amines appear to reach their target sites on
muscle and on neuromuscular junctions by diffusion or
circulation in body fluids, rather than by conventional
synapses, but they may be transported to release sites near

or within the muscle by neurosecretory axons that accompany motor neurons to the muscle.

In Manduca, in virtually all motor nerves that have been examined, a population of very small axons that are loosely wrapped with glial cells lie as a collar outside of the neurolemma that binds the larger motor and sensory axons. Small axons in that position also ramify with the motor axon amongst the muscle fibers (Fig. 7), as has also been observed by Hoyle et al. (1980) for the DUMETi. Small varicosities containing a few dense cored vesicles of various types, as well as clear cored vesicles, may be seen in these collar axons all along their lengths. But obvious release sites are rare and postsynaptic specializations on the muscle have not been reported. The only example with the anatomical correlates of a presynaptic release site occurred in a series of varicosities clustered where the motor nerve first entered the third axillary muscle. Varicosities from one of the collar axons each contained several discrete spots of electron dense fuzzy material surrounded by clear vesicles; dense-cored vesicles were nearby (Fig. 8).

Small axons lying outside the main nerve sheath in the "collar" position have been observed in other insect nerves, and some have been identified as processes of peripheral neurosecretory neurons (e.g. Fifield, Finlayson 1978). In Manduca, examination of the peripheral or "collar" axons in the nerve to the dorsal longitudinal flight muscle has indicated that these axons could have cell bodies lying in several different locations, and that they may contain several types of dense-cored vesicles (Wasserman, Rheuben, in preparation). These cell bodies were located in cobalt backfills of the nerve to the dorsal longitudinal muscle (IIN1b) and of branches of the transverse nerve, viewed either as whole mounts or sectioned and intensified using Timm's method. There are two cell bodies lying on the nerve branch linking the transverse nerve and nerve IIN1b, each containing a different type of vesicle. In addition, cells from both the prothoracic and mesothoracic ganglia appear to contribute axons to the collar. No axon from a DUM cell could be identified, but one could still be present. Three distinct types of vesicles were found in the "collar" axons of nerve IIN1b (Wasserman, 1982). These observations, if nothing else, imply that the role of modulators on skeletal muscle may be complex, if the anatomical diversity of innervation of this one muscle

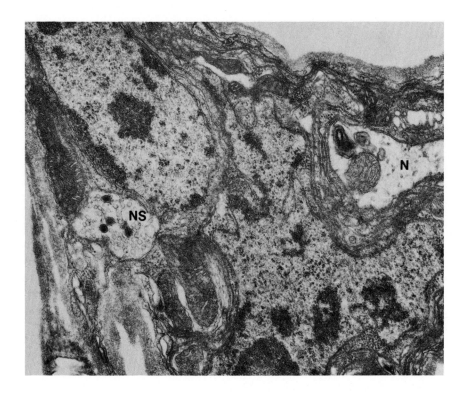

Figure 7. Section through a nerve twig lying within the
subalar muscle. In addition to the fine branches of the
motor axon (N), there is a branch of a small axon lying
at the edge of the glial sheath. This axon contains both
clear and dense-cored vesicles (NS). In our preparations
and in those previously described (Hoyle et al. 1980) these
are characteristic positions for the two types of axons.
However, not all twigs containing the motor axon also contain the profile of a neurosecretory type. 30,400X

indicates anything about the function. One of these axons
may contain octopamine, as indicated by the muscle's
response to exogenously applied octopamine, and as indicated by the presence of octopamine in the hemolymph (Klaassen, in preparation). The identities or effects of other
modulators in this system remain as yet uninvestigated.

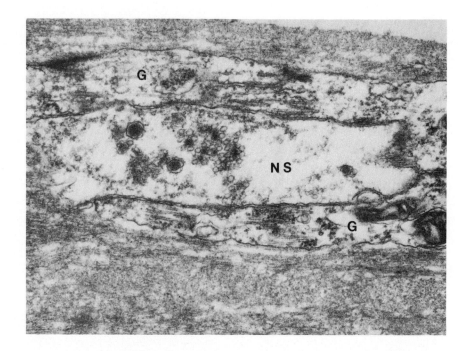

Figure 8. Grazing longitudinal section through the vari-
cosity of a "collar" axon. The rest of the nerve is not
included in the section. The section was obtained near
where the main nerve, containing motor axons to the upper,
middle and lower portions of the middle third axillary
muscle, approached the main muscle mass. G, glial cell;
NS, neurosecretory axon. 42,800X.

CONJECTURES

        While conclusive evidence is lacking for the mecha-
nisms underlying the differences in the time course and
amplitude of fast, slow, and immature junctions, several
anatomical specializations are available for conjecture.
The slow and immature junctions have in common smaller
areas of postsynaptic receptors and either slightly (immature) or substantially (slow) shorter junctions. Both of
these structural features could theoretically limit peak
amplitude of the e.j.p., but the very small postsynaptic
patches of the immature terminal are more likely to have

that effect. Similar observations have been made about the
initial contacts formed by neurons in culture with Xenopus
muscle (Kidokoro, Yeh 1982). Assuming quantal content is
normal at early stages, small numbers of receptors and the
confined nature of the synaptic cleft could lead to saturation and prolonged transmitter action. In addition, the
electrical coupling that we see in Manduca at that stage
would prolong the electrical response. Not suggested by
structural studies but equally important to the time course
is the maturity of any mechanism that would be responsible
for the uptake of glutamate from the cleft, or any matura-
tional changes in the properties of the receptor-ion
channel complex. The time course of the e.j.p. of the
adult slow terminals, on the other hand, might be due to
asynchronous release from distributed terminals that are
likely only to be releasing a few quanta at a time, because
of their shorter total length of active zone. The asynchronous nature would then arise from the conduction
properties of the intramuscular axonal branches.

The modulatory effect of neurosecretions on neuromuscular junctions is likely to be an integral part of
their functioning. In Manduca, octopamine, in the levels
that we measured, could serve to boost the fast e.j.p.'s
above threshold for contraction at the end of development
to make sure emergence and initial flight could occur.
Judging from the diverse morphological types of neurosecretory axons present, however, additional modulatory or
trophic effects on muscle and nerve terminals remain to be
described.

REFERENCES

Battelle BA, Kravitz EA (1978). Targets of octopamine
    action in the lobster. Cyclic nucleotide changes and
    physiological effects in hemolymph, heart and exo-
    skeletal muscle. J Pharmacol Exp Ther 205:438.
Bowman WC, Nott MW (1969). Actions of sympathomimetic
    amines and their antagonists on skeletal muscle.
    Pharmacol Rev 31:27.
Dudel J (1965). Facilitory effects of 5-HT on the cray-
    fish neuromuscular junction. Arch Exp Path Pharmacol
    249:515.

Enyeart J (1981). Cyclic AMP, 5-HT, and the modulation of transmitter release at the crayfish neuromuscular junction. J Neurobiol 12:505.

Evans PD (1980). Biogenic amines in the insect central nervous system. In Berridge MJ, Treherne JE, Wigglesworth VB (eds): "Advances in Insect Physiology", Volume 15, New York: Academic Press, p 317.

Evans PD, O'Shea M (1978). The identification of an octopaminergic neurone and the modulation of a myogenic rhythm in the locust. J Exp Biol 73:235.

Evans PD, Siegler MVS (1982). Octopamine mediated relaxation of maintained and catch tension in locust skeletal muscle. J Physiol 324:93.

Fifield SM, Finlayson LH (1978). Peripheral neurons and peripheral neurosecretion in the stick insect, Carausius morosus. Proc R Soc Lond B 200:63.

Florey E, Rathmayer M (1978). The effects of octopamine and other amines on the heart and on neuromuscular transmission in decapod crustaceans: further evidence for a role as neurohormone. Comp Biochem Physiol 61C:229.

Glusman S, Kravitz EA (1982). The action of serotonin on excitatory nerve terminals in lobster nerve-muscle preparations. J Physiol 325:223.

Hagiwara S (1953). Neuro-muscular transmission in insects. Jap J Physiol 3(4):284.

Hagiwara S, Watanabe A (1954). Action potential of insect muscle examined with intra-cellular electrode. Jap J Physiol 4(1):65.

Hill RH, Govind CK (1981). Comparison of fast and slow synaptic terminals in lobster muscle. Cell Tissue Res 221:303.

Hoyle G, Colquhoun W, Williams M (1980). Fine structure of an octopaminergic neuron and its terminals. J Neurobiol 11(1):103.

Hoyle G, Dagan D, Moberly B, Colquhoun W (1974). Dorsal unpaired median neurons make neurosecretory endings on skeletal muscle. J Exp Biol 187:159.

Jacobs JR, Atwood HL (1981). Long term facilitation of tension in crustacean muscle and its modulation by temperature, activity and circulating amines. J Comp Physiol 144:335.

Jenkinson DH, Stamenovic BA, Whitaker BDL (1968). The effect of noradrenaline on the end-plate potential in twitch fibres of the frog. J Physiol 195:743.

Kidokoro Y, Yeh E (1982). Initial synaptic tranmission at the growth cone in <u>Xenopus</u> nerve-muscle culture. Proc Nat Acad Sci USA 79:6727.

Lingle C (1981). The modulatory action of dopamine on crustacean foregut neuromuscular preparations. J Exp Biol 94:285.

Livingstone M, Schaeffer SF, Kravitz EA (1981). Biochemistry and ultrastructure of serotonergic nerve endings in the lobster: serotonin and octopamine are contained in different nerve endings. J Neurobiol 12:27.

O'Shea M, Bishop CA (1982). Neuropeptide proctolin associated with an identified skeletal motoneuron. J Neurosci 2(9):1242.

O'Shea M, Evans PD (1979). Potentiation of neuromuscular transmission by an octopaminergic neurone in the locust. J Exp Biol 79:169.

Rheuben MB (1982). Ultrastructural differences between slow and fast neuromuscular junctions of <u>Manduca</u>. Soc Neurosci Abstr 8:279.

Rheuben MB, Kammer AE (1981). Membrane structure and physiology of an immature synapse. J Neurocytol 10:557.

Rheuben MB, Kammer AE (1982). Structural and electrotonic connections between developing moth muscle fibers. J Neurobiol 13(6):559.

Rheuben MB, Reese TS (1978). Three-dimensional structure and membrane specializations of moth excitatory neuromuscular synapse. J Ultrastruct Res 65:95.

Swann JW, Sinback CN, Carpenter DO (1978). Dopamine-induced muscle contractions and modulation of neuromuscular transmission in <u>Aplysia</u>. Brain Res 157:167.

Titmus MJ (1981). Ultrastructure of identified fast excitatory, slow excitatory and inhibitory neuromuscular junctions in the locust. J Neurocytol 10:363.

Wasserman AJ (1982). Central and peripheral neurosecretory contributions to an insect flight motor nerve. Ph.D. thesis, The Pennsylvania State University.

Supported by NIH Career Development Award NS-00301 and grant NS-17132 to M.B.R. and NSF grant BNS 75-18569 to A.E.K.

M. Henkart; A. Kammer; M. Rheuben; J. Wallace

The Physiology of Excitable Cells, pages 411–421
© 1983 Alan R. Liss, Inc., 150 Fifth Avenue, New York, NY 10011

STUDY OF SYNAPTIC TRANSMISSION USING A TEMPERATURE-SENSITIVE
DROSOPHILA MUTANT

Kazuo Ikeda and J. H. Koenig

City of Hope Research Institute
1450 East Duarte Road
Duarte, CA 91010

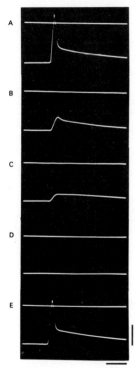

One approach to study of the nervous
system is to perturb the system and observe the effect this has on function.
One can do this in various ways: chemically, electrically, or surgically. In
Drosophila it is possible to do this
genetically by the use of mutants.

The ongoing study of synaptic transmission in our laboratory was initiated by
a collaborative work with Hagiwara (Ikeda
et al., 1976) which demonstrated that the
temperature-sensitive mutant, shibire[ts1]
(shi), isolated as a paralytic mutant in
Suzuki's laboratory (Hall et al, 1972),
affects the synaptic transmission mechanism. In this paper, it was shown that in
the dorsal longitudinal flight muscles
(DLM) of shi flies, the excitatory junction potential (e.j.p.), which is normal-
appearing at 18°C, gradually diminishes in
amplitude as the temperature is brought
above about 27°C, until it is almost completely abolished at 29°C (Fig. 1). This

Figure 1. Intracellular recording from DLM in a shi fly.
A. 19°C. Full sized action potential. B. 28°C. Junction
potential with a small regenerative component. C. 29°C.
Junction potential of reduced amplitude. D. 30°C. Transmission is almost completely blocked. E. 19°C. Response
recovers completely when temperature is lowered. Bars:
10 msec, 50 mV.

effect is completely reversible if the temperature is lowered. It was further shown in this paper that nerve conduction and the response of the muscle membrane to direct stimulation were normal at 29°C, pinpointing the site of this defect to the transmission mechanism itself.

Much has been learned about the nature of the shi defect since this early collaborative work with Hagiwara. It has now been demonstrated that this block in transmission is due to a presynaptic defect, rather than a postsynaptic one. This was demonstrated in two ways. Firstly, it was shown that the sensitivity of the subsynaptic receptor to 1-glutamate, the putative transmitter for this synapse (Ikeda, 1980), was not decreased at 29°C in shi flies. Since the synapses on the DLM lie deep in invaginations of the muscle membrane, it was necessary to apply the gluta- mate to the bath rather than iontophoretically. [Since it appears that no extrajunctional receptors exist on this muscle (Ikeda, 1980), this application demonstrates the response of the subsynaptic receptors to 1-glutamate.] When 1-glutamate was applied, a transient depolarization, the amplitude of which was dependent on the concentration of glutamate, was observed. It was shown that at 29°C, when transmission is almost completely blocked in shi, the depolarizing response to 1-glutamate was of the same magnitude as that seen at 18°C (Fig. 2). Thus, the dimi- nished e.j.p. does not appear to be due to a decrease in postsynaptic sensitivity to the transmitter. A second way in which it was shown that the shi effect is presynaptic involves the use of mosaic flies. Bilateral mosaics, which were made up of shi tissue on one side and normal tissue on

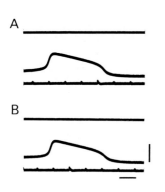

the other, were created using an unstable ring X-chromosome [In(1) w^{vC}] (Lindsley and Grell, 1968). Taking advantage of the fact that DLM fibers 5-6 (nomenclature: Mihalyi, 1936) are innervated by a motor neuron which lies contrala- terally (Ikeda and Koenig, 1982), it was possible to observe the effect on transmission when a shi

Figure 2. Examples of response of DLM Fiber 3 to bath-applied 1- glutamate in a shi fly at 18°C (A) and 29°C (B). Bars: 1 sec, 50 mV.

muscle fiber is innervated by a normal motor neuron and a
normal muscle fiber is innervated by a shi motor neuron.
We consistently observed the transmission block at 29°C
only when the motor neuron was mutant. When a shi muscle
fiber was innervated by a normal motor neuron, no effect
was observed. Thus, the defect must lie in the presynaptic
terminal (Koenig and Ikeda, 1983).

Another observation which has been made is that at
29°C in shi flies, vesicle depletion occurs at various
types of synapses (Poodry and Edgar, 1979, tibial NMJ;
Koenig et al., 1983, DLM; Kosaka and Ikeda, 1983a, CNS,
sensory and NMJ's). It has now been shown that in the DLM,
there is a correlation between the loss of the e.j.p., a
reduction in the frequency of spontaneously released
miniature e.j.p.'s (m.e.j.p.'s) and vesicle depletion
(Koenig et al., 1983). In shi when the e.j.p.'s of the DLM
are normal, many m.e.j.p.'s are observed (Fig. 3a).
However, at 29°C, when the e.j.p. is greatly diminished,
very few m.e.j.p.'s are seen (Fig. 3B). [At 29°C, normal
synapses show a very high rate of spontaneous release

A

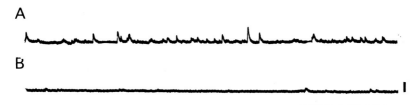

B

Figure 3.   Examples of spontaneously released m.e.j.p.'s in
the DLM of a shi fly at 25°C (A) and 29°C (B). Bars: 1
sec, 1mV.

(Koenig et al., 1983).] Morphologically, it could be shown
that in shi, the DLM synapses were almost completely
depleted of synaptic vesicles when the e.j.p. and
m.e.j.p.'s were thus affected (Fig. 4). These observations
suggest that the cause of the reduction in e.j.p. amplitude
and m.e.j.p. frequency is a reduction in the number of
vesicles available for release. Since muscle fiber activity was monitored continuously while the temperature was
being raised in these experiments, it was confirmed that no
excessive release of transmitter occurred which would normally deplete these synapses (Koenig et al., 1983). This
suggests that the depletion might be due to a block in the

If indeed the process of endocytosis is blocked, then
this abnormality might exist in non-nervous tissue as well.
Therefore, the garland cells of Drosophila, which are
believed to be very active in endocytosis via coated
vesicles (Wigglesworth, 1972), were observed electron-
microscopically. These cells, which lie in a cluster near
the junction of the ventriculus with the esophagus, are
believed to function in the segregation and storage of
waste products (Wigglesworth, 1972). They are characterized by a cortical layer which includes numerous
labyrinthine channels, coated profiles, and tubular elements. These structures can be seen in the specimen shown
in Fig. 6A, which has been impregnated with tannic acid to
distinguish those structures which are continuous with the
extracellular space from those which are not.

When compared to the garland cells of flies at 18°C,
(shi or wild-type), it was observed that the garland cells
of shi at 29°C demonstrated 3 major changes which were not
observed in wild-type at this temperature. These were:
(1) The surface area of the plasma membrane increased
drastically due to an increase in the length and complexity
of the labyrinthine channels. This can be seen in the

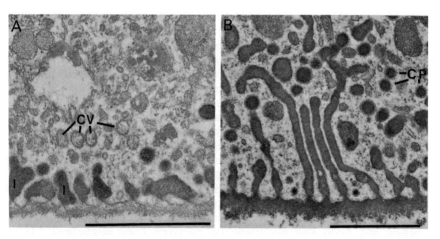

Figure 6. Tannic acid impregnated specimens showing cortical layer of garland cell in shi. A. 18°C. Normal-
appearing cortical layer. B. 29°C. Cortical layer with
elongated labyrinthine channels. CV: coated vesicles; CP:
coated pits; 1:labyrinthin channels; Bars: 1.0 μm X 37,500
(A); 1.0 μm X 27,500 (B).

tannic acid impregnated specimen shown in Fig. 6B. (2) The
number of coated pits increased, while the number of coated
vesicles and tubular structures decreased. Numerous
(tannic acid impregnated) coated pits were observed lined
up along the plasma membrane (Fig. 7), while very few
coated vesicles (unimpregnated) were found (also see Fig.
6). (3) Compared to 19°C, uptake of horseradish peroxidase
(HRP) was greatly decreased at 29°C in shi flies, while
wild-type flies showed some increase in HRP uptake at this
temperature.

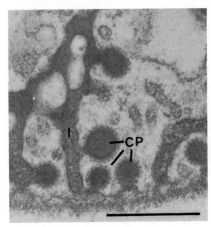

Figure 7. Cortical layer of
garland cell in a shi fly at
29°C. Note many tannic acid
impregnated coated pits budding
off of the outer plasma
membrane and labyrinthine channels. Bar: 0.5 μm X 56,000.

These observations suggest that indeed the process of
endocytosis is blocked in the garland cells of shi flies at
29°C. Furthermore, there is a striking similarity between
the effect which is observed in the garland cells and that
observed in the presynaptic terminals of neurons. Thus,
there is a great increase in membranous structures in both
preparations, an increase in pit formation, and a decrease
in the number of vesicles observed. These similarities
suggest that the neurons are affected in much the same way
as the garland cells. The evidence all suggests that the
process of endocytosis is blocked at 29°C in shi flies.

DISCUSSION

Our purpose in studying the mutant, shi, is to gain an
understanding of the mechanism of synaptic transmission.
At the present time, controversy exists concerning this
mechanism. The vesicle hypothesis, proposed in 1955 by
Katz to explain the quantal nature of transmitter release,

has been well accepted for many years. This hypothesis proposes that transmitter is stored in synaptic vesicles and is released by a process of exocytosis. The demonstration that transmitter indeed exists in vesicles, gave strong support to this hypothesis. Furthermore, treatments which cause massive transmitter release (e.g., strenuous stimulation, application of various toxins, depolarization with K^+ or carbachol) seem to cause depletion of vesicles, and structures which are claimed to be endo-cytotic pits are sometimes observed. However, these effects were not observed with less severe treatments [for evidence on the vesicle hypothesis, see reviews by Ceccarelli and Hurlbut (1980), Isreal et al. (1979), Zimmerman (1979)].

In recent years, evidence for nonvesicular release of transmitter has been presented. Thus, it has been shown that cytoplasmic ACh is preferentially released over vesicular ACh and that transmission is blocked by the hydrolysis of cytoplasmic ACh [for evidence for nonvesicular release, see review by Tauc (1982)]. The mechanism by which the cytoplasmic transmitter is released is thought to be through channels in the presynaptic membrane. Thus, the role of synaptic vesicles in the transmission process has been thrown into doubt.

The observations on the effect of the shi gene on transmission are quite pertinent to this question. The data suggest that the mutant gene affects the process of endocytosis, apparently slowing down or blocking the pinch-off mechanism. If this is true, then a block in the endocytosis of synaptic vesicles affects synaptic transmission in the following ways:
 1) If release of transmitter occurs (i.e., moderate activity) while the temperature is raised above 28°C, transmission gradually becomes blocked by 29°C. However, if release is blocked while raising the temperature, transmission is normal at 29°C. (Activity of about 0.5 Hz will cause it to gradually diminish at this temperature, however). Thus, the release mechanism itself seems to be normal at 29°C., but the availability of transmitter (or transmitter in releasable form) seems to gradually decrease as transmitter is released. Thus, a block in endocytosis at the presynaptic terminal appears to cause a block in the production of new (releasable) transmitter.
 2) When moderate release of transmitter occurs, vesicle depletion, an increase in membranous structures,

and pit formation occur. These structural changes are
strikingly similar to those which have been reported to
occur as a result of massive transmitter release (Heuser et
al. (1979). When release is blocked, these morphological
changes do not occur. Thus, a block in the endocytosis
process causes vesicle depletion if moderate transmitter
release occurs.

The observations mentioned above are explained well by
the vesicle hypothesis. Thus, if transmitter is released
by exocytosis of synaptic vesicles, it would be expected
that if exocytosis proceeds normally, a block in the
recycling of these vesicles should cause vesicle depletion
and thus block transmission by reducing the amount of
transmitter which is in releasable form. Without exocytosis (transmitter release), transmission would not be
blocked at 29°C.

If transmitter release is nonvesicular, then the role
of the synaptic vesicles is unknown. The observations on
the shi effect show that moderate activity causes the
disappearance of the synaptic vesicles from the terminal,
and, in addition, the appearance of what appear to be endo-
cytotic pits and various membranous structures. This is
the first morphological evidence showing that exocytosis is
occurring with only moderate transmitter release. (Other
methods which show these effects are biologically quite
severe.) Thus, one function of these vesicles appears to
involve exocytosis. Furthermore, by blocking the formation
of these vesicles, the production of releasable transmitter
substance seems to be affected. Thus, the observations
suggest that the vesicles are somehow involved in the production of transmitter in releasable form.

Interpretation of these results is not simple.
Although the vesicle hypothesis explains the data well, the
evidence for nonvesicular release cannot be ignored.
However, any hypothesis concerning the role of the vesicles
in synaptic tranmission should be able to explain these
observations. Obviously, further experimental data is
necessary before a conclusion can be drawn. Hopefully,
further experimentation on this mutant, which reversibly
blocks transmission by such a biologically benign method as
raising the temperature, will contribute greatly to our
understanding of the transmission mechanism.

We wish to thank Dr. Toshio Kosaka for providing the EM photographs which appear in this article. The EM work was done by Dr. Kosaka while on leave from the University of Tokyo. The research was supported by USPHS NIH grant NS-18856.

REFERENCES

Hall L, Grigliatti TA, Poodry CA, Suzuki DT, Junker A (1972). Three temperature-sensitive loci causing paralysis in Drosophila melanogaster. Can J Genet Cytol 14:728.

Heuser JE, Reese TS (1973). Evidence for recycling of synaptic vesicle membrane during transmitter release at the frog neuromuscular junction. J Cell Biol 57:315.

Heuser JE, Reese TS, Dennis MJ, Jan F, Jan L, Evans L (1979). Synaptic vesicle exocytosis captured by quick freezing and correlated with quantal transmitter release. J Cell Biol 81:275.

Ikeda K (1980). Neuromuscular physiology. In Ashburner M, Wright TRF (eds): "Genetics and Biology of Drosophila," Vol. 2, Academic Press, pp. 369-405.

Ikeda K, Koenig JH (1982). Morphological identification of the 5 motor neurons innervating the dorsal longitudinal flight muscle of Drosophila. Soc Neurosci Abstracts, Vol 8, p. 737.

Ikeda K, Ozawa S, Hagiwara S (1976). Synaptic transmission reversibly conditioned by a single-gene mutation in Drosophila melanogaster. Nature 259:489.

Israel M, Dunant Y, Manaranche R (1979). The present status of the vesicular hypothesis. Progr Neurobiol 13:237.

Koenig JH, Ikeda K (1983). Evidence for a presynaptic blockage of transmission in a temperature sensitive mutant of Drosophila. (Submitted)

Koenig JH, Saito K, Ikeda K (1983). Reversible control of synaptic transmission in a single-gene mutant of Drosophila melanogaster. J Cell Biol. (In press)

Kosaka T, Ikeda K (1983a). Possible temperature-dependent blockage of synaptic vesicle recycling induced by a single gene mutation in Drosophila. J Neurobiol. (In press)

Kosaka T, Ikeda K (1983b). Reversible blockage of membrane retrieval and endocytosis in the garland cell of the temperature-sensitive mutant of Drosophila melanogaster, shibire[tsi]. (Submitted)

Lindsley DL, Grell EH (1968). Genetic variations of
  Drosophila melanogaster. Carnegie Insti Wash Publ No.
  627.
Mihalyi F (1936). Untersuchungen uber Anatomie und
  Mechanik der Flugorgane an der Stubenfliege. Arb ung
  Biol Forsch Inst 8:106.
Poodry CA, Edgar L (1979). Reversible alterations in the
  neuromuscular junctions of Drosophila melanogaster
  bearing a temperature-sensitive mutation, shibire. J Cell
  Biol 81:520.
Salkoff L, Kelly L (1978). Temperature-induced seizure and
  and frequency-dependent neuromuscular block in a ts
  mutant of Drosophila. Nature 173:156.
Tauc L (1982). Nonvesicular release of neurotransmitter.
  Physiol Rev 62(3):857.
Wigglesworth VB (1972). The principles of insect phy-
  siology. 7th Edition. London: Chapman and Hall.

The Physiology of Excitable Cells, pages 423–433
© 1983 Alan R. Liss, Inc., 150 Fifth Avenue, New York, NY 10011

CORRELATED NERVE AND MUSCLE DIFFERENTIATION IN THE BULLFROG
CUTANEOUS PECTORIS

Diana C. Linden* and Michael S. Letinsky

Department of Physiology, Jerry Lewis Neuromus-
cular Research Center, University of California,
Los Angeles 90024, and *Biology Department,
Occidental College, Los Angeles, California 90041

Understanding the mechanisms responsible for the functional development of the highly ordered neuronal circuitry
seen in mature animals has been a challenging area of neuroscience for decades. Due to the obvious problems of small
size, complexity, and limited accessibility of neurons in
the central nervous system, many important studies have
used the developing nervous systems of submammalian vertebrates and invertebrates. The experimental advantage of
such systems has been applied beautifully, for example, by
Hagiwara and his co-workers to study the properties of
developing ionic channels (Hagiwara and Miyasaki, 1977).
Such studies are not only fundamental to our understanding
of neuronal development, but have suggested that specific
ionic conductances may play a role in the regulation of
synaptic interactions during development. To explore this
possibility we have turned to the vertebrate neuromuscular
junction where the formation of synaptic contacts can be
studied at the single cell level.

The development of functional neuromuscular contacts
involves an intimate and maintained relationship between a
developing motor nerve cell and its target muscle cell.
The initial cellular interactions between exploring neurons
and uninnervated muscle have been studied in tissue culture
(Kidokoro et al., 1980; Weldon and Cohen, 1979). Our research is directed toward studying the course of development
of nerve, muscle, and neuromuscular synaptic properties in
vivo. Others have investigated particular aspects of neuromuscular development in a variety of preparations: the

number of muscle fibers have been counted in rat muscles (Harris, 1981; Slater, 1982; reviewed by Dennis, 1981), the number of nerve fibers in ventral roots in chick have also been determined (Jacobson, 1978), and even the number and location of motor neuron cell bodies to individual muscles at different stages of chick and amphibian development have been reported (Jacobson, 1978; Grinnell and Herrera, 1981 for reviews). However, there are few reports in the literature which correlate the interaction during development of an individual muscle and the nerve innervating it (e.g., Slater, 1982). We have undertaken such a study using a multi-faceted approach to investigate the development of the frog cutaneous pectoris (CP) nerve and muscle. Such an analysis should provide a comprehensive picture of the important events in neuromuscular differentiation, and will establish an experimental foundation for further studies on mechanisms involved in de novo synaptogenesis.

There are several fundamental questions which we have set out to answer: (1) What is the time course of maturation of the CP muscle; at what stage of differentiation is the muscle when innervation occurs; how does the muscle develop relative to motor neuron numbers, synapse formation, and elimination? Specifically, (2) When is the adult complement of CP axons (myelinated and unmyelinated) in contact with the muscle and when this occurs has the muscle matured similarly; does the number of axons change during maturation of neuromuscular synaptic specializations such as distribution of acetylcholine receptors (AChR), acetylcholinesterase (AChE) and functional transmission? (3) How do these properties change as the tadpole metamorphoses into a frog? (4) During the period of most rapid elimination of multiple elimination (7-21 days post-metamorphosis; Morrison-Graham, 1981) are any of the above nerve or muscle characteristics altered?

The Cutaneous Pectoris Nerve-Muscle Preparation

For these developmental studies we used the CP muscle from bullfrog (Rana catesbeiana) tadpoles and recently metamorphosed froglets. This muscle has been used widely in electrophysiological and anatomical studies exploring the factors involved in motor nerve regeneration (Letinsky et al., 1976; McMahan et al., 1981), neuromuscular sprouting (reviewed by Brown et al., 1981), and motor neuron synapse elimination (reviewed by Grinnell and Herrera, 1981).

Several attributes of this muscle make it ideal for correlated histological and physiological research. The CP muscle forms a broad trapezoidal sheet, only 3-5 muscle fibers thick in the adult; each fiber has a single synaptic site which transiently becomes multiply innervated during development (Letinsky and Morrison-Graham, 1980). The CP can be removed as a whole mount even when it is quite immature (tadpole stages VIII-IX; front legs emerge at XX and metamorphosis is at stage XXV; see Letinsky and Morrison-Graham, 1980). Thus we can study nerve-muscle interactions at any stage of development from initial neuromuscular junction formation through synapse elimination and into the adult.

Muscle Development and Muscle Fiber Maturation

Muscle fibers are formed as a result of fusion of mononucleate myoblasts. At the earliest stages of muscle development, the muscle primordium is composed exclusively of mononucleate muscle cells and connective tissue cells. Multinucleate myotubes form as many myoblasts fuse together.

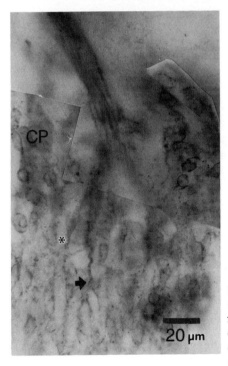

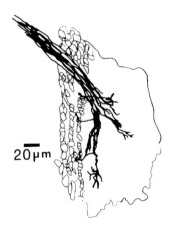

Figure 1. Stage XI NBT/AChE stained CP. The muscle is mainly composed of myoblasts organized in rows. Arrow indicates small nerve trunk, * small myotube.

At the earliest stage of differentiation studied, stage IX, the CP nerve was present in the presumptive CP muscle, which consisted of an unorganized mass of small globular cells (presumably myoblasts). At stage XI the muscle was largely made up of myoblasts which appeared to line up in rows throughout the muscle. However, interspersed amongst these rows were small myotubes; the entire CP muscle was not wider than 150 μm at this time (see Figure 1). Most axons ran together in bundles, with few straying from the main nerve trunks. No specialized nerve contacts or AChE staining were present in whole mounts processed for nitroblue tetrazolium (NBT)/AChE staining (Letinsky and DeCino, 1980); and no muscle contractions were evoked by indirect nerve stimulation. In addition, AChRs were not seen in the CP after rhodamine conjugated alpha-bungarotoxin (R-αBGT) staining, although neuromuscular junctions in the more mature surrounding muscles stained intensely.

The CP was mainly composed of myotubes by stage XII (Figure 2). Fine axons branched throughout the muscle and in many places neuromuscular contacts (small bulbous endings) were present. No AChE reaction product was present anywhere in the muscle (neither at the myotendinous junctions nor at the motor nerve endings). No contraction was evoked with nerve stimulation; however, extremely small weakly fluorescent patches of R-αBGT staining were localized beneath the main nerve trunks. These were completely bleached after one or at the most two photographs were taken. No extrajunctional R-αBGT staining (indicating AChR accumulations) was apparent. At stages XIII-XIV very slow nerve-evoked muscle contractions were occasionally seen in a few myotubes. Not all preparations at this age contracted in response to nerve stimulation.

When viewed as a whole mount the stage XVI muscles appeared well-developed and resembled the adult CP (Figure 4). However, only about 15% of the adult number of muscle cells, mainly myotubes and myoblasts (counted in two different electron microscopic cross sections of the muscle) were present in the CP (Figure 5). Ultrastructurally the muscle was relatively immature (Figure 4). Many muscle cells were small, scarcely larger in diameter than the nucleus, and had no or few contractile filaments in the sarcoplasm. Although the muscle cells appeared relatively undifferentiated, nerve stimulation did elicit a slow contraction. Neuromuscular contacts were often quite complex, with many axons innerva-

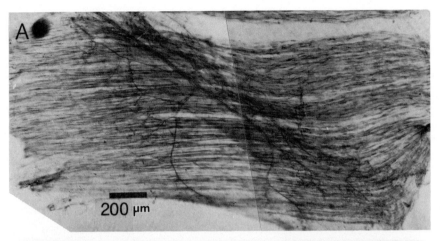

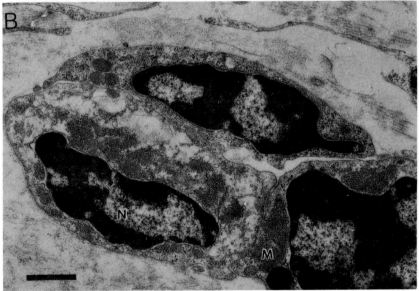

Figure 4. Stage XVI CP. (A) NBT/AChE stained whole mount.
(B) Cross section of muscle fibers from middle region of
the muscle. Note varying amounts of contractile myofila-
ments M; nucleus N. Notice the adult-like appearance of
the whole mount (A) and the relative ultrastructural immatur-
ity of individual muscle fibers (B). Scale is 1 μm.

ting and overrunning each synaptic site (Letinsky and Morrison-Graham, 1980). Subneural AChE staining was prominent.
Further muscle maturation resulted in increased numbers of
muscle fibers as well as the addition of myofibrils in each
muscle cell. By stage XVIII many muscle fibers resembled
those of the adult, however myoblasts and myotubes with
sparse amounts of contractile filaments were still present.
Some myoblasts and undifferentiated cells were closely apposed
to well differentiated myotubes. At this stage 35% of the
adult number of muscle fibers were present. AChRs were
localized in loose aggregates beneath the nerve terminals.
In adjacent muscles (e.g., LP in Figure 2) the adult pattern
of AChR staining with subneural striations was prominent.
By stage XX-XXI, the full complement of muscle fibers (Figure 5) was present in the CP. This number remained relatively constant thereafter throughout the life of the animal.
This means that during metamorphosis and during the period
of elimination of multiple innervation (7-21 days post-metamorphosis), the number of muscle fibers was stable.

Cutaneous Pectoris Nerve Development

The CP nerve was present at the earliest stages of muscle
development when the muscle was composed entirely of undifferentiated myoblasts. By stage XII, when myotubes were
first identified ultrastructurally, the nerve trunk appeared
to be well formed, although no axons were myelinated in the
nerve cross sections 100-200 μm from the nerve's entry into
the muscle (Figure 3). The total number of axons in the CP
nerve at this time (118) was similar to that in adult(33 $\pm$ 6
myelinated, 80 $\pm$ 40 unmyelinated; see Figure 5.) The diameters of these axons were bimodally distributed. The population of the large unmyelinated axons from stage XII CP nerve
(about 30 nerve fibers) was 1.1 $\pm$ 0.2 μm in diameter;
this is more than twice the mean diameter of all the axons
(0.48 $\pm$ 0.37 μm) in the CP nerve at this time. These large
axons probably correspond to the sensory and motor axons
which will become myelinated by stage XVI (see below).
Therefore, most CP motor and sensory axons are present in the
CP muscle as early as stage XII. The mean diameter of the
smaller nerve fibers was 0.29 $\pm$ 0.1 μm, which is 73% of the
adult unmyelinated fiber mean diameter.

By the time that neuromuscular junctions are functional
(stage XVI), 60% of the adult number of myelinated axons
were present. The total number of axons (myelinated plus

unmyelinated) was similar to that of stage XII, as well as that of the adult. The mean diameter of unmyelinated axons

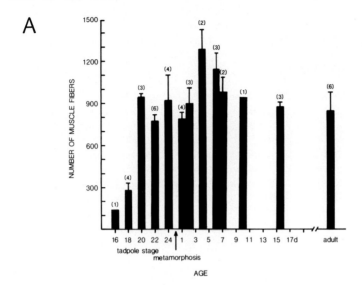

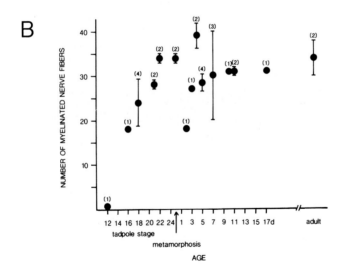

Figure 5. (A) Mean number of muscle fibers (± S.D.) at different stages of development. (B) Mean number of myelinated axons (± S.D.) at different stages of development. No myelinated axons were present at stage XII. Parentheses enclose numbers of preparations.

was the same as that in the adult (stage XVI = 0.39 ± 0.3 μm; adult = 0.40 ± 0.1 μm), but the mean diameter of the axon itself of the myelinated axons was less than 30% of the adult. At stage XVIII, 80% of the adult number of myelinated axons were present; however, there was an increase in the number of unmyelinated axons (143 ± 38) which was almost twice that of the adult (80 ± 40).

The adult number of muscle fibers developed by stage XXI; also, the adult number of myelinated axons was present. Therefore, before and throughout the period of rapid elimination of multiple innervation in the CP muscle (7-21 days post-metamorphosis) the adult number of axons was present and remained constant. We can conclude that the large motor and sensory axons innervating the muscle are not lost during synapse elimination. This supports the hypothesis that entire motor neurons are not lost in the formation of singly-innervated synapses. It is most probable that intramuscular nerve twigs retract (reviewed by Grinnell and Herrera, 1981).

In summary, at the earliest stages of development, when the CP consisted of an unorganized mass of myoblasts, the CP axons were already present. The adult number of CP axons, though still unmyelinated, was present about the time when myotubes developed. Well developed neuromuscular AChRs, AChE, and nerve-evoked muscle contractions were present when 60% of the CP axons were myelinated and 30% of the muscle fibers were present. Thereafter maturation consisted of further myelination of axons and growth in number and size of muscle fibers. Before metamorphosis the CP muscle, motor nerve, and neuromuscular junction were well-developed and functional. The CP was adult-like in most respects, with the important exception that all synapses were extensively polyneuronally innervated; most neuromuscular junctions become singly innervated about 7-21 days post-metamorphosis. Throughout synapse elimination muscle and CP nerve characteristics do not change.

Acknowledgments

We thank B. Woloske for expert technical assistance. This work was supported by USPHS grant NS13470.

References

Brown MC, Holland RL, Hopkins WG (1981). Motor nerve sprout-
    ing. Ann Rev Neurosci 4:17.
Dennis MJ (1981). Development of the neuromuscular junction:
    Inductive interactions between cells. Ann Rev Neurosci
    4:43.
Grinnell AD, Herrera AA (1981). Specificity and plasticity
    of neuromuscular contacts: Long-term regulation of
    motoneuron function. Prog in Neurobiol 17:203.
Hagiwara S, Miyasaki S (1977). Changes in excitability of
    the cell membrane during 'differentiation without cleav-
    age' in the egg of the annelid, *Chaetopterus pergamen-
    taceus*. J Physiol 272:197.
Harris AJ (1981). Embryonic growth and innervation of rat
    skeletal muscles. I. Neural regulation of muscle fibre
    numbers. Phil Trans Roy Soc Lond B 29:-257.
Jacobson M (1978). "Developmental Neurobiology", 2nd ed.
    New York: Plenum Press.
Kidokoro Y, Anderson MJ, Gruener R (1980). Changes in syn-
    aptic potential properties during acetylcholine receptor
    accumulation and neurospecific interactions in *Xenopus*
    nerve-muscle cell culture. Dev Biol 78:464.
Letinsky MS, DeCino P (1980). Histological staining of
    pre- and postsynaptic components of amphibian neuro-
    muscular junctions. J Neurocytol 9:305.
Letinsky MS, Fischbeck KH, McMahan UJ (1976). Precision
    of reinnervation of original postsynaptic sites in frog
    muscle after a nerve crush. J Neurocytol 5:691.
Letinsky MS, Morrison-Graham K (1980). Structure of devel-
    oping frog neuromuscular junctions. J Neurocytol 9:321.
McMahan UJ, Edington DR, Kuffler DP (1981). Factors that
    influence regeneration of the neuromuscular junction.
    J Exp Biol 89: 31-42.
Morrison-Graham K (1981). Synapse elimination at the de-
    veloping frog neuromuscular junction. Ph.D. thesis,
    UCLA Department of Neuroscience.
Slater CR (1982). Postnatal maturation of nerve-muscle
    junctions in hindlimb muscles of the mouse. Dev Biol
    94:11.
Weldon PR, Cohen MW (1979). Development of synaptic ultra-
    structure at neuromuscular contacts in an amphibian cell
    culture system. J Neurocytol 8:239.

Top: W.J. Moody, Jr.; S. Hagiwara
Bottom: Y. Kidokoro; M. Letinsky; D. Junge

The Physiology of Excitable Cells, pages 435–449
© 1983 Alan R. Liss, Inc., 150 Fifth Avenue, New York, NY 10011

THE REGULATION OF SYNAPTIC STRENGTH AT ANURAN
NEUROMUSCULAR JUNCTIONS

A.D. Grinnell, A.A. Herrera*, B.M. Nudell,
L.O. Trussell, A.J. D'Alonzo, and P. A. Pawson

Jerry Lewis Neuromuscular Research Center, UCLA
School of Medicine, Los Angeles, CA 90024 (*Dept.
of Biology, U.S.C., Los Angeles, CA 90007)

Frog neuromuscular junctions differ greatly in size
and strength--by up to ten-fold in total terminal branch
length and up to 25-fold in endplate potential (EPP) amplitude in the sartorius muscle (Grinnell, Herrera 1981). One
might suppose that all junctions are sufficiently strong to
excite the postsynaptic muscle fiber, and that differences
in strength in curare or Mg^{2+}-blocked preparations are of
no functional importance; but this is not the case. Even
in normal Ringer, as many as 20% of the fibers in the sartorius receive no inputs that are suprathreshold on single
stimulation of the innervating axons even though there are
two or three endplates on each fiber (Grinnell, Herrera
1980). Moreover, synapses change in number, size, and
strength in response to peripheral conditions and the
animal's physiological state: terminals enlarge and additional endplates are added as muscle fibers grow (Kuno et
al., 1971; Bennett, Pettigrew 1975; Nudell, Grinnell 1983);
they grow and retract as a function of seasonal change or
hormonal state (Wernig et al., 1980; Mallart et al., 1980;
Grinnell, Herrera 1981); they sprout when synaptic function
is blocked or adjacent fibers are denervated (cf. Brown et
al., 1981); and they can be functionally displaced by competition with other synaptic inputs (cf. Mark, 1980;
Grinnell, Herrera 1981) even when these are several millimeters away on the same fibers (Grinnell et al., 1979).
These indications of plasticity imply the existence of a
set of regulatory factors, that, for any given fiber and
its synaptic inputs, determines the properties of those
synapses. We have been interested in trying to identify

the important regulatory factors, and in analyzing how they
interact to achieve the equilibrium observed.

Motor unit size and intrinsic differences between motoneurons

Junctions in the frog cutaneous pectoris (c.p.) muscle
are, on the average, much stronger and more uniform in
safety margin than junctions in the sartorius muscle.
Although the junctions in both cases appear to be similar
in size and shape, c.p. junctions release two to three
times as much transmitter, whether expressed as total EPP
quantal content or release per unit length of terminal
(Grinnell, Herrera 1980). The disparity in synaptic
strength could be due to differences between the two populations of motoneurons and/or to some difference in the
pattern of innervation of the two muscles. C.p. muscle
fibers have only one endplate, while most sartorius muscle
fibers are innervated at two or three different sites.
Moreover, the two muscles have nearly the same number of
fibers and are innervated by approximately the same number
of motoneurons, so it can be calculated that each sartorius
motoneuron must, on the average, form two to three times as
many junctions as c.p. motoneurons (Grinnell, Herrera 1980).
One can imagine that the sartorius junctions are weaker
either because of competitive influences from other junctions on the same fiber, or because each sartorius motoneuron, in trying to maintain many more synapses, is over-
extended, i.e., the amount of some critical substance
supplied by the soma is limiting. We have attempted to
distinguish between the effects of competition and synaptic
numbers by experiments in which we removed half of the sartorius muscle fibers, crushed the nerve, and allowed the
full complement of axons to reinnervate the remaining half
muscle. Each muscle fiber became reinnervated selectively
at old endplates, with no evidence of new ectopic endplates. Hence competitive interaction should have been as
great as in the original condition, or potentially greater,
if there is increased polyneuronal innervation at reinnervated junctions (Rotshenker, McMahan 1976). On the other
hand, the motor units in the reinnervated half muscles were
all scaled down in size. Thus each motoneuron presumably
formed fewer junctions in the half muscle than originally,
in the intact muscle. If competitive interaction were
responsible for the relative weakness of sartorius

junctions, this should have been relatively little changed
by the manipulation. If the number of synapses maintained
is more important, one might predict the junctions in the
half muscle would be stronger. In fact, they were found to
release two to three times more transmitter per unit terminal length than junctions in reinnervated whole or un-
operated sartorius muscles (Herrera, Grinnell 1980). This
finding implies that the effectiveness of synapses does,
in fact, depend on the supply of some essential substance
from the cell body, and that changes in the number of
synapses (or synaptic area) a given neuron must maintain
can sharply affect its release properties. This is consistent with findings suggesting that junctions formed by
motor axon sprouts are relatively weak (Bennett, Raftos
1977); and easily displaced by regenerating nerves that
are maintaining less than a full complement of endings
(Cass et al., 1973; Wigston 1980; Haimann et al., 1981b).
The enhanced release from terminals in half muscles gradually returns to normal in about six months, indicating
that a new equilibrium is eventually established (Herrera &
Grinnell, unpublished).

It is well known that motor units in frog muscles can
differ greatly in size. From the evidence presented above,
it might be predicted that large motor units would tend to
have weaker junctions than smaller motor units. In fact,
just the opposite correlation obtains in the normal sartorius (Grinnell, Trussell 1983). Motor units range in
size from 0.1 to 40% of the fibers contracting to a single
stimulus of the whole nerve. The large motor units thus
represent 200 or more fibers, each innervated at least
once by a suprathreshold synapse, while the smallest motor
units appear to involve no more than one or two fibers in
the twitch. Yet when the single axons driving such motor
units are isolated and tetanized, the small motor units
are seen to include up to 100 times as many fibers as contracted in the twitch. Axons innervating the smallest
motor units, as judged by twitch, in fact innervate a much
larger number of fibers, but with subthreshold inputs.
This is not true of the largest motor units, in which a
tetanus drives relatively few fibers that did not contract
in the twitch. This difference shows up clearly in tests
of the effect of changes in calcium concentration in the
Ringer (Figure 1). An analysis of the properties of many
motor units reveals that there is an inverse linear
relationship between the number of fibers innervated by an

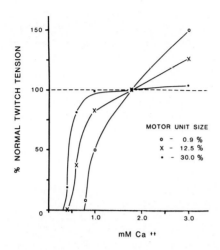

Figure 1. Motor unit twitch tension in Ringer solutions of different calcium concentration. Tension is expressed as the percent of twitch tension found in normal frog Ringer. Shown are three motor units from one muscle. Motor unit size is expressed as a percentage of the tension evoked by whole nerve stimulation.

axon and the percentage of junctions formed by that axon that are subthreshold in normal Ringer (Grinnell, Trussell 1983).

That the smallest motor units also have the highest percentage of subthreshold junctions appears to contradict the findings with the half muscles; but the two sets of data in fact describe very different situations. Taken together, they suggest that motoneurons emerge from the period of development endowed with certain intrinsic synapse maintaining capabilities, and that these differ for different neurons. A reduction in the number of synapses any neuron must maintain may result in an increase in efficacy for the remaining terminals. (It is possible that the smallest, weakest motor units are selectively helped by a reduction in number of synapses formed, but evidence for this is not compelling (Herrera, Grinnell 1980)).

We can conclude that in later stages of development and in the mature animal motoneurons differ in their ability to maintain synapses and in the total amount of transmitter they can release. The bases for those differences in capability are unknown and are a central interest of our laboratory. They might represent true intrinsic differences, the result of genetic variability in the population; or they may result from developmental processes, e.g., the timing of peripheral differentiation, or differences in activity pattern driven by central generators. It would seem reasonable, for example, to postulate that the first terminals to form on a fiber tend to form relatively large, strong synapses and be successful in the eventual synapse elimination process, and that "strong" motoneurons are those that have a slight developmental edge--that grow into the muscle first or are capable of forming synapses more rapidly at early stages. Axons reaching the muscle later may mainly add components to junctions already innervated at least once, and consequently, for unknown reasons, be less likely to prevail during the process of synapse elimination. On the other hand, most junctions become singly innervated by the process of synapse elimination, and it is not clear why, under these conditions, the junctions formed by neurons with large peripheral fields should be generally stronger or more uniform in strength than those formed by neurons innervating fewer endplates. It would be of interest to know, for example, whether a motoneuron acquires a large peripheral field because it has strong intrinsic growth and synapse maintaining capabilities, or whether it acquires a large axon and greater synaptic maintaining capability because, by chance, it happened to innervate more fibers and receive more positive feedback than other neurons early in development. In the case of sartorius muscle, of course, the presence of other endplates on the fiber could significantly affect the properties of any given junction even after the elimination of polyneuronal inputs to that junction (see below).

Competitive interaction between synapses, and the role of peripheral regulation

Important as the intrinsic differences between motoneurons may be, it is clear that there is another important level of control of synaptic structure and function: peripheral feedback regulation. Different terminals formed by the same axon do not necessarily innervate fibers of the same size, and they can differ dramatically in the amount of transmitter released per unit length of terminal (unpublished observations). These differences probably are the result of local peripheral influences.

At least one important regulatory influence is the size of the target cell itself. It has long been recognized that as muscle fibers grow, the terminals innervating them increase in size, as does the amount of transmitter they release, compensating for the decrease in muscle fiber input

impedance (Kuno et al., 1971). However, the correlation between terminal size and quantal content is a very "noisy"
one, with a correlation coefficient of only about 0.4. At
any one terminal size, for example, the quantal content may
differ by more than 10-fold, and terminals with the same
quantal contents can vary by more than three-fold in size
(Kuno et al., 1971).

We have recently shown that much of this variability
can be explained by the existence of an inverse correlation
between terminal branch length and release per unit length
for terminals on muscle fibers of the same size (input resistance) in the singly innervated c.p. muscle (Figure 2)
(Nudell, Grinnell 1982). This might represent a regulatory
process in which relatively short terminals, in order to
compensate for their small size, are somehow induced to release more transmitter, or a process in which terminals differ intrinsically in their release per unit length, and ones
with lower release efficacy are somehow induced to grow
longer to increase their total output. The nature of the
feedback regulating either of these processes is not obvious, but presumably represents a mechanism designed to

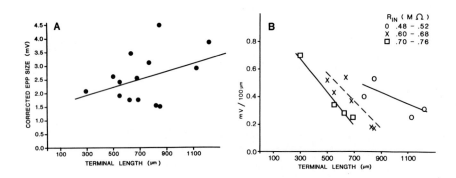

Figure 2. A, Transmitter release, judged by EPP size corrected for resting potential, nonlinear summation, and input
resistance, versus terminal length for 14 endplates in one
curarized c.p. muscle. The correlation coefficient for the
regression line is 0.4. B, Lines of best fit for data from
the cells in A with release per unit terminal length plotted
against terminal length. Endplates are grouped on the basis
of muscle fiber input resistance.

insure adequate synaptic input to the muscle fiber. Because
the shortest terminals tend to be absolutely stronger
(larger EPP and quantal content) than the longest terminals
in the fiber group, it seems likely that the regulatory
mechanism acts principally by inducing greater growth in intrinsically weaker terminals. This conclusion is also suggested by the fact that the terminals on larger muscle fibers in the same muscle tend to be larger, but again show an
inverse correlation between length and release per unit
length, with approximately the same range of release per
unit length values as terminals in populations of smaller
fibers (Figure 2).

Given the evidence above that there is some form of
feedback from muscle fibers that influences the growth and/or
strength of terminals innervating them, it is of interest to
try to assess the role of this regulatory mechanism in cases
where two or more synaptic inputs innervate the same fiber.
To this end, we have examined the structure and function of
junctions in the Xenopus pectoralis muscle (Nudell, Grinnell
1983). These fibers gradually acquire second endplates as
they grow and mature after metamorphosis. At the time we
have studied them physiologically, almost 95% of the fibers
are innervated at two endplates. Those fibers that remain
singly innervated when most others already have two endplates tend to be of smaller diameter. Moreover, the single
endplates usually are centrally located on the fiber and
very strong, often with a larger EPP than the summed EPPs of
the two endplates on doubly innervated fibers of equivalent
input resistance (see also Haimann et al., 1981). When
there are two endplates, they are almost always separated by
20-30% of the muscle fiber length or more (2-3 mm in a
typical experimental animal).

On fibers with two endplates, there is a striking similarity in length of the two junctions: nearly 70% of the
junctions are within 20% of the length of the other junction
on the same fiber, whereas random comparison between junctions on fibers of the same input resistance shows much less
similarity (Nudell, Grinnell 1983). There is no corresponding tendency for paired junctions to be of similar strength.
Thus there seems to be something expressed uniformly throughout the length of a given muscle fiber that strongly influences the length to which terminals will grow, with the
result that both junctions, even when they differ greatly in
strength, are of similar length. Obviously, under these

conditions, there can be no tight correlation between the
length and release per unit length of any given terminal,
such as was obtained in the c.p. muscle (Nudell, Grinnell
1982). However, if one compares the length to which either
terminal grows (or the total terminal length of both
junctions) with the summed synaptic output of both junctions (or mean release per unit length), a good correlation
is seen (Figure 3).

From these data have emerged a general model, similar
in many respects to that of Jansen et al. (1978; see also
Kuffler et al., 1980), for regulation of synaptic number,
position, size, and strength (Nudell, Grinnell 1983). The
extent of longitudinal terminal branch growth appears to be
determined to a significant degree by feedback from the target muscle fibers. This feedback we presume to depend on the
level of a terminal growth promoting substance that is inserted into, or secreted through, the fiber membrane in
approximately equal amounts along the fiber's length. Moreover, more of this substance is produced as muscle fibers
grow, explaining the tendency for larger fibers to have
larger endplates. As fibers grow beyond a certain size, the
same substance is likely to be responsible for the attrac-
tion and acceptance of additional endplates, and it may be
responsible for inducing sprouting from nearby axons in
cases of partial denervation. In addition, however, the
synthesis or expression of this terminal growth promoting
substance appears to be inversely proportional to the total
synaptic input to the fiber. Very strong terminals are
seemingly best able to keep fibers refractory to further innervation. Moreover, each junction (perhaps in proportion
to its strength and/or length, but we have no good evidence
for this) clearly inhibits additional synapse formation
close to it. Junctions are usually separated by 20-40% of
the fiber's length. It may be, therefore, that the central
location of surviving single junctions contributes to their
ability to keep the readily innervatible portion (the central
60%) of the fiber refractory to further innervation.

The mechanism by which synapses control a muscle fiber's
expression of terminal growth promoting substance is of interest. In view of the convincing demonstration that extrajunctional AChR and a variety of other muscle fiber properties are controlled by muscle electrical or contractile
activity (Lømo 1976), it is tempting to suggest that the
level of muscle fiber activity also controls terminal growth.

However, this does not appear a satisfactory explanation for
two reasons: (a) the good inverse correlation between total
transmitter release and terminal length requires that the
EPP amplitudes for both junctions be taken into account. Yet

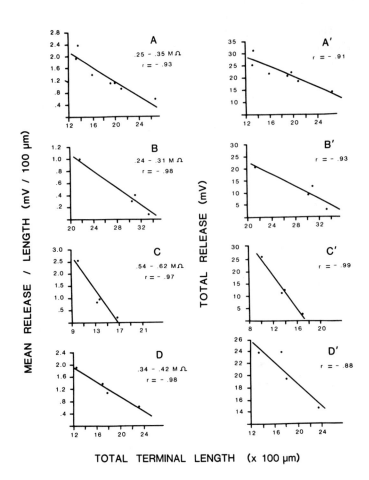

Figure 3. Relationship between transmitter release and
total terminal length for the paired terminals on four
groups of muscle fibers, each composed of fibers of similar
input resistance (R_{in}). EPPs were corrected for resting
potential and normalized to the mean R_{in} of each group.
r= correlation coefficient for the line of best fit
(Nudell, Grinnell 1983).

there is no obvious reason why postsynaptic activity, a
threshold phenomenon, should be directly related to the
summed EPP amplitude of the two junctions, when one or both
might have a large safety factor, others might be near or
below threshold. And (b), we have found that in more than
half the doubly innervated fibers, both endplates are inner--
vated by the same axon, in which case addition of the second
input, which appears to satisfy the fiber's need for innervation, does not affect the muscle fiber's activity level
(Nudell, Grinnell 1983).

It is noteworthy also that this model helps account for
at least two forms of competitive interaction between terminals. By virtue of local depletion or inactivation of a
muscle fiber's terminal growth promoting substance, a terminal prevents other junctions from being added nearby on
the fiber. Secondly, since terminal length is determined by
summed synaptic strength, an intrinsically weak terminal
(low release per unit length), coupled with a strong terminal, will not grow to nearly the length, or acquire as large
an EPP amplitude, as it would if coupled with a weaker synapse. Thus the size and effective strength of a junction is
negatively influenced by its paired input in proportion to
the strength of that input.

What accounts for differences in release per unit length?

In the structure-function correlations described above,
synaptic efficacy is often expressed as transmitter released per unit length. This differs in average sartorius
and c.p. junctions, it varies inversely with terminal length
in muscle fibers of the same size in the c.p., and it probably varies as a function of the motoneuron from which a
given terminal is derived. It can be interpreted as "intrinsic" terminal release efficacy. However, we have little
information about what differences between terminals account
for the differences in release per unit length. We don't
even know if normalization of release to length is an appropriate procedure. Although it is generally assumed that
release is equally probable from all parts of the terminal,
this has not been proved; and at least one report claims the
contrary: that in many toad sartorius junctions, at low
$[Ca^{++}]$, 1-3% of the terminal length releases as much as 30-
50% of the total transmitter output, i.e., most release is
from a single "hot spot" (Bennett, Lavidis 1979). We are

now investigating the uniformity of release from frog motor
nerve terminals by determining the numbers of single quantum
EPPs occuring at different points between intracellular
electrodes placed at both ends of identified junctions, and
so far find no deviations greater than 6% from the levels
predicted on the basis of uniform release (D'Alonzo,
Grinnell 1982). This needs more extensive investigation,
however.

Assuming that release along the length of a terminal
branch is uniform in any given terminal, it is important to
decide what is different about two terminals that release
vastly different amounts of transmitter per unit length.
We have compared the morphology and physiology of randomly
selected terminals of the c.p. and sartorius muscles, the
former of which, on the average, release 2.4 times more
transmitter per unit length (Grinnell, Herrera 1980). We
find no correlation between apparent diameter, as judged by
light microscopy, and synaptic efficacy. Also, there is no
difference in spacing of active zones, assuming that they
lie opposite postsynaptic folds (Peper et al., 1974). At
the EM level, no correlated differences are seen in vesicle
density, mitochondrial number, or terminal diameter, although a barely significant difference was seen in the degree of glial interposition between the nerve terminal and
the postsynaptic surface, with greater interposition in the
case of sartorius terminals (Grinnell, Herrera 1980). In a
more exhaustive ultrastructural study of a number of identified, physiologically characterized, endplates that differed by over 10-fold in release per unit length, a good
positive correlation was found between release per unit
length and the average width of close apposition between
pre- and postsynaptic structures (Herrera, Grinnell 1982).
The relationship was a very steep one, with the release increasing almost as the square of synaptic area. An even
better, steeper correlation was found between release/unit
length and the ratio of apposition width/terminal perimeter.
However, since both apposition width and width/perimeter
ratios are largely determined by the degree of Schwann cell
interposition between the terminal and the muscle fiber, it
is quite possible that differences in this parameter are
the result, rather than the cause, of differing release levels. Comparable differences in glial interposition have
been correlated with differing levels of neurohormone release from pituitary cells (Tweedle 1983). In short, we
have an interesting morphological correlate of transmitter

release efficacy, but are not yet convinced that it represents a causal explanation for differences in transmitter release per unit length of terminal. It seems likely that fundamental physiological differences are responsible and considerable effort is being directed at identifying them.

Physiologically, terminals that differ in release efficacy show a similar, approximately fourth power, relationship between quantal content and external $[Ca^{++}]$, and show an equivalent dependence on $[Ca^{++}]$ for evoked and spontaneous transmitter release in response to imposed depolarizations (Grinnell, Herrera 1980). The differences in release are found in curarized preparations as well as in low Ca^{++}, high Mg^{++} Ringer. Moreover, there is no evidence for partial action potential invasion in the weaker terminals. Miniature EPP frequency and EPP quantal content are closely proportional, in both strong and weak junctions. In general, our data suggest that calcium, once it gets into the terminal, acts in the same way in all cases to cause transmitter release. However, strong terminals seem to be characterized either by a larger than normal chronic Ca^{++} leak or by reduced internal Ca^{++} buffering, resulting in an elevated active Ca^{++} level. It is interesting to note that in patients with myasthenic syndrome reduced transmitter release is correlated with fewer and disorganized aggregates of the presynaptic membrane particles thought to represent calcium channels (Fukunaga et al, 1982).

Contralateral effects

While considering factors that regulate synaptic strength, mention should be made of two, probably related, effects of denervation of a frog muscle. Rotshenker and his colleagues have demonstrated very elegantly that unilateral denervation of the c.p. causes the homologous contralateral axons to sprout and polyneuronally innervate fibers (Rotshenker 1979;1982); and we have demonstrated that contralateral sartorius nerve terminals release increased amounts of transmitter (Herrera, Grinnell 1981). The sprouting response, which has been studied quite thoroughly, appears to be specific to the contralateral homologous muscle, and clearly depends on centripetal axonal transport of one or more substances from the periphery that are lacking when the muscle is missing or inactive, or axonal transport is blocked (Rotshenker 1982). When sartorius denervation

results in enhanced EPP quantal contents on the contralateral
side, much less sprouting is seen than is reported for the
c.p. (Herrera, Grinnell 1981). However, in some frogs, perhaps because of their endocrine state, release from sartorius
terminals is unusually high, comparable to that normally seen
from c.p. terminals. In these sartorius muscles, it is our
impression that there is much less change in release efficacy
upon contralateral denervation but nearly as much terminal
sprouting as is seen in the c.p. (Pawson unpublished). These
observations suggest a possible relationship between synaptic
efficacy and sprouting. The effect of contralateral denervation may be first to increase transmitter release efficacy
and, when that reaches a "ceiling level" characteristic of
each particular terminal or motoneuron, to induce the axon
terminals to sprout. Moreover, Rotshenker (personal communication) has noted that there appears to be a "ceiling effect"
for sprouting; in certain frogs, where polyneuronal innervation of c.p. junctions is already high, denervation does
not significantly increase it.

The importance of retrograde transport of some sub-
stance(s) for these contralateral effects draws attention to
the importance of the cell body, and to interactions between
motoneurons in regulation terminal release and growth. It
is possible that the strong sprouting response of intact
axons in partially denervated muscles, which is clearly dependent on local peripheral sprouting signals (Brown et al.
1981), is partially mediated centrally as well, and again may
involve an increase in release efficacy as well as axonal
growth.

Conclusion. Our understanding of the regulation of synaptic
strength at neuromuscular junctions is still in its infancy.
However, these initial studies of the structure and function
of identified terminals already permit us to conclude (a)
that motoneurons differ intrinsically in their transmitter
release and synapse maintaining capabilities, (b) that
changes in release efficacy can result from motoneuron responses to axotomy of contralateral homologues, and (c) that
superimposed on these central controls are important peripheral mechanisms that influence the number, location, size
and strength of terminals on the basis of muscle fiber size
and the total synaptic input to the fiber.

References
Bennett MR, Lavidis NA (1979). The effect of calcium ions

on the secretion of quanta evoked by an impulse at nerve terminal release sites. J Gen Physiol 74:429.

Bennett MR, Pettigrew AG (1975). The formation of synapses in amphibian striated muscle during development. J Physiol 252:203.

Bennett MR, Raftos J (1977). The formation and regression of synapses during the reinnervation of axolotl striated muscles. J Physiol 265:261.

Brown MC, Holland RL, Hopkins WG (1981). Motor nerve sprouting. Ann Rev Neurosci 4:17.

Cass DT, Sutton TJ, Mark RF (1973). Competition between nerves for functional connexions with axolotl muscles. Nature 243:201.

D'Alonzo AJ, Grinnell AD (1982). Uniformity of transmitter release along the length of frog motor nerve terminals. Soc Neurosci Abs 8:492.

Fukunaga H, Engel AG, Osame M, Lambert EH (1982). Paucity and disorganization of presynaptic membrane active zones in the Lambert-Eaton myasthenic syndrome. Muscle & Nerve 5:686.

Grinnell AD, Herrera AA (1980). Physiological regulation of synaptic effectiveness at frog neuromuscular junctions. J Physiol 307:301.

Grinnell AD, Herrera AA (1981). Specificity and plasticity of neuromuscular connections: long-term regulation of motoneuron function. Prog in Neurobiol 17:203.

Grinnell AD, Letinsky MS, Rheuben MB (1979). Competitive interaction between foreign nerves innervating frog skeletal muscle. J Physiol 289:241.

Grinnell AD, Trussell LO (1983). Synaptic strength as a function of motor unit size in the normal frog sartorius. J Physiol (in press).

Haimann C, Mallart A, Tomás I Ferré J, Zilber-Gachelin NF (1981). Interaction between motor axons from two different nerves reinnervating the pectoral muscle of Xenopus laevis. J. Physiol 310:257.

Herrera AA, Grinnell AD (1980). Transmitter release from frog motor nerve terminals depends on motor unit size. Nature 287:649.

Herrera AA, Grinnell AD (1981). Contralateral denervation causes enhanced transmitter release from frog motor nerve terminals. Nature 291:495.

Herrera AA, Grinnell AD (1982). Synaptic apposition, glial obstruction and synaptic strength at frog neuromuscular junctions. Soc Neurosci Abs 8:492.

Jansen JKS, Thompson W, Kuffler DP (1978). The formation and maintenance of synaptic connections as illustrated by studies of the neuromuscular junction. Prog Brain Res 48:3.

Kuffler DP, Thompson W, Jansen JKS (1980). The fate of foreign endplates in cross-innervated rat soleus muscle. Proc R Soc Lond B 208:189.

Kuno M, Turkanis SA, Weakley JN (1971). Correlation between nerve terminal size and transmitter release at the neuromuscular junction of the frog. J Physiol 213:545.

Lømo T (1976). The role of activity in the control of membrane and contractile properties of skeletal muscle. In Thesleff S (ed): "Motor Innervation of Muscle," New York: Academic Press, p 289.

Mallart A, Angaut-Petit D, Zilber-Gachelin NF, Tomás I Ferré J, Haimann C (1980). Synaptic efficacy and turnover of endings in pauci-innervated muscle fibres of *Xenopus laevis*. In Taxi J (ed): "Ontogenesis and Functional Mechanisms of Peripheral Synapses." Amsterdam: Elsevier Press, p 213.

Mark RF (1980). Synaptic repression at neuromuscular junctions. Physiol Rev. 60:355.

Nudell BM, Grinnell AD (1982). Inverse relationship between transmitter release and terminal length in synapses on frog muscle fibers of uniform input resistance. J Neurosci 2:216.

Nudell BM, Grinnell AD (1983). Regulation of synaptic position, size, and strength in anuran skeletal muscle. J Neurosci 3:161.

Peper K, Dreyer F, Sandri C, Akert K, Moor H (1974). Structure and ultrastructure of the frog motor endplate. Cell & Tissue Res 149:437.

Rotshenker S (1979). Synapse formation in intact innervated cutaneous-pectoris muscles of the frog following denervation of the opposite muscle. J Physiol (Lond) 292:535.

Rotshenker S, McMahan UJ (1976). Altered patterns of innervation in frog muscle after denervation. J Neurocytol 5:719.

Rotshenker S (1982). Transneuronal and peripheral mechanisms for the induction of motor neuron sprouting. J Neurosci 2:1359.

Tweedle CD (1983). Ultrastructural changes in the neurohypophysis during increased hormone release. Prog in Brain Res (in press).

Wernig A, Pécot-Dechavassine M, Stover H (1980). Sprouting and regression of the nerve at the frog neuromuscular junction in normal conditions and after prolonged paralysis with curare. J Neurocytol 9:277.

Top: A.D. Grinnell; K. Negishi
Bottom: Satoko Hagiwara; S. Hagiwara; F. Lee

The Physiology of Excitable Cells, pages 451–460
© 1983 Alan R. Liss, Inc., 150 Fifth Avenue, New York, NY 10011

ON THE DIFFERENTIATION OF ELECTRIC ORGANS IN THE ABSENCE OF
CENTRAL CONNECTIONS OR PERIPHERAL INNERVATION

Thomas Szabo, Frank Kirschbaum

Dept. de Neurophysiol. Sensorielle, Lab. Physiol.
Nerveuse, F 91190 Gif sur Yvette, and Zoologisches
Inst., Univ. zu Köln, Weyertal 119, 5 Köln 41 GFR

In the last five or six decades many experiments have
been done to investigate the extent to which the nervous
system is indispensable for the development of an effector
organ. In this respect, the phenomenon of amphibian limb
regeneration is a good model for studying the trophic
function of the neuron. Singer, in 1952, showed that an
amputated limb does not regenerate if the spinal nerves are
transsected at the same time. Similarly, if the limb is
denervated during early stages of regeneration, the regenerating limb breaks down and is resorbed. Sidman and Singer
(1951) showed that the trophic function is apparently not
linked to the function of impulse induction, since regeneration will also proceed in the presence of sensory nerves,
or even implanted sensory ganglia (Kamarin and Singer,
1959).

A quite different result is obtained if denervation is
done during development. Harrison (1904) has shown that the
embryonic development of limbs can occur in the absence of
innervation and that such aneurogenic limbs also regenerate,
despite the absence of a nerve supply (see also Yntema,
1979). However, to date no definitive conclusions have
been drawn concerning the way in which nerves exert
their trophic influence during regeneration (Guth, 1965).

The limb, which was taken as a model in all of these
experiments, is a complex organ. Thus, the possibility of
working on an effector organ without sensory innervation--
such as the electric organ of weakly electric fish--was
very attractive.

The major aim of this study was to determine whether the nervous system exerts a trophic function on electrocyte development. We asked the following questions: 1) Are central connections indispensable for differentiation of motoneurons and/or for the differentiation of the effector organ innervated by them? 2) Does regeneration occur in the teleost spinal cord? 3) Does electric organ (EO) differentiation occur in the absence of its innervation?

The experiments were carried out on the weakly electric fish, Pollimyrus isidori (Fig. 1). This species presents several experimental advantages:

1) It can be bred in our laboratory (Kirschbaum 1975, 1977) and therefore was available at any developmental stage.

2) The adult electric organ (AEO) is innervated by a small pool of motoneurons (AEMN) restricted to two metameric segments of the caudal spinal cord. The motoneuron pool as well as the electric organ is located in the caudal peduncle, far away from the brain.

3) The electromotoneurons (EMN) receive exclusively one type of afferent terminals, originating from the medullary center. They receive no inputs from the spinal cord. The caudal position of the EMN pool permits the interruption by spinal section of the specific descending pathway which activates EMN and EO.

4) The same descending path is also connected to the larval EMNs (LEMN), which activate the electric organ (LEO), and gives information about the functioning of the pacemaker center (P). The larval electric organ discharge (LEOD), to which the adult electric organ discharge (AEOD) is coupled with a fixed time relationship, serves as an indicator for the detection of the first appearance of the AEOD during ontogenetic development and makes control of the presence or absence of the AEOD relatively easy.

5) Finally, the AEO of this fish has a very restricted nerve/electrocyte synaptic junctional area, convenient for localization in electron microscopy, and hence well suited for detection of early synaptic contacts during development.

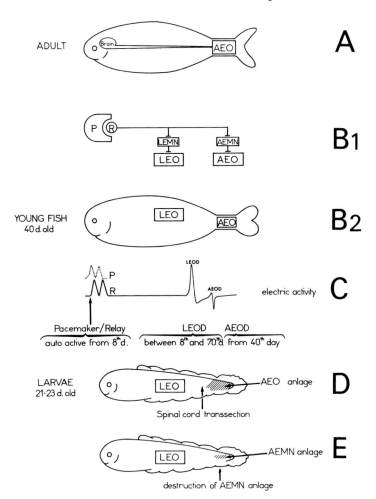

Fig. 1, A and B. Schematic representation of adult (AEO) and larval (LEO) electric organ location in adult (A) and young fish (B2) - B1. Schematic drawing of AEO and LEO innervation: P, pacemaker, R, medullary relay, LEMN and AEMN, larval and adult electromotoneurons. - C. Electric activity in young (40 days old) fish. P and R, "pacemaker" and relay nucleus activity; LEOD, larval and AEOD, adult electric organ discharge. - D and E. 21-23 days old larval fish with adult electric organ (AEO) anlarge (D) and adult electromotoneuron (AEMN) anlage (E). Arrows indicate the level of spinal transsection (D) and that of AEMN anlage destruction (E).

All larvae displayed a LEOD which persisted after the operation. The transsection was made in different fish on day 21, 23 or 27, during the period when the AEO starts to differentiate.

After 2 or 3 months survival, when the fish were a minimum of 16 mm in length, at which size AEO activity was observed in normal development (Westby and Kirschbaum, 1978), the electrical activity was recorded and the animals sacrificed.

Light microscopical observations of the histologically processed larvae with exclusively larval EOD's showed that 1) the EO was "normally" developed: like the control (Fig. 2D and F), it showed a laminated structure with each electrocyte displaying its typical stalk (Fig. 2A); 2) the EMN's as well as electrocyte innervation exhibited normal histological structure (Fig. 2B and C), identical to that found in the control (Fig. 2E and F); 3) the spinal cord at the level of the section was not regenerated.

It may be concluded that the AEO differentiates in the absence of central connections, i.e., in the absence of descending pacemaker activity.

In a second group of fish we tested whether the electrocytes of AEO differentiate in the absence of AEMN's, i.e., AEMN's in the absence of innervation. For these experiments the same developmental stage (23 or 27 days) was chosen. The EMN anlage was destroyed with a fine needle by lateral approach (Fig. 1E). At the penetration point, of course, a small part of the electrocyte-anlage was also destroyed.

After 2 or 3 months survival, the young fish showed only LEOD activity. The animals were sacrificed and processed histologically for light microscopy.

Light microscopical study of these preparations resulted in the following observations:
1. A rather normally constituted adult electric organ with parallel stacks of electrocytes, was developed in the caudal peduncle (Fig. 3A); however, the number of electrocytes was about 25% less than in a normally developed organ.
2. In the absence of any innervation, each electrocyte

possessed a stalk of usual configuration (Fig. 3B)
3. EMNs were absent from the spinal cord.

We conclude that electrocyte differentiation as well
as EO development occurs also in the absence of spinal innervation. In spite of the absence of EMNs, the postsynaptic area, that is to say the stalk, was, however, formed.

In a third group of fish, we totally amputated the
tail at one of three different levels (1, 2 and 3 in Fig. 3)
or totally destroyed the caudal spinal cord after

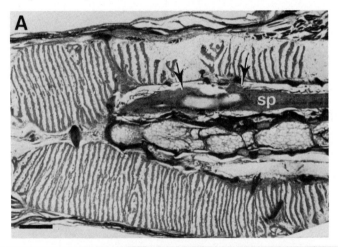

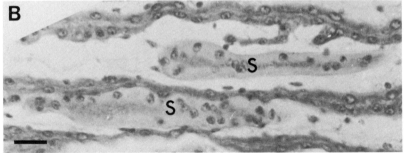

Fig. 3 A. Microphotograph of an adult electric organ where
the EMN anlage (arrows) was destroyed 23 days after spawning. 2 months survival after operation. sp, spinal cord.
B, well developed electrocyte stalk (S) in the EO shown in
Fig. 3A. Note the lack of innervation. Bar, 150 µm in A
and 10 µm in B.

amputation of the caudal fin.

Figure 4 shows that the development depends on the
level of the amputation. A section at level 3 interferes
only with the development of the caudal fin. A section at
level 2, posterior to the EMN pool, does not prevent
either caudal peduncle differentiation or development of
the EO or EMNS. In contrast, a section anterior to the
EMN pool (level 1), prevents differentiation of the caudal
peduncle. A few electrocytes were differentiated anterior
to the section but they were not active in the absence of
EMN.

According to these experiments one can draw the general

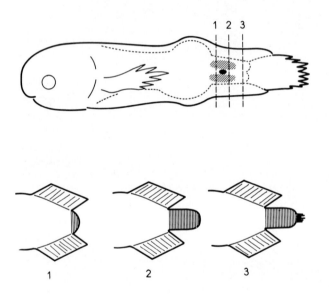

Fig. 4. Above: Schematic drawing showing amputation levels
1,2, and 3 in the 23-27 day old larvae. Dashed area, adult
electric organ anlage. Solid black: AEMN anlage.
Below: Schematic representation of the developed caudal
peduncle after section at level 1,2 and 3. Vertically
dashed area: electric organ.

conclusion that the development of the electric organ
is not impeded by the interruption of descending specific
or unspecific central connections of the EMNs. Our findings also suggest the conclusion that EO differentiation
does not depend on the presence of specific neuronal
elements.

These results raise several questions. Is the silent,
inactive EO functional; that is, are the EMNs and the
electrocytes excitable? Is there synaptic transmission
from EMN to electrocyte? What are the histochemical
properties of the organ?

These operations have been made at a stage of development at which the descending central pathway has already
established connections with the EMNs. In other words, if
these connections provide a signal for EMN differentiation,
it is reasonable to ask if and how this differentiation
would occur in the case of a spinal section made at an
earlier stage of development. Finally, a longer survival
period may answer the question, whether a nonfunctional
AEOD will persist indefinitely, and whether in the absence
of an AEOD, or in the presence of a nonfunctional AEO, the
LEO can permanently substitute for the AEO.

Acknowledgements. This investigation was supported by the
European Science Foundation, the Deutsche Forschungsgemein-
schaft and the Fondation pour la Recerche Médicale
Francaise.

Denizot JP, Kirschbaum F, Westby GWM, Tsuji S (1982). On
the development of the adult electric organ in mormyrid
fish *Pollimyrus isidori* (with special focus on the
innervation). J Neurocytol 11:913.
Guth L (1969). "Trophic" effects of vertebrate neurons.
Neurosc Res Prog Bull 7:1.
Harrison RG (1904). An experimental study of the relation
of the nervous system to the developing musculature in
the embryo of the frog. Amer J Anat 3:197.
Kamrin AA, Singer M (1959). The growth influence of
spinal ganglia implanted into the denervated forelimb
regenerate of the newt, *Triturus*. J Morphol 104:415.
Kirschbaum F (1975). Environmental factors control the

periodical reproduction of tropical electric fish.
Experientia 31:1159.

Kirschbaum F (1977). Electric organ ontogeny: distinct
larval organ precedes the adult organ in weakly electric
fish. Naturwiss 64:387.

Kirschbaum F (1981). Ontogeny of both larval electric or-
gan and electromotoneurones in *Pollimyrus isidori*
(Mormyridae, Teleostei). In Szabo T, Czeh G (eds):
"Sensory physiology of aquatic lower vertebrates,"
London: Pergamon Press , p 129.

Sidman RL, Singer M (1951). Stimulation of forelimb regeneration in the newt, *Triturus viridescens*, by a sensory
nerve supply isolated from the central nervous system.
Amer J Physiol 165:257.

Singer M (1952). The influence of the nerve in regeneration of the amphibian extremity. Quart Rev Biol 27:169.

Westby GWM, Kirschbaum F (1977). Emergence and development
of the electric organ discharge in the mormyrid fish
Pollimyrus isidori. I. The larval discharge. J comp
Physiol 122:251.

Westby GWM, Kirschbaum F (1978). Emergence and development
of the electric organ discharge in the mormyrid fish
Pollimyrus isidori. II. Replacement of the larval by
the adult discharge. J comp Physiol 127:45.

Yntema CL (1979). Regulation in sparsely innervated and
aneurogenic forelimbs of amblystoma larvae. J Exptl
Zool 140:101.

The Physiology of Excitable Cells, pages 461–473
© 1983 Alan R. Liss, Inc., 150 Fifth Avenue, New York, NY 10011

ADRENAL MEDULLARY TUMOR CELLS (CLONE PC-12) CHEMICALLY FUSED
AND GROWN IN CULTURE: A MODEL SYSTEM FOR STUDYING NEURONAL
DEVELOPMENT

Susanne Huttner, Raj Kapur, Kathleen Morrison-
Graham, and Paul O'Lague
Jerry Lewis Neuromuscular Research Center and
Biology Department, University of California
Los Angeles, California 90024

Isolation of nerve growth factor (NGF) has been invaluable in the creation of in vitro model systems for studying
developing sympathetic neurons and their sensitivity to
environmental signals. Such systems have recently provided
evidence for an unsuspected amount of plasticity in the
developmental choices open to neural crest derivatives,
especially sympathetic neurons and adrenal medullary cells.
In addition they have provided information that may be
relevant to early developmental events (for reviews of NGF
and sympathetic neuron development see: Black 1978;
Bradshaw 1978; Greene, Shooter 1980; Thoenen, Barde 1980;
Vinores, Guroff 1980).

One system that is proving useful for studying early
events is a clone of cells (PC-12) isolated by Greene and
Tischler (1976) from a rat pheochromocytoma, an adrenal
medullary tumor. Cells of this clone exhibit a broad spectrum of properties characteristic of sympathetic neurons,
including sensitivity to NGF (Greene, Tischler 1976). However, in contrast to developing sympathetic neurons they do
not depend upon NGF for their survival. In this paper we
describe some morphological and electrophysiological experiments on PC-12 cells that have been chemically fused and
grown in culture in the presence or absence of NGF (for
methods see: O'Lague, Huttner 1980). Fusion produced large
cells (up to 350 µm in diameter) and facilitated electrophysiological recordings. This permitted several novel
experiments such as intracellular recordings from growing
tips of neurite-like processes. Preliminary reports of some
of these experiments have appeared (Huttner 1980; Huttner,

O'Lague 1981, 1982; Kapur et al. 1982; O'Lague, Huttner 1980).

MORPHOLOGY OF FUSED PC-12 CELLS

Following exposure to the general fusigen polyethyelene glycol, individual cells fused forming multinucleate, syncytial cells (O'Lague, Huttner 1980; see also Davidson et al. 1976). These multinucleate cells were easily distinguished from aggregates of single cells. As seen in Figure 1A, a phase-contrast micrograph of a fused cell, the nuclei of fused cells were collectively surrounded by the outline of and phase-halo produced by the giant cell's plasma membrane In the transmission electron micrograph shown in Figure 2 of an

Fig. 1. Phase-contrast micrographs of fused PC-12 cells grown in cell culture. A) Multinucleate PC-12 cell (M). Several unfused cells are present (arrow). B) Multinucleate PC-12 cell (M) three days after plating with NGF. Note large growth cone (arrow).

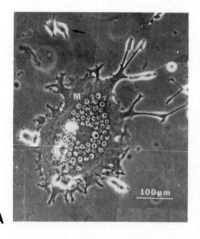
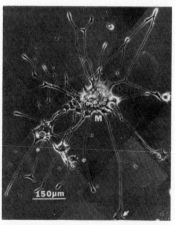

A   B

Fig. 2. (Facing page) Electron micrograph of a multinucleate PC-12 cell (identified as fused in the phase-contrast microscope), three days after plating without NGF. Section was taken 10 μm from the dish surface.

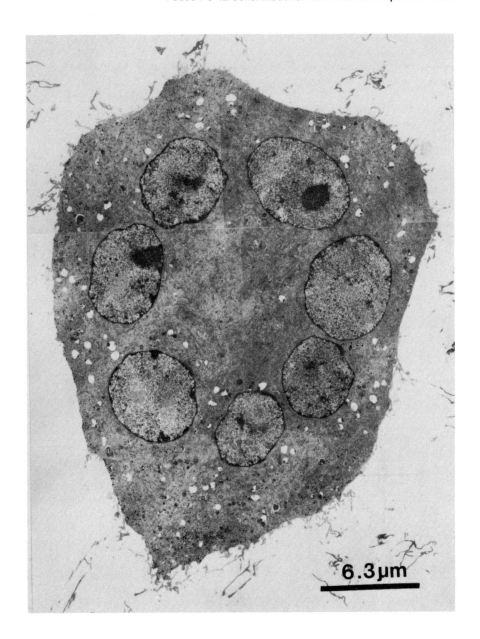

6.3 μm

individual cell, previously identified in the phase-contrast microscope as fused, all nuclei were encased within one plasmalemma. In general the cytoplasmic constitutents of fused and unfused cells were qualitatively similar (Morrison-Graham and O'Lague, in preparation). For example, chromaffin-like granules, a characteristic of unfused PC-12 cells, were abundant in fused cells and ranged in diameter from 40-170 nm in diameter in both groups of cells. When grown in the presence of NGF (7s) fused cells extended neurite-like processes that ended in tips resembling growth cones (Figure 1B, arrow; O'Lague and Huttner, 1980). An unexpected finding was that the growth cones of multinucleate cells were often many times larger than those of unfused cells. This difference allowed us to test directly their excitability and general membrane properties (see below). Ultrastructurally the processes and growth cones appeared similar to both those of unfused cells (Luckenbill-Edds et al, 1979) and to those of sympathetic neurons grown in culture (Bunge, 1973; Landis, 1978). The only striking ultrastructural difference noted was quantitative. For example, the neurite-like processes and growth cones of fused cells contained a larger number of microtubules than those of unfused cells yet their organization within the cytoplasm was similar (Morrison-Graham and O'Lague, in preparation).

We also unexpectedly found that the growth characteristics of PC-12 cells grown with NGF were strikingly different from those of unfused cells. In general fused cells extended more processes per cell and extended them at a greater rate than unfused cells (Huttner, 1980; Kapur et al, 1982). The basis of this difference is unknown, however there was a direct relationship between the number of processes extended, the number of nuclei, and the cell volume which, taken together, suggests effects due to altered surface-to-volume geometry or to altered gene dosage, among other possibilities (Huttner, 1980; Kapur et al, 1982).

ELECTROPHYSIOLOGICAL PROPERTIES OF THE CELL SOMA

Dichter et al (1977) found that PC-12 cells grown in the absence of NGF were electrically inexcitable but, after a two-week incubation with NGF, did exhibit action potentials sensitive to tetrodotoxin (TTX), a Na-channel blocker. In addition, they suggested that all cells, whether grown with or without NGF, exhibited conventional outward (delayed) rectification (see below). Tests for the presence or regulation of this or other ionic conductance mechanisms were not reported.

As a model system, PC-12 cells would be expected to exhibit other ionic mechanisms because rat sympathetic neurons grown in culture are known to possess a rich and complex repertoire of electrophysiological conductance mechanisms, including in addition to Na-dependent action potentials and outward rectification (McAfee, Yarowsky 1979; O'Lague et al. 1978), a Ca-action potential mechanism and Ca-activated K-conductance (McAfee, Yarowsky 1979; O'Lague et al. 1978), an inward (anomalous) rectification conductance (Christ, Nishi 1973), and early in their development an anode-break K-conductance (Pope et al. 1979).

Intracellular recordings with two microelectrodes were made during continuous perfusion from large PC-12 cells grown with or without NGF (Huttner 1980; O'Lague, Huttner 1980). In both growth conditions resting potentials ranged from -50 to -70 mV. Similar to the findings of Dichter et al. (1977) for unfused cells, we found that fused cells responded to NGF with the gradual appearance over several weeks of a Na-dependent, TTX-sensitive action potential mechanism. In addition we detected in both conditions other conductance mechanisms similar to those found in sympathetic neurons (Huttner 1980; O'Lague, Huttner 1980); a detailed manuscript of this and the following is in preparation).

In the presence of external tetraethylammonium (TEA: usually 20mM), a conventional blocker of outward rectification (Hodgkin, Huxley 1952), we could evoke an overshooting action potential (Figure 3) in cells grown with or without NGF. This action potential was identified as a Ca-potential by a variety of criteria (see reviews by Hagiwara 1975; Hagiwara, Byerly 1981) including: 1) it was dependent upon external Ca^{++} (Figure 3A-E): 2) it was blocked by Cd^{++} (Figure 3F-G), Co^{++}, or Mn^{++}; and 3) it was generated in low external Ca^{++} when Sr^{++} or Ba^{++} was present. This action potential, in addition, was insensitive to TTX or to reductions in external Na^+.

The presence of other conductance mechanisms was revealed in steady-state V-I plots by regions of decreased input resistance (Figure 4). The decrease associated with depolarization was produced by two distinct K-dependent conductances.

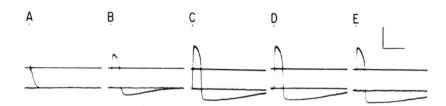

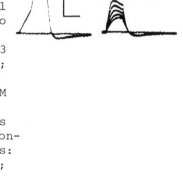

Fig. 3. Dependence of the action potential and afterhyperpolarization evoked in TEA (18.5 mM) on external Ca^{++} and the sensitivity of both to Cd^{++}; recorded in fused PC-12 cell grown without NGF (40 days, A-E; 23 days F-G). A) 0.1 mM Ca, 2.7 mM Mg; B) 0.5 mM Ca, 2.3 mM Mg; C) 1.0 mM Ca, 1.8 mM Mg; D) 2.0 mM Ca, 0.8 mM Mg; E) 2.8 mM Ca; F) 2.8 mM Ca; G) 2.8 mM Ca, 0.75 mM Cd. Mg^{++} was used to maintain divalent cation concentration constant. Vertical bars: A-E= 40 mV, 2 nA; F-G= 20 mV, 2 nA; Horizontal bar: 100 msec.

Application of TEA significantly increased the resistance in this region, an indication of outward rectification. TEA plus Co^{++} further increased membrane resistance towards linearity. Mn^{++}, Mg^{++}, or in normal external Ca^{++}, Cd^{++} mimicked the effect of Co^{++}. These results indicated the presence of a Ca-activated K-conductance mechanism (Meech, 1978). Both conductances were dependent upon external K^{+} in a manner predicted by the Nernst relationship (Huttner, 1980). It is unlikely that the effects of divalent cations could be explained by shifts in surface charge because other voltage-dependent conductances were not affected (see below).

Further evidence for a Ca-activated K-conductance was obtained in studies of the long after-hyperpolarization which followed the Ca-action potential. This after-hyperpolarization had a reversal potential of approximately -85 mV (a value

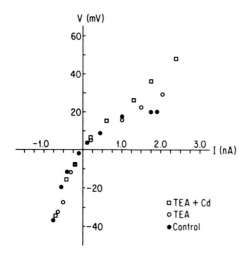

Fig. 4. Steady-state voltage-current (V-I) characteristics of a fused PC-12 cell (three days after plating without NGF); external Ca^{++} was reduced to 0.1 mM and Mg^{++} was raised to 2.8 mM to eliminate evoked action potentials. Control, closed circles; 20 mM TEA, open circles; 20 mM TEA plus 2.8 mM Co^{++} (instead of Mg^{++}), squares.

close to that of sympathetic neurons (McAfee and Yarowsky, 1979). Its duration depended upon external Ca^{++} in the same manner as the amplitude of the Ca-action potential, that is, both increased with increasing Ca^{++} (compare Figures 3B and 3C).

Another conductance mechanism present was associated with the hyperpolarizing region of the steady-state V-I curve (Figure 4). Generally a region of decreased input resistance occurred at potentials more hyperpolarized than about -85 mV with respect to zero potential and depended upon external K^{+} (Huttner, 1980). This conductance was insensitive to Co^{++}, Mn^{++}, or TEA (at concentrations which blocked the outward rectification). Voltage responses associated with this conductance had no reversal potential. This behavior resembled inward (anomalous) rectification observed in other cells (e.g., Adrian, 1969; Hagiwara and Yoshii, 1979).

In addition to those mentioned above, we observed yet another conductance which could be distinguished from the others on the basis of its voltage-dependence, pharmacology, and time course (Huttner, 1980; O'Lague and Huttner, 1980). Upon termination of a relatively large amplitude (usually greater than 30 mV with respect to rest) hyperpolarization, the membrane potential did not return to the original resting level with the time constant of the membrane. Instead it gradually decayed to rest over a period ranging from 100 msec to a few seconds (Figure 5). This hyperpolarization was accompanied by a decreased membrane resistance (Figure 5), was dependent upon external K^+, had a reversal potential of approximately -85 mV, and was insensitive to TEA (at concentrations which blocked the outward rectification conductance) and to Ca-channel blockers. Taken together these characteristics separate this mechanism from the other three. This conductance closely resembles the anode-break conductance seen in a wide variety of neurons (Connor, Stevens 1971; Hagiwara, Saito 1959; Neher 1971) including sympathetic neurons (Pope et al. 1979).

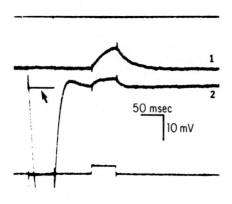

Fig. 5. The anode-break hyperpolarizing response of a fused PC-12 cell, two days in culture without NGF. Top trace: two superimposed current traces, one evoking the hyperpolarizing response (trace 2), the other, control (trace 1). Voltage responses to square pulses of depolarizing current (both traces 1 and 2) are shown with (trace 2) or without (trace 1) the preceding hyperpolarizing current pulse (arrow).

The presence of a spectrum of channels is characteristic of neurons in general (e.g., Adams et al, 1980) and developmental changes have been observed in channel populations (e.g., Spitzer, 1979). However little is known about their development or about factors that regulate their expression. In the PC-12 cells NGF clearly plays an important role in the appearance of Na-action potential channels but all of the other conductance mechanisms are present in the presence or absence of NGF. Possible quantitative changes with NGF were not studied.

ELECTROPHYSIOLOGICAL PROPERTIES OF THE GROWTH CONE

The morphology and ultrastructure of growth cones has been described at length (e.g., Johnston and Wessels, 1980;

Pfenninger, 1978) yet until recently little was known about
their physiology. This lack of information at the physiological level is undoubtedly due to the general inaccessibility
caused by their small size. However advances in the use of
voltage-sensitive dyes enabled Grinvald and colleagues
Anglister et al, 1982; Grinvald and Farber, 1981; Grinvald
et al, 1981) to carry out an elegant series of in vitro
experiments in which they found strong evidence for the
presence of active Ca-currents in the growth cones of neuroblastoma cells. In our studies we have exploited the large
size of growth cones of fused PC-12 cells to test the membrane
properties directly by making intracellular recordings
(Huttner and O'Lague, 1982, and manuscript in preparation).

Although the growth cones were quite fragile we obtained
resting potentials ranging from -50 to -65 mV with a single
microelectrode. The membranes of the growth cones were
inexcitable in control solutions (Figure 6a) and the shape
of voltage responses to injected current resembled that seen
in the cell body (see above). Excitability at the growth
cone could be unmasked, as in the case of the cell body, by
blocking the outward rectification with TEA. We applied TEA
in a localized and controlled fashion using a micropipette
(5-10 μm in diameter and filled with 20 mM TEA). We found
that action potentials resembling the Ca-action potentials
recorded in the cell body could be evoked by depolarizing
current passed into the growth cone. This occurred only when
the TEA pipette was immediately above the growth cone (Figure
6b). These potentials were greatly attenuated when the pipette was moved more than approximately 10 μm away. Similar
to the Ca-action potentials of the cell body, these action
potentials were insensitive to TTX but were blocked by Cd^{++}.
In addition, we found that the growth cone action potentials
were followed by a long-lasting after-hyperpolarization which
resembled that produced by the Ca-activated K-conductance in
the cell body (see above). Taken together these results indicate that the growth cone membrane contains several conductance
mechanisms similar to those seen in recordings from the cell
body.

For the growth cone experiments reported here we chose
cells grown in NGF for fewer than five days because beyond
this time processes formed complex networks. At that early
time Na-action potentials are usually not seen in the cell
body. In the small number of recordings that we made we found
no evidence of a Na-action potential mechanism, yet in all

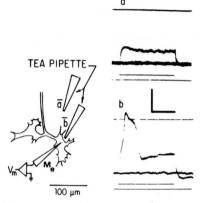

Fig. 6. Intracellular recording from a growth cone. Left: Schematic diagram showing the recording arrangement. Intracellular microelectrode (M_e) for passing current and recording. Positions of the micropipette for locally applying TEA (20mM), one directly above the growth cone (b) and another at a distance (a). Right: Voltage responses (middle traces in a and b) to current pulses (lower traces), when (a) the TEA pipette is at a̅ and (b) when the pipette is at b̅. Upper traces indicate zero potential. Note middle trace in (b) is a superposition of responses with and without the current pulse. Bars: Horizontal= 50 msec; Vertical= 30 mV, 2.5 nA.

cases the cells were grown in NGF (see above). It will be of interest to determine whether the growth cone acquires a Na-action potential at a time when the cell body exhibits one.

In summary, we have attempted to demonstrate further the usefulness of PC-12 cells as a model system for studying aspects of neuronal differentiation. The fused PC-12 cells have enabled us to characterize a wide range of electrophysiological properties in various parts of individual cells. Similar to postnatal sympathetic neurons, PC-12 cells exhibit a spectrum of ionic mechanisms for K^+, Na^+, and Ca^{++}, as well as other neuron-like properties. Some of these properties are dependent upon NGF, a presumed trophic factor for sympathetic neurons. We have also found evidence for the presence of Ca-channels within or very near to the growing membrane of growth cones, a finding consistent with that of Grinvald and Farber (1981). Uncovering possible roles for these channels in the regulation of neurite extension and other developmentally related functions associated with growth cones now appears feasible.

ACKNOWLEDGMENTS

This work was supported by NINCDS (NS 12901) and UCLA MDA
Center Grant (Project 18); SLH was supported by an MDA post-
doctoral fellowship and an NIH training grant fellowship;
KMG was supported by a Giannini Foundation fellowship.
Expert technical assistance was supplied by Puntipa Kwanyuen
and John Watson (tissue culture) and Herman Kabe (photography).

REFERENCES

Adams, DJ, Smith, SJ, and Thompson, SH (1980). Ionic currents
  in molluscan soma. Ann Rev Neurosci 3:141.
Adrian, RH (1969). Rectification in muscle membrane. Prog
  Biophys Molec Biol 19:341.
Anglister, L, Farber, IC, Shahar, A, and Grinvald, A (1982).
  Localization of voltage-sensitive calcium channels along
  developing neurites: Their possible role in regulating
  neurite elongation. Dev Biol 94:351.
Black, IB (1978). Regulation of autonomic development. Ann
  Rev Neurosci 1:183.
Bradshaw, RA (1978). Nerve growth factor. Ann Rev Biochem
  47.
Bunge, MB (1973). Fine structure of nerve fibers and growth
  cones of isolated sympathetic neurons in culture. J Cell
  Biol 56:713.
Christ, DD, and Nishi, S (1973). Anomalous rectification of
  mammalian sympathetic ganglion cells. Exper Neurol 40:806.
Connor, JA, and Stevens, CF (1971). Voltage clamp studies of
  a transient outward current in gastropod neural somata.
  J Physiol 213:21.
Davidson, RL, O'Malley, KA, and Wheeler, TB (1976). Poly-
  ethylene glycol-induced mammalian cell hybridization:
  Effect of polyethylene glycol molecular weight and concen-
  tration. Som Cell Genet 2:271.
Dichter, MA, Tischler, AS, and Greene, LA (1977). Nerve
  growth factor-induced increase in electrical excitability
  and acetylcholine sensitivity of a rat pheochromocytoma
  cell line. Nature 268:561.
Greene, LA, and Shooter, EM (1980). The nerve growth factor:
  Biochemistry, synthesis, and mechanism of action. Ann
  Rev Neurosci 3:353.
Greene, LA, and Tischler, AS (1976). Establishment of a
  noradrenergic clonal line of rat pheochromocytoma cells
  which respond to nerve growth factor. Proc Natl Acad Sci
  73:2424.

Grinvald, A, and Farber, IC (1981). Optical recording of calcium action potentials from growth cones of cultured neurons with a laser microbeam. Science 212:1164.

Grinvald, A, Ross, WN, and Farber, I (1981). Simultaneous optical measurements of electrical activity from multiple sites on processes of cultured neurons. Proc Natl Acad Sci 78:3245.

Hagiwara, S (1975). Ca-dependent action potential. In Eisenmann, G (ed): "Membranes: A Series of Advances", Vol 3, New York: Dekker, p. 359.

Hagiwara S, and Byerly, L (1981). Calcium channel. Ann Rev Neurosci 4:69.

Hagiwara, S, and Saito, N (1959). Voltage-current relations in nerve cell membranes of onchidium verruculatum. J Physiol 148:161.

Hagiwara, S, and Yoshi, M (1979). Effects of internal potassium and sodium on the anomalous rectification of the starfish egg as examined by internal perfusion. J Physiol 292:251.

Hodgkin, AL, and Huxley, AF (1952). A quantitative description of membrane current and its application to conduction and excitation in nerve. J Physiol 117:500.

Huttner, SL (1980). An in vitro study of the effects of nerve growth factor on the electrophysiological and morphological properties of giant cells produced from the neuron-like clone PC-12 by chemically-induced cell fusion. Doctoral dissertation, UCLA.

Huttner, SL, and O'Lague, PH (1981). Electrophysiological studies of giant pheochromocytoma (PC-12) cells grown in cell culture. Intern Biophys Cong Abstr M-H-6.

Huttner, SL, and O'Lague, PH (1982). Excitability of growth cones of multinucleate PC-12 cells. Soc Neurosci Abstr 33.9.

Johnston, RN, and Wessels, NK (1980). Regulation of the elongating nerve fiber. Curr Topics Dev Biol 16:165.

Kapur, R, Huttner, S, and O'Lague, P (1982). Morphological and growth characteristics of cells from the neuron-like clone PC-12: Chemically induced cell fusion and effects of nerve growth factor. Soc Neurosci Abstr 83.5.

Landis, SC (1978). Growth cones of cultured sympathetic neurons contain adrenergic vesicles. J Cell Biol 78:R8.

Luckenbill-Edds, L, Van Horn, C, and Greene, LA (1979). Fine structure of initial outgrowth of processes induced in a pheochromocytoma cell line (PC-12) by nerve growth factor. J Neurocytol 8:493.

McAfee, DA, Yarowsky, PJ (1979). Calcium-dependent potentials in the mammalian sympathetic neuron. J Physiol 290:507.

Meech, RW (1978). Calcium-dependent potassium activation in nervous tissues. Ann Rev Biophys Bioeng 7:1.

Neher, E (1971). Two fast transient current components during voltage clamp on snail neurones. J gen Physiol 58:36.

O'Lague, PH, Potter, DD, and Furshpan, EJ (1978). Studies on rat sympathetic neurons developing in cell culture. Dev Biol 67:384.

O'Lague, PH, and Huttner, SL (1980). Physiological and morphological studies of rat pheochromocytoma cells (PC12) chemically fused and grown in culture. Proc Natl Acad Sci 77:1701.

Patterson, PH (1978). Environmental determinants of autonomic neurotransmitter functions. Ann Rev Neurosci 1:1.

Pfenninger, KH (1978). Organization of neuronal membranes. Ann Rev Neurosci 1:445.

Pope, B, Grinnell, AD, and O'Lague, PH (1979). Embryonic chick sympathetic neurons grown in culture have action potential mechanisms for Na, Ca, and K and several other K conductance mechanisms. Soc Neurosci Abstr 63.3.

Spitzer, NC (1979). Ion channels in development. Ann Rev Neurosci 2:363.

Thoenen, H, and Barde, YA (1980). Physiology of nerve growth factor. Physiol Rev 60:1284.

Vinores, S, and Guroff, G (1980). Nerve growth factor: mechanism of action. Ann Rev Biophys Bioeng 9:223.

P. O'Lague; L.A. Jaffe

The Physiology of Excitable Cells, pages 475–489

STEADY TRANSCELLULAR ION CURRENTS

Richard Nuccitelli

Department of Zoology
University of California
Davis, CA 95616

In this volume on the physiology of excitable cells
voltage-sensitive ion channels and the action potentials
resulting from them are at center stage, and little attention has been given to the passive ion fluxes which con-
stantly flow through these membrane ion channels. Decades
ago, tracer studies revealed these steady ion fluxes through
the plasma membrane, but it is only in the past several
years that we have been able to measure the small extracellular voltages generated by this ion flow which allows us
to map the spatial distribution of these ion movements.
These studies have led to the somewhat surprising conclusion
that cells do not generally display a uniform distribution
of ion leaks, but instead separate active channel types
resulting in distinct inward and outward current regions.
This generates a steady transcellular ion current which is
usually closely correlated with cellular function. These
steady currents reflect the distribution of ion channels
and can influence cellular function by perturbing intracellular ion concentrations and generating both intracellular and extracellular voltage gradients.

I will first briefly describe the measurement technique
and provide an overview of the transcellular current patterns
observed thus far, and will then discuss specific examples
in which such currents are playing a causal role in a
specific cellular function.

THE VIBRATING PROBE TECHNIQUE FOR MEASURING STEADY TRANS-
CELLULAR CURRENTS

Such ion currents must flow through the extracellular
medium of fixed resistivity and in so doing will generate a
small voltage gradient. By measuring this voltage gradient,
one could directly calculate the current density generating
it. This can be seen clearly from Ohm's Law:

$$I = E/\rho = -(1/\rho) \; \vec{\nabla} \; V = -(1/\rho) \; \frac{\partial V}{\partial r} \; \hat{a}_r \simeq -(1/\rho) \; \frac{\Delta V}{\Delta r} \; \hat{a}_r$$

where I is current density, E is the electric field strength,
ρ is the medium resistivity, and ΔV is the voltage gradient
measured over the small distance, Δr, and $\hat{a}_r$ is the unit
vector in the radial direction. The smaller the Δr used,
the better the approximation to the true current density.
Therefore, the challenge is to measure extracellular voltage gradients generated by an ion flux of less than 1 $\mu A/cm^2$
over very small distances. The magnitude of this challenge
can be appreciated when the calculation of the expected
voltage gradient generated over a 20 μm displacement along a
radial vector beginning 25 μm outside the cell's plasma
membrane is made, resulting in a value of 6 nV in seawater
and 30 nV in serum. Standard 3M KCl-filled microelectrodes
have resistances on the order of 10^6 ohms and a resolution
limited by their noise values of 10^{-5} volts. They are
therefore not sensitive enough to detect these steady
voltages of 10^{-8} volts. Lionel Jaffe and I overcame this
difficulty nine years ago by developing a much more sensitive electrode system called the vibrating probe (Jaffe and
Nuccitelli, 1974). This was accomplished by lowering the
electrode's resistance 1000-fold (by filling the glass
microelectrode with metal and plating a platinum sphere at
its tip) and by signal averaging and filtering using a
phase-sensitive lock-in amplifier. The probe impedance is
on the order of 10^3 ohms and the system's sensitivity is 10^{-8}
volts. These changes resulted in a 100- to 1000-fold im-
provement in resolution over the standard microelectrode.
This is quite comparable to the increase in resolution
obtained when one goes from the light microscope to the
electron microscope, so that now we can detect the nanovolt
gradients generated by the steady ion fluxes flowing through
single cells. Moreover, the spatial resolution of the
vibrating probe technique is about 10 μm so that the spatial
distribution of these membrane ion fluxes can now be revealed.

This instrument is now commercially available (Vibrating Probe Co., Davis, CA) and is being used in fifteen laboratories around the world. The main difference between this technique and the popular patch-clamp technique is temporal resolution. The probe averages over time intervals on the order of seconds in order to obtain the desired sensitivity of 10^{-8} volts, while the patch clamp can detect events on the order of milliseconds in length. Moreover, the probe detects the natural slowly-varying currents crossing the membrane in specific regions whereas the patch-clamp fixes the voltage outside the patch of membrane under study and measures the current necessary to do this. Nevertheless, both the vibrating probe technique and classical voltage clamp techniques can provide similar insights into membrane channels. A recent example of this is based on studies of Ca^{2+}-activated Cl^- channel in the frog egg. Miledi (1982) recently concluded from voltage clamp studies that there is a Cl^- channel in the Xenopus egg which is opened by Ca^{2+} influx which occurs when the membrane potential is more positive than -20 mV. Exactly the same type of Ca^{2+}-activated Cl^- channel was described by Robinson (1979) in the Xenopus oöcyte using the vibrating probe, and in addition the spatial distribution of the active channels was revealed by the probe technique.

TRANSCELLULAR ION CURRENTS ARE OBSERVED IN CELLS THROUGH-OUT THE PLANT AND ANIMAL KINGDOMS

During the past nine years, a wide variety of cell types in both plants and animals have been studied using the vibrating probe. In every case, steady transcellular ion currents have been found, usually closely correlated with the axis of polarity. It would appear, therefore, that most cells do not have a uniform distribution of ion channels and pumps, but segregate channel types to varying degrees. Figure 1 summarizes most of the single-cell studies to date.

Plant Cells

Water mold. One of the lowest plant cells to be studied is the water mold, Blastocladiella emersonii (Stump, et al, 1980). In growing cells, positive current on the order of 1 μA/cm^2 enters the rhizoid and leaves from the thallus. There is some evidence that H^+ carries part of this current.

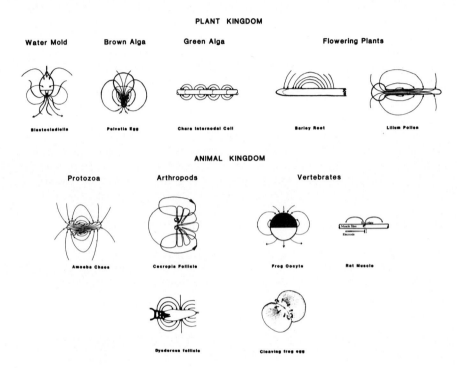

Fig. 1. Transcellular ion current patterns measured in plant and animal cells.

Sporulation is associated with a reversal of this current pattern, and positive current begins entering the thallus region from which the spores will be released. This cell, therefore, drives a transcellular current through itself along its axis of polarity, and reverses its direction according to the stage of the life cycle.

  Pelvetia egg. The egg of the brown alga, Pelvetia fastigiata, also drives a current along its axis of polarity (Nuccitelli and Jaffe, 1974, 1975, 1976; Nuccitelli, 1978) and evidence will be presented below that this current, which is carried by Ca^{2+} influx and Cl^- efflux, is a controlling factor in axis determination.

Chara internodal cells. The freshwater green alga, _Chara corallina_, takes in HCO_3^- as its carbon source, fixes the CO_2 and sends out the remaining OH^-. The regions of HCO_3^- influx (outward positive current) and OH^- efflux (inward positive current) are separated to form neighboring bands of current pictured in Fig. 1 (Lucas and Nuccitelli, 1980). Thus, even in this cell with very little growth, the ion channels or transporters are spatially separated. Both the inward and outward currents are light dependent, falling off within seconds of light removal, and the magnitude of these transport currents is much larger than that of the elongation currents, reaching peaks of 80 $\mu A/cm^2$.

Barley roots. Growing roots and root hairs of barley seedlings, _Hordeum vulgare_ L., drive steady currents through themselves. Current of about 2 $\mu A/cm^2$ in magnitude enters both the main elongation zone of the root and the growing tips of elongating root hairs (Weinsenseel, et al, 1979). Both the inward and outward currents appear to be carried largely by H^+, OH^- or HCO_3^-.

Easter Lily pollen grain. The higher flowering plant, _Lilium longiflorum_, exhibits transcellular currents in its germinating pollen grain (Weinsenseel, et al, 1975; Weinsenseel and Jaffe, 1976). Positive current of about 4 $\mu A/cm^2$ enters the ungerminated grain's prospective growth site and leaves its opposite end. After the grain germinates to form the pollen tube, current enters along most of the tube and leaves the grain as diagrammed in Fig. 1. This is another example of a tip-growing plant cell which drives an ion current into the region of vesicle secretion and along its axis of polarity.

Animal Cells

Transcellular ion currents have been measured in a wide range of animal cells as well. As in most of the plant cells mentioned, the current pattern is closely correlated with the axis of polarity.

Amoebae. The amoeba, _Chaos chaos_, drives a steady current of 0.1 $\mu A/cm^2$ into its tail or uroid zone and out its pseudopods (Nuccitelli, et al, 1977). This current pattern changes as the amoeba's morphology changes, and there is a striking correlation between the region of highest inward current density and tail formation. Spontaneous cytoplasmic streaming polarity reversals occur frequently in Chaos, and these are always preceded by a change in the transcellular current such that the region of largest inward current becomes the new tail. Ca^{2+} influx appears to carry some of the inward current.

Silk moth follicle. The nurse cell-oöcyte complex of the silk moth, Hyalophora cecropia, drives an ion current into the nurse cell end and out the oöcyte end (Woodruff and Telfer, 1974; Jaffe and Woodruff, 1979). This current is important for the polarized transport of protein along the cytoplasmic bridge connecting the nurse cells with the oöcyte (Woodruff and Telfer, 1973), and there is strong evidence that this current drives the electrophoretic transport of protein from nurse cells to oöcyte.

Cotton bug ovariole. The morphology of the ovariole of the cotton bug, Dysdercus intermedius, is quite different from the silk moth ovariole because nurse strands between the trophic syncytium and oöcytes can measure several millimeters in length. The basic current pattern, however, is analogous to that found in Cecropia with current entering over the tropharium and leaving the small follicles (Dittman, et al, 1981). This current is also about an order of magnitude larger than that measured in Cecropia. In another telotrophic hemipteran, Rhodnius prolixus, an intercellular voltage gradient of up to 10 mV with the oöcyte more positive has been measured (Telfer, et al, 1981). The magnitude of this gradient is very sensitive to juvenile hormone. Therefore, in these three insect systems, the transcellular current pattern is strongly correlated with the axis of polarity and may be directly involved in the polarized transport of protein between cells.

Frog oöcyte. The immature oöcyte of the frog, Xenopus laevis drives 1 $\mu A/cm^2$ into the animal hemisphere and out of the vegetal hemisphere prior to complete maturation (Robinson, 1979). Here, again, the current pattern is along the main axis of polarity and the inward current exhibits Ca^{2+} influx and Cl^- efflux components. This current pattern is found in all oöcytes; however, when progesterone or any of several other maturation-producing agents are applied, the current decreases to nearly zero. Thus, the mature egg appears to be in a more quiescent state waiting for fertilization to reactivate it.

Cleaving frog egg. The mature, unfertilized Xenopus egg has no detectable transcellular current until fertilization triggers an inward current pulse which enters the egg at the site of the sperm-egg fusion and spreads over the egg in a ring-shaped wave during about 3 min. after fertilization. The fertilized egg then becomes quiescent until first cleavage. About 10 min. after first cleavage has begun, new, unpigmented membrane begins to appear in the cleavage furrow. At this same time, outward current is measured near this unpigmented membrane (Kline, et al, 1983). From this point on through at least the 32-cell stage, each blastomere drives a current through itself, into the old membrane and out of the newly-inserted membrane. These blastomeres form an epithelium with the blastocoel on one side and the external medium on the other side, and the outward current region forms the basolateral margins of this epithelial layer. Therefore, an apical-basal transcellular current is associated with the apical-basal polarity of these epithelial cells. Lionel Jaffe (1981b) has proposed a model of epithelial cell polarity control by transcellular ion currents.

Mammalian muscle fiber. The rat skeletal muscle fiber drives a steady current of 10 $\mu A/cm^2$ out of the end plate region and into either side of the end plate (Betz, et al, 1980). This current was initially thought to represent the electrogenic component of the Na^+-K^+ ATPase, but more recent studies suggest that it is mainly due to a Cl^- leak (Betz and Caldwell, personal communication). When acetylcholine is added extracellularly, this outward end plate current immediately reverses and increases dramatically as Na^+ rushes in through the open acetylcholine receptor ion channels. Therefore, even in the fully differentiated muscle fiber, there is a segregation of ion channels leading to a steady transcellular current which may act as a guiding factor in muscle innervation.

These investigations of the steady ion current patterns mainly around single cells have led to two generalizations:

1. Active ion channels are generally not uniformly distributed in the cell's plasma membrane, but are spatially separated resulting in transcellular ion currents. 2. These ion current patterns are often closely correlated with the axis of cell polarity in both plant and animal cells. Based on these steady ion current measurements made in a

wide variety of cell types from both lower and higher plants
and animals, it is clear that these transcellular ion
currents are signals of cell polarity. However, the far
more interesting question is whether these currents can
affect, influence or control cell polarity. Does the plasma
membrane directly influence cell function by separating ion
channel types and driving these ion currents through cells?

DO THESE TRANSCELLULAR ION CURRENTS PLAY A CAUSAL ROLE IN
THE CONTROL OF CELL POLARITY?

Transcellular ion currents can have the most direct
effect on the cell by either changing local ion concentrations or by generating voltage gradients. Furthermore,
these effects could act on the inside or outside of the cell's
plasma membrane. The remainder of this review will concentrate on a few well-studied cases which indicate that
the transcellular currents can affect cell polarity. See
also Jaffe (1979; 1981 a,b) and Jaffe and Nuccitelli (1977).

Intracellular Ion Concentration Gradients

Tip-growing plant cells provide the strongest evidence
that steady transcellular currents can generate ion concentration gradients which, in turn, can influence the
cell's polarity. The most extensively studied system is the
germinating egg of the brown alga, Pelvetia, and this current
pattern is quite similar to that found in the germinating
pollen grain of the higher plant, the Easter Lily.

The Pelvetia egg has no predetermined axis of polarity
and can germinate a rhizoid at any point on its surface.
This process involves the localized secretion of wall-
softening enzymes and the increase in turgor pressure to 5
atmospheres which is accomplished by pumping in K^+ and Cl^-.
The cell then bulges (germinates) at the weakest region
where wall softening has occurred and the originally apolar
zygote becomes differentiated into two different regions
and then cells. The establishment of this axis of secretion
can be influenced by a variety of environmental vectors
including light, temperature, and pH (Jaffe, 1968, Table I).
The response to unilateral light is to germinate on the side
opposite to the light. This would be useful in nature since
the rhizoid outgrowth will form the holdfast for the plant
and should form opposite the sun. One can use this light

response to orient the germination site while studying the transcellular ion current pattern associated with it using the vibrating probe as shown in Fig. 2a (Nuccitelli, 1978).

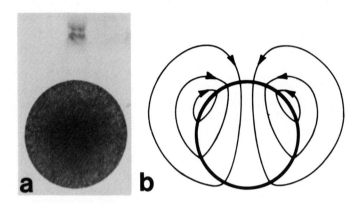

Fig. 2. Current measurements around the Pelvetia egg. a) Photomicrograph of the vibrating probe near a 3-hr-old Pelvetia egg. The egg's diameter is 100 μm. b) The electrical current pattern generated by the germinating egg just before germination, which will occur at the top of the figure.

Currents are detected around this egg as early as 30 min. after fertilization and tend to enter at the dark hemisphere. The early spatial current pattern is unstable and shifts position, often with more than one inward current region. However, current enters mainly on the side where germination will occur and is usually largest at the prospective cortical clearing region where the rhizoid forms. The current pattern observed during the two-hour period prior to germination is more stable and is shown in Fig. 2b. The site of inward current always predicts the germination site, even when the axis is reversed by light direction reversal.

In order to understand how this current might be affecting the egg, one must first know which ions are carrying the current. Therefore, these current measurements were next carried out in artificial sea waters in which various ion substitutions were made. These studies in-dicated that Ca^{2+} influx and Cl^- efflux carried the inward current with K^+ efflux as the probable outward current carrier. Robinson and Jaffe (1975) directly measured the

$^{45}Ca^{2+}$ tracer flux through the egg by separating the rhizoid and thallus ends by fitting eggs tightly into holes in a nickel screen, and orienting them with unilateral light. By measuring both $^{45}Ca^{2+}$ influx and efflux at either end they found that 2 picoamps of Ca^{2+} current traversed the egg. This means that about 10% of the total transcellular current is carried by Ca^{2+} influx at this stage. While only a small fraction of the current is due to Ca^{2+} influx, the subsequent intracellular free ion concentration change is certainly much larger for Ca^{2+} than for the major current carrier, Cl^-. This is because the free Ca^{2+} concentration in all cells is extremely low (about 10^{-7} M) compared to Cl^-_i, so that even a small Ca^{2+} influx will result in a very large concentration gradient.

Up to this point, we have a plasma membrane-driven current carried partly by Ca^{2+} entering the prospective germination region to generate a Ca^{2+} gradient within the cytoplasm. If such a Ca^{2+} gradient were important for the establishment of an axis of polarity, manipulations of the gradient should affect cell polarization. The best test for causality is to impose a Ca^{2+} gradient on the ungerminated embryo. This critical experiment was done by Robinson and Cone (1980) using a gradient of the calcium ionophore, A23187. They found that 60 to 80 percent of the eggs lying within one egg diameter of a glass fiber coated with ionophore formed a rhizoid on the hemisphere nearer the fiber. This is strong evidence that imposed Ca^{2+} gradients can orient the axis of polarity and supports the hypothesis that the natural transcellular Ca^{2+} current is an effector of cell polarity in the Pelvetia egg. Moreover, a wide variety of other tip-growing plant cells exhibit high Ca^{2+} levels at their growing tips (L.A. Jaffe, et al, 1975; Reiss and Herth, 1978, 1979) and those which have been studied all drive an ion current into that region (Weisenseel, et al, 1975, 1979).

Intracellular Voltage Gradients

The second way that these steady transcellular currents might influence cellular function is through the intracellular voltage gradient which they generate. The best example of this phenomenon is found in the developing follicle of the silk moth which exhibits polarized transport of protein from the nurse cells to the oöcyte across cytoplasmic bridges. These cells drive a steady current through

the bridge (Jaffe and Woodruff, 1979) to generate a 5-10 mV
difference between the nurse cells and oöcyte which electro-
phoreses negatively charged proteins and RNA into the oöcyte.
Woodruff and Telfer (1973, 1974, 1980) have elegantly
demonstrated that protein transport across the cytoplasmic
bridge is strictly a function of the protein's net charge.
The positively charged protein, lysozyme, can be made to
reverse its transport direction by simply reversing its
charge by methylcarboxylation, and uncharged proteins (at
pH 7) will move in either direction across the bridge.
This is the strongest case thus for intracellular transport
by electrophoresis, and an excellent example of a causal
role for transcellular ion currents.

Another example is the steady dark current of 70 pA
measured in retinal rods by Hagins et al, (1970). They
estimate that this current might generate a 2 mV cytoplasmic
voltage drop across the neck region of the rod.

As a third example, Zeuthen (1977) has measured an
intracellular voltage gradient of 0.6 mV/mm in epithelial
cells of Necterus gall bladder. This might be generated by
the transcellular Na^+ current which enters the mucosal side
and leaves the basal side of these cells.

Extracellular Voltage Gradients

In some cases cells can naturally find themselves in
steady electric fields generated by transcellular currents
in other cells. The best example of this is the epithelial
cell sheet (such as frog skin) which has been well studied
in many systems and is known to generate substantial voltages of up to 100 mV across itself. This voltage will
drive current out of regions of low resistance where a cut
in the cell layer has occurred (Barker, et al, 1982) or
where active cell growth is occurring (Jaffe and Stern,
1979). Cells near these regions of current flow may well
experience electric fields on the order of 100 mV/mm (Barker,
et al, 1982). Fields of this magnitude can influence the
distribution of membrane proteins by lateral electrophoresis
(Jaffe, 1977; Poo and Robinson, 1977; Poo, 1981), and can
perturb the transmembrane voltage. They can also influence
macrophage cell motility (Orida and Feldman, 1982). Carol
Erickson and I have recently been investigating the possibility that such fields might initiate or direct embryonic cell

movements by placing quail somitic fibroblasts on glass coverslips in steady electric fields. We have observed a striking sensitivity to fields as small as a few millivolts per cell width. Cells begin to orient with their long axes perpendicular to the field lines (along isopotential lines), and then move toward the negative pole by crawling sideways (Fig. 3).

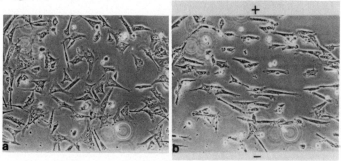

Fig. 3. Quail somitic fibroblasts before and after exposure to electric field. (a) Cultured on glass in F12 medium for 48 hours. (b) Same cells after a 90 min exposure to a 600 mV/mm field.

This response is most striking at field strengths of about 600 mV/mm (12 mV per cell width), but can be detected in fields as low as 50 mV/mm (1 mV/cell width).

In summary, it is clear that extracellular voltage gradients can influence embryonic cell motility, and others have reported directed neurite outgrowth in similar small fields (Jaffe and Poo, 1979; Patel and Poo, 1982; Hinkle, et al, 1981; Freeman et al, 1981). Therefore it is possible that transcellular ion currents may influence development by guiding embryonic cell movements and nerve outgrowth direction.

The vibrating probe has opened up a new dimension of cellular organization that is providing many insights into the mechanisms of cell polarity control via transcellular currents.

REFERENCES

Barker AT, Jaffe, LF, Vanable JW Jr (1982). The glabrous epidermis of cavies contains a powerful battery. Am J Physiol 242:R358.

Betz WJ, Caldwell JH, Ribchester RR, Robinson KR, Stump RF (1980). Endogenous electric field around muscle fibers depends on the Na^+-K^+ pump. Nature 287:235-237.

Cooper MS, Keller RE (1982). Electrical currents induce perpendicular orientation and cathode-directed migration of amphibian neural crest cells in culture. J Cell Biol 95:323a.

Dittman F, Ehni R, Engels W (1981). Bioelectric aspects of the Hemipteran Teleotrophic ovariole (Dysdercus intermedius. Roux's Arch Develop Biol 190:221-225.

Freeman JA, Weiss JM, Snipes GJ, Mayes B, Norden JJ (1981). Growth cones of goldfish retinal neurites generate DC currents and orient in an electric field. Soc Neurosci Abstr 7:550.

Hagins WA, Penn RD, Yoshikami S (1970). Dark currents and photocurrent in retinal rods. Biophys J 10:380-412.

Hinkle L, McCaig CD, Robinson KR (1981). The direction of growth of differentiating neurones and myoblasts from frog embryos in an applied electric field. J Physiol 314:121-135.

Jaffe LA, Weisenseel MH, Jaffe LF (1975). Calcium accumulation within the growing tips of pollen tubes. J Cell Biol 67:488-492.

Jaffe, LF (1968). Localization in the developing Fucus egg and the general role of localizing currents. Advan Morphogen 7:295-328.

Jaffe LF (1977). Electrophoresis along cell membranes. Nature 265:600-602.

Jaffe, LF (1979). Control of development by ionic currents. In Cone RA, Dowling JE (eds): "Membrane Transduction Mechanisms." New York:Raven Press, pp. 199-231.

Jaffe LF (1981a). Control of development by steady ionic currents. Fed Proc 40:125-127.

Jaffe LF (1981b). The role of ionic currents in establishing developmental pattern. Phil Trans R Soc Lond B 295: 553-566.

Jaffe LF, Nuccitelli R (1974). An ultrasensitive vibrating probe for measuring steady extracellular currents. J Cell Biol 63:614-628.

Jaffe LF, Nuccitelli R (1977). Electrical controls of development. Annu Rev Biophys Bioeng 6:445-476.

Jaffe LF, Poo M-m (1979). Neurites grow faster towards the cathode than the anode in a steady field. J Exp Zool 209:115-128.

Jaffe LF, Stern CD (1979). Strong electrical currents leave the primitive streak of chick embryos. Science 206:569-571.

Jaffe LF, Woodruff RI (1979). Large electrical currents traverse developing Cecropia follicles. Proc Natl Acad Sci 76:1328-1332.

Kline D, Robinson KR, Nuccitelli R (1983). Ion currents and membrane asymmetries in the cleaving Xenopus egg. J Cell Biol, in press.

Lucas WJ, Nuccitelli R (1980). HCO_3^- and OH^- transport across the plasmalemma of Chara. Planta 150:120-131.

Miledi R (1982). A calcium-dependent transient outward current in Xenopus laevis oöcytes. Proc R Soc Lond B 215:491.

Nuccitelli R (1978). Oöplasmic segregation and secretion in the Pelvetia egg is accompanied by a membrane-generated electrical current. Develop Biol 62:13-33.

Nuccitelli R, Jaffe LF (1974). Spontaneous current pulses through developing fucoid eggs. Proc Natl Acad Sci (USA) 71:4855-4859.

Nuccitelli R, Jaffe LF (1975). The pulse current pattern generated by developing fucoid eggs. J Cell Biol 64: 636-643.

Nuccitelli R, Jaffe LF (1976). The ionic components of the current pulses generated by developing fucoid eggs. Develop Biol 49:518-531.

Nuccitelli R, Poo M-m, Jaffe LF (1977). Relations between ameboid movement and membrane-controlled electrical currents. J Gen Physiol 69:743-763.

Orida N, Feldman JD (1982). Directional protrusive pseudo-podial activity and motility in macrophages induced by extracellular electric fields. Cell Motility 2:243.

Patel N, Poo M-m (1982). Orientation of neurite growth by extracellular electric fields. J Neurosci 2:483-496.

Poo M-m (1981). In situ electrophoresis of membrane components. Annu Rev Biophys Bioeng 10:245-276.

Poo M-m, Robinson KR (1977). Electrophoresis of concanavalin A receptors along embryonic muscle cell membrane. Nature 265:602-605.

Reiss HD, Herth W (1978). Visualization of the Ca^{2+} gradient in growing pollen tubes of Lilium longiflorum with chlorotetracycline fluorescence. Protoplasma 97:373-377.

Reiss HD, Herth W (1979). Calcium gradients in tip growing plant cells visualized by chlorotetracycline fluorescence. Planta 146:615-621.

Robinson KR (1979). Electrical currents through full-grown and maturing Xenopus oöcytes. Proc Natl Acad Sci (USA) 76:837-841.

Robinson KR, Cone R (1980). Polarization of fucoid eggs by a calcium ionophore gradient. Science 207:77-78.

Robinson KR, Jaffe LF (1975). Polarizing fucoid eggs drive a calcium current through themselves. Science 187:70-72.

Stump RF, Robinson KR (1982). Directed movement of Xenopus embryonic cells in a electric field. J Cell Biol 95:331a.

Stump RF, Robinson KR, Harold RL, Harold FM (1980). Endogenous electrical currents in the water mold Blastocladiella emersonii during growth and sporulation. Proc Natl Acad Sci (USA) 77:6673-6677.

Telfer WJ, Woodruff RI, Huebner E (1981). Electrical polarity and cellular differentiation in Meroistic ovaries. Amer Zool 21:675-686.

Weisenseel MH, Jaffe LF (1976). The major growth current through lily pollen tubes enters as K^+ and leaves as H^+. Planta 133:1-7.

Weisenseel MH, Nuccitelli R, Jaffe LF (1975). Large electrical currents traverse growing pollen tubes. J Cell Biol 66:556-567.

Weisenseel MH, Dorn A, Jaffe LF (1979). Natural H^+ currents traverse growing roots and root hairs of Barley (Hordeum vulgare L.). Plant Physiol 64:512-518.

Woodruff RI, Telfer WH (1973). Polarized intercellular bridges in ovarian follicles of the Cecropia moth. J Cell Biol 58:172-188.

Woodruff RI, Telfer WH (1974). Electrical properties of ovarian cells linked by intercellular bridges. Ann NY Acad Sci 238:408-419.

Woodruff RI, Telfer WH (1980). Electrophoresis of proteins in intercellular bridges. Nature 286:84-86.

Zeuthen T (1977). Intracellular gradients of electrical potential in the epithelial cells of the Necturus gall bladder. J Mem Biol 33:281-309.

S. Hagiwara with <u>Dosidicus</u> <u>gigas</u>, Chile, 1967

SECTION V
SENSORY AND CNS PHYSIOLOGY

The Physiology of Excitable Cells, pages 493–496
© 1983 Alan R. Liss, Inc., 150 Fifth Avenue, New York, NY 10011

INTRODUCTION

Closely related to Hagiwara's work on action potentials
and synaptic mechanisms has been a recurrent interest in
sensory transduction, synaptic integration, information processing, and neurophysiological correlates of behavior.
Hagi's Ph.D. thesis, written in a hospital bed as he recovered from lung surgery, was a pioneering statistical
analysis of information coding in impulse trains from
stretch receptors (Hagiwara 1950; Hagiwara 1954). He went
on soon thereafter to make the first single fiber recordings
from single taste fibers in cats with Y. Zotterman and M. J.
Cohen,(Cohen, Hagiwara, Zotterman 1954). In the early
1960's, before embarking on his brilliant studies of calcium
channels, he and his collaborators published an important
series of papers on electroreceptors and the various coding
mechanisms found in different electric fish (Hagiwara,
Morita 1963; Hagiwara, Szabo, Enger 1965a; Hagiwara, Szabo,
Enger 1965b; Szabo, Hagiwara 1967). Subsequently, while
pursuing his analysis of ion permeation channels in a variety
of cells, Hagi turned his attention to transduction in the
barnacle photoreceptor, using this large cell to achieve the
first voltage-clamp analysis of a light-evoked membrane conductance change (Brown, Meech, Koike, Hagiwara 1969; Brown,
Hagiwara, Koike, Meech 1970; Brown, Hagiwara, Koike, Meech
1971; Koike, Brown, Hagiwara 1971). With various associates,
he helped elucidate the role of Ca^{++} in mechano- sensory
transduction in vetebrate hair cells (Sand, Ozawa, Hagiwara
1975; Sand 1975); he analyzed neuromuscular coordination of
wing beat in hummingbirds (Hagiwara, Chichibu, Simpson 1968),
and on an Alpha Helix expedition to New Guinea, he even
left his imprint on the study of neural correlates of echolocation in bats (Grinnell, Hagiwara 1972a,b). In each instance the questions he has asked have been fundamental, the
preparation and procedures adopted for answering them innovative and productive.

In keeping with these interests, and reflecting the
general emphasis within the field on an understanding of
sensory function at the cellular level, several of the papers
included in this section deal with transduction and synaptic
mechanisms early in the sensory pathways. C. EDWARDS

assesses what has been learned about the channels opened in the frog spindle and crayfish stretch receptors by mechanical stimulation, and goes on to consider the role of calcium fluxes in mediating receptor adaptation. Remarkable progress has been made in recent years in elucidating the mechanical and electrical properties of vertebrate hair cells (see Hudspeth 1983). In this volume YOSHIOKA, YANAGISAWA, and KATSUKI describe experiments suggesting that phosphotidic acid may serve as a calcium ionophore and help mediate transduction in the hair cells. S. CHICHIBU, using a combination of scanning electron microscopy and electrophysiology, analyzes the relationship between the type of mechano-sensory hairs on the crayfish antenna, and their sensitivity and receptive field properties.

Several papers are devoted to visual receptors and pathways, as well. The identity of the transmitters released by different receptor cells has been a long standing problem. New techniques for identifying possible neural transmitters and their synthetic enzymes now make it possible to make reasonable judgements in many cases. For example, in the barnacle photoreceptor, introduced in Hagiwara's lab, H. KOIKE finds that although ACh appears to be involved as a transmitter at some point in the pathway to the supraesophageal ganglion, the photoreceptor itself probably releases GABA as its neurotransmitter. In the retinas of certain vertebrates, such techniques have revealed a previously unrecognized population of dopaminergic interplexiform cells that project to horizontal and bipolar cells and that have reciprocal synapses with amacrine cells, which apparently provide their input (Dowling, Ehinger 1978). NEGISHI, TERANISHI and SATO report that this system serves the important function of regulating the degree of coupling of horizontal cells by gap junctions. This greatly influences the spatial localization of responses to a light stimulus. And in another study of the vertebrate retina, this time in turtles, MARCHIAFAVA, WEILER & STRETTOI have analyzed response properties of different bipolar, amacrine, and ganglion cells injected with horseradish peroxidase (HRP) to permit correlation of morphology with physiology. The result is the identification of two discrete retinal circuits, one of which involves inputs from bipolar cells only, the other from both bipolars and amacrines to ganglion cells.

Unquestionably, one of the most exciting developments in all of CNS physiology has been the demonstration that

visual cortical circuitry is subject to "functional vali-
dation" on the basis of experiences during a critical postnatal period (for review, see Wiesel 1982). It is exceedingly important to know what is happening at the synaptic level
when some inputs are retained and strengthened, others are
functionally lost, as a function of use. TOYAMA ET AL may
have made an important contribution toward this end by developing a visual cortical brain slice preparation in which
they can demonstrate changes in synaptic circuitry following
electrical stimulation of the white matter.

Regulation of visceral drives, such as feeding behavior,
has long been a subject of physiological study. The subject
is so complex, however, involving so many hormones and regulatory factors, that progress in understanding mechanisms
of control has been slow. In this section OMURA ET AL add
important new data about the effects of many of these putative regulatory factors on neurons of two parts of the
hypothalamus thought to represent feeding and satiety
centers.

Finally, fittingly, and with characteristic flair,
Hagi's early and close associate, T.H. BULLOCK, ponders the
problems of understanding whole animal behavior, in this
case the slothfulness of the sloth, in terms of the information gained by studying the properties of single cells.

References

Brown HM, Hagiwara S, Koike H, Meech RW (1970). Membrane
   properties of a barnacle photoreceptor examined by the
   voltage clamp technique. J Physiol Lond 208:385.
Brown HM, Hagiwara S, Koike H, Meech RW (1971). Electrical
   characteristics of a barnacle photoreceptor. Fed Proc 30:
   69.
Brown HM, Meech R, Koike H, Hagiwara S (1969). Current-
   voltage regulations during illumination: photoreceptor
   membrane of a barnacle. Science 166:240.
Cohen MJ, Hagiwara S, Zotterman Y (1954). The response
   spectrum of taste fibres in the cat: a single fibre
   analysis. Acta Physiol Scand 33:316.
Dowling JE, Ehinger B (1978). The interplexiform cell
   system. I, Synapses of the dopaminergic neurons of
   the goldfish retina. Proc Roy Soc Lond B 201:7.
Grinnell A, Hagiwara S (1972a). Adaptations of the auditory

nervous system for echolocation: Studies of New Guinea bats. Z vergl Physiol 76:41.

Grinnell A, Hagiwara S (1972b). Studies of auditory neurophysiology in nonecholocating bats, and adaptation for echolocation in one genus *Rousettus*. Z Vergl Physiol 76:82.

Hagiwara S (1950). On the fluctuation of the inteval of rhythmic excitation. II. Analysis on impulses from stretch receptor of a frog muscle. Report Physiol Sci Inst Tokyo U. 4:28.

Hagiwara S (1954). Analysis of interval fluctuation of the sensory nerve impulses. Jap J Physiol 4:234.

Hagiwara S, Chichibu S, Simpson N (1968). Neuromuscular mechanisms of wing beat in hummingbirds. Z vergl Physiol 60:209.

Hagiwara S, Morita H (1963). Coding mechanisms of electroreceptor fibers in some electric fish. J Neurophysiol 26:551.

Hagiwara S, Szabo T, Enger PS (1965). Physiological properties of electroreceptors in the electrical eel, *Electrophorus electricus*. J Neurophysiol 28:775.

Hagiwara S, Szabo T, Enger PS (1965). Electroreceptor mechanisms in a high-frequency weakly electric fish, *Sternarchus albifrons*. J Neurophysiol 28:784.

Hudspeth AJ (1983). Mechanoelectrical transduction by hair cells in the acoustico lateralis sensory system. Ann Rev Neurosci 6:187.

Koike H, Brown HM, Hagiwara S (1971). Hyperpolarization of a barnacle photoreceptor membrane following illumination. J Gen Physiol 57:723.

Sand O (1975). Effects of different ionic environments on the mechano-sensitivity of lateral line organs in the mudpuppy. J Comp Physiol Sec A 102:27.

Sand O, Ozawa S, Hagiwara S (1975). Electrical and mechanical stimulation of hair cells in the mudpuppy. J Comp Physiol. A 102:13.

Szabo T, Hagiwara S (1967). A latency-change mechanism involved in sensory coding of electric fish (Mormyrids). Physiol & Behavior 2:331.

Wiesel TN (1982). Postnatal development of the visual cortex and the influence of environment. Nature 299:583.

The Physiology of Excitable Cells, pages 497–503

THE IONIC MECHANISMS UNDERLYING THE RECEPTOR POTENTIAL IN
MECHANORECEPTORS

Charles Edwards

Neurobiology Research Center and
Department of Biological Sciences
State University of New York at Albany
1400 Washington Avenue
Albany, New York, 12222, U.S.A.

Stretch or mechanical distortion of a mechanoreceptor
produces a depolarizing generator potential which initiates
the action potentials. Katz (1950) first separated the two
potentials in the frog spindle by using procaine to block
the action potentials. The generator potential rises
rapidly with the onset of a maintained stretch (dynamic
phase of the response), and decreases rapidly (adaptation)
to a more or less constant level (static phase); there
follows a slow phase of adaptation which seems to be
mechanical in origin (Swerup, Rydqvist, Ottoson 1983). The
release of tension is usually followed by a small transient
hyperpolarization (off effect) (frog spindle, Katz 1950;
crayfish stretch receptor, Eyzaguirre, Kuffler 1955). I
shall review the results of studies on the ionic basis of
the receptor potential and of the mechanisms of adaptation
and the off effect. The discussion will focus on the
crustacean stretch receptor and the frog muscle spindle,
because there are more data on these two organs than on
others.

IONIC BASIS OF THE RECEPTOR POTENTIAL

The principal ion moving through the channel
responsible for the receptor potential is Na. In the
Pacinian corpuscle, Diamond, Gray & Inman (1958) found that
the size of the receptor potential was decreased by about
90% by the substitution of Na by choline. The first
systematic study of the ionic basis of the receptor
potential showed that the response to stretch of the

crayfish stretch receptor was reduced after the replacement
of Na by hydrazinium, choline, tetramethylammonium,
tetraethylammonium or tris (Obara 1968). A small but
significant permeability to arginine was also found (Klie,
Wellhöner 1973). The absence of an effect by replacement
of Cl by acetate implies that anions are not involved
(Obara 1968). The finding that these substitutions reduce
the receptor potential by the same amount at all lengths
(Obara 1968) suggests that mechanical distortion opens a
large, non-selective cation channel in an all-or-none
manner. An increase of stimulus intensity opens more
channels. The alternate possibility, that channels open
wider with increasing distortion, is precluded by the data;
if this were the case, the replacement of Na by a larger
cation would shift the generator potential-stretch curve to
greater stretches without changing the slope.

The ideal way to investigate channel selectivity is by
measurement of the reversal potential with voltage clamp,
and the results of studies with this technique on the
crayfish stretch receptor have confirmed that stretch opens
a large, non-selective cation channel. The channel admits
univalent cations as large as arginine; the ratio of its
permeability to that of Na is something less than 0.25
(Brown, Ottoson, Rydqvist 1978, see also Edwards 1982).
Divalent cations such as Ca, Mg, Sr and Ba are quite
permeant (Edwards, Ottoson, Rydqvist, Swerup 1981). In the
muscle spindle the amplitude of the receptor potential is
also much reduced by the replacement of Na by choline
(frog, Ottoson 1964), tris, glucosamine, tetraethylammonium
or spermidine (cat, Hunt, Wilkinson, Fukami 1978). Voltage
clamp has not been applied to this organ, but the evidence
suggests that the size of the channel responsible for the
receptor potential is similar to that in the crayfish
stretch receptor.

MECHANISM OF ADAPTATION

Given that the rising phase of the potential is, in
part at least, the result of the opening of a large cation
channel, what mechanisms might be responsible for
adaptation? There is much evidence that Ca plays a role in
adaptation, some of which will be reviewed here.

There are two likely pathways for Ca entry. First,

in the crayfish stretch receptor Ca enters through the channel responsible for the receptor potential (see above). Ca entry probably also occurs in the frog spindle since the size of the receptor potential in the absence of Na varies with the Ca concentration (Hunt, Wilkinson, Fukami 1978). Second, in the spindle there is good evidence for the presence of a voltage activated Ca conductance. In the presence of tetrodotoxin, stretch of the spindle produces a small spike whose amplitude is related to the concentration of Ca in the bath (Ito, Komatsu 1979). A Ca spike can also be elicited by electrical stimulation (Ito, Komatsu, Fujitsuka 1981) and it appears to originate in the non-myelinated filaments of the nerve fiber (Ito, Komatsu, Kaneko 1980). There is no direct evidence for a voltage dependent Ca conductance in the membrane of the crustacean stretch receptor. The ionic conductances of the lobster stretch receptor have been investigated, and no Ca channel was found (Gestrelius, Grampp, Sjölin 1981); however the depolarizing potentials used may not have been large enough to demonstrate the existence of the voltage dependent Ca conductance. There is some indirect evidence for such a channel, since the shape of the receptor potential in the tetrodotoxin blocked receptor is modified by D600, a Ca channel blocker (Ottoson, Swerup 1981).

Given two pathways for entry of Ca, what are the effects of changes of external and internal Ca on the two phases of the receptor potential? In isotonic Ca the dynamic phase of the receptor potential of the crayfish stretch receptor is increased and the static phase is virtually eliminated. On the other hand, when Ca entry is reduced either by block with D600 or by using a solution without Ca and K, the peak amplitude of the dynamic response is reduced and it barely exceeds that of the static phase (Ottoson, Swerup 1981). Consistent with these results are the findings that intracellular injection of Ca reduces the static phase, and injection of EGTA increases it (Ottoson, Swerup 1982a). In the spindle, both removal of Ca from the bath and Ca channel blockers, such as Mn, Mg and verapamil, depress the dynamic phase with only small effects on the static phase (Ito, Komatsu, Katsuta 1981). Thus the magnitude of the dynamic phase is dependent on the entry of Ca and the amplitude of the static phase is reduced by increases in internal Ca.

While much of the initial depolarization is due to Na

entry, the evidence reviewed above suggests that the depolarizing receptor potential transiently activates the voltage dependent Ca conductance. The increase in internal Ca that results from this and from the receptor potential plays a role in adaptation. One possible mechanism for the Ca effect would be that it could block, from the inside, the channels responsible for the receptor potential; a similar mechanism has been proposed to explain part of the desensitization to acetylcholine found at the neuromuscular junction (Chesnut 1983). There is no evidence to support this hypothesis at present, and it is likely that such evidence would be difficult to obtain. The increase in internal Ca can produce adaptation by way of the Ca activated K conductance, which is present in the membrane of the lobster stretch receptor (Gestrelius, Grampp, Sjölin 1981). In the spindle, the effects of quinine, a blocker of this conductance, suggest it is present there also (Ito, Komatsu 1980). The activation of the Ca activated K channel would hyperpolarize the membrane and this would account for at least part of the adaptation. Indeed, in the frog muscle spindle, Ba, which is usually quite permeant through the voltage dependent Ca channel, but does not activate the Ca activated K conductance, reduces the amount of adaptation, as measured by the frequency of action potentials (Ito, Komatsu, Kaneko, Katsuta 1981).

The depolarization produced by the receptor potential, the voltage-activated Ca conductance, and the action potential if present should activate the voltage dependent K conductance which is responsible for the falling phase of the action potential. In the crayfish stretch receptor injection of TEA, which blocks this conductance in the giant axon of the squid (Armstrong, Binstock 1965), increases the amplitude of the static phase of the stretch induced current measured with potential clamp more than it affects the dynamic phase (Ottoson, Swerup 1982b).

Therefore during the static phase of the receptor potential, there are inward currents due to the open receptor potential channels with a possible contribution from the voltage activated Ca conductance; counteracting this would be outward currents flowing through either or both of the K channels, i. e. the voltage dependent and the Ca activated conductances. The increased K conductance is probably responsible for the transient hyperpolarization (off effect) following termination of the stretch.

LATENCY OF CHANNEL OPENING AND A POSSIBLE MECHANISM

Mechanical distortion opens channels in an all-or-none manner with a latency of milliseconds (Katz 1949; Eyzaguirre, Kuffler 1955). Examination of the ultrastructure of the crayfish stretch receptor with the electron microscope suggests that stretch deforms dendritic tips. These are cylindrical processes whose diameter is fairly uniform and which lack mitochondria. They are embedded in a mass of connective tissue and are not surrounded by sheath cells (Tao-Cheng, Hirosawa, Nakajima 1981). A possible mechanism for the opening of the channels is suggested by some recent results with patch clamp. A channel has been found in the membrane of chick myotubes in culture which is permeable to both Na and K ($P_{Na}/P_K \simeq 0.3$). Pressure increases the occurrence of bursts and the probability of the channel's being open (F. Guharay, F. Sachs, personal communication).

In summary, the mechanical distortion of the dendrites of a stretch receptor neuron opens non-selective cation channels. The depolarization, due largely to Na entry, as well as some slight entry of Ca, may activate the voltage dependent Ca channel. These effects may activate either or both of two K channels: one voltage dependent, and one Ca dependent. The K conductances produce hyperpolarization and therefore adaptation as well as the off response after the termination of the stretch.

ACKNOWLEDGEMENT

It is a pleasure to dedicate this paper to Susumu Hagiwara on his 60th birthday. My adventures with the crayfish stretch receptor started with our joint experiments some 26 years ago. I am grateful to John Schmidt for his comments on the manuscript and to David Ottoson, Frederick Sachs and Christer Swerup for sharing their thoughts and unpublished data. My research has been supported by grants from the NIH (NS 07681).

REFERENCES

Armstrong CM, Binstock L (1965). Anomalous rectification in the squid giant axon injected with tetraethylammonium chloride. J gen Physiol 48:859.

Brown HM, Ottoson D, Rydqvist B (1978). Crayfish stretch receptor: an investigation with voltage-clamp and ion-sensitive electrodes. J Physiol 284:155.

Chesnut T (1983). Two component desensitization at the neuromuscular junction of the frog. J Physiol 336:229.

Diamond J, Gray JAB, Inman DR (1958). The relation between receptor potential and the concentration of sodium ions. J Physiol 142:382.

Edwards C (1982). The selectivity of ion channels in nerve and muscle. Neurosci 7:1335.

Edwards C, Ottoson D, Rydqvist B, Swerup C (1981). The permeability of the transducer membrane of the crayfish stretch receptor to calcium and to other divalent cations. Neurosci 6:1455.

Eyzaguirre C, Kuffler SW (1955). Processes of excitation in the dendrites and in the soma of single isolated sensory nerve cells of the lobster and crayfish. J gen Physiol 39:87.

Gestrelius S, Grampp W, Sjölin L (1981). Subthreshold and near-threshold membrane currents in lobster stretch receptor neurones. J Physiol 310:191.

Hunt CC, Wilkinson RS, Fukami Y (1978). Ionic basis of the receptor potential in primary endings of mammalian muscle spindles. J Gen Physiol 71:683.

Ito F, Komatsu Y (1979). Calcium-dependent regenerative responses in the afferent nerve terminal of the frog muscle spindle. Brain Res 175:160.

Ito F, Komatsu Y (1980). Rectangular fluctuations in potential of the afferent nerve terminal during depolarization in the frog muscle spindle. Neurosci Lett 16:1.

Ito F, Komatsu Y, Fujitsuka N (1981). Calcium spike induced by electrical stimulation to the sensory nerve terminal of the frog muscle spindle. Neurosci Lett 27:135.

Ito F, Komatsu Y, Kaneko N (1980) Site of origin of calcium spike in frog muscle spindle. Brain Res 202:459.

Ito F, Komatsu Y, Katsuta N (1981). Effects of calcium blockers on the discharge pattern of frog muscle spindle. Brain Res 218:383.

Ito F, Komatsu Y, Kaneko N, Katsuta N (1981). Control of

the variability of the afferent discharge rate in frog muscle spindle by potassium blockers. Brain Res 216:199.

Katz B (1950). Depolarization of sensory terminals and the initiation of impulses in the muscle spindle. J Physiol 111:261.

Klie JW, Wellhöner HH (1973) Voltage clamp studies on the stretch response in the neuron of the slowly adapting crayfish stretch receptor. Pflugers Arch 342:93.

Obara S (1968). Effects of some organic cations on generator potential of crayfish stretch receptor. J gen Physiol 52:363.

Ottoson D (1964). The effect of sodium deficiency on the response of the isolated muscle spindle. J Physiol 171:109.

Ottoson D, Swerup C (1981). Ionic basis of adaptation in crustacean stretch receptors. J Physiol 317:26P.

Ottoson D, Swerup C (1982a). Studies on the role of calcium in adaptation of the crustacean stretch receptor. Effects of intracellular injection of calcium, EGTA and TEA. Brain Res 244:337.

Ottoson D, Swerup C (1982b). Effects of TEA-injection on outward potassium current in crustacean stretch receptor. Acta physiol Scand Suppl 508:56.

Swerup C, Rydqvist B, Ottoson D (1983). Time characteristics and potential dependence of early and late adaptation in the crustacean stretch receptor. Acta physiol Scand in press.

Tao-Cheng J, Hirosawa K, Nakajima Y (1981). Ultrastructure of the crayfish stretch receptor in relation to its function. J Comp Neurol 200:1.

The Physiology of Excitable Cells, pages 505–514
© 1983 Alan R. Liss, Inc., 150 Fifth Avenue, New York, NY 10011

A MOLECULAR ASPECT OF MECHANICAL TRANSDUCTION IN HAIR CELLS

Tohru Yoshioka[1], Keiji Yanagisawa[2] and
Yasuji Katsuki[3]

1) Yokohama City University, Yokohama, Japan
2) Tsurumi University, Tsurumi, Japan and
3) National Institute for Physiological Science,
   Okazaki, Japan

The transduction process of vertebrate hair cells is extremely difficult to elucidate because of their complex geometry and small size. Stimulation with sound is thought to deform the membranes of sensory cells, leading to change of permeability to Na^+ and K^+. However, membrane components mediating these changes have not yet been identified (Katsuki, 1982).

The toxic action of aminoglycosides has been ascertained by clinical study, and biochemical changes produced by aminoglycoside treatment of hair cell tissue were reported by Hawkins (1970) and Schacht (1976). Schacht et al. concentrated their study on the interaction of neomycin and polyphosphoinositide, and suggested that polyphosphoinositide may act, in vivo, as a receptor for neomycin.

Recently Yoshioka et al. (1982) found that phosphatidic acid (PA), which is classified as an acidic lipid, as well as polyphosphoinositide, may be important in visual transduction, and they suggest that PA may serve as a Ca ionophore. In the present report we discuss the effect of neomycin on PA formation in the inner ear hair cell and the interaction of PA with Ca, in vivo and in vitro. We also describe the production of monoclonal antibody to PA.

THE EFFECT OF AMINOGLYCOSIDE ON PA FORMATION IN THE INNER EAR HAIR CELL OF GUINEA PIG

The effect of neomycin on PA formation in the cochlea tissue was previously reported by Yanagisawa et al. (1982).

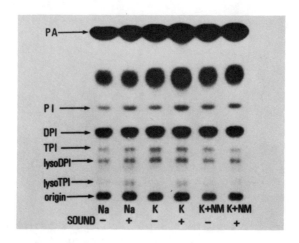

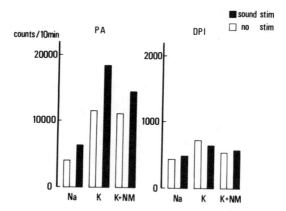

Fig. 1 Upper: Autoradiographs of 32_p incorporated phospholipids in incubation under various conditions. Sodium rich and potassium rich solutions expressed as Na and K respectively. K+NM means neomycin added to K rich Ringer solution. Application of sound stimulation is expressed as + or − corresponding to the presence or absence of sound stimulation.

   Lower: Radioactivity of 32_p incorporated in each band of autoradiograph.
(See Fig. 5 legend for abbreviations.)

In the present study the cochlea was dissected from the inner ear of guinea pig and the tissues quickly removed. Since the amount of PA is very small, a tracer technique was employed to detect the phospholipid. The tissue was incubated in a reaction mixture containing 20 µCi of [γ-32p]ATP. High sodium and high potassium reaction mixtures, phosphate buffered to pH 7.0, were used as incubation solutions, the composition of the former was 130 mM NaCl, 10 mM NaHCO$_3$, 4 mM KCl, 1 mM MgCl$_2$, 1.5 mM CaCl$_2$ and 1 mM NaH$_2$PO$_4$. In the latter, Na and K were exchanged completely. Lipids were extracted with an acidic solvent and separated by thin layer chromatography. Labelled phospholipid was detected by autoradiography followed by counting the radioactivity of the labelled band after scratching from silica gel plate (Fig. 1).

The effect of sound stimulation was determined by applying high frequency sound at 2 second intervals (1 sec 90 dB pulses swept 1-4 KHz) to living guinea pigs for about 2 hours. The neomycin effect was investigated by adding 110 µM neomycin to the reaction mixture. Incubation was carried out for 30 min at room temperature.

As shown in Fig. 1, both potassium ion and sound stimulation increase the amount of labeled PA. Since PA is formed from diglycerides (DG) by phosphorylation, using ATP with the aid of DG kinase, the increase of the radioactivity in PA might be due to the activation of DG kinase by potassium ions and mechanical stimulation by sound. The mechanism of facilitation of the label in PA is unclear at present, but it is probably related to the change in the intracellular concentration of calcium. Neomycin suppressed the increase in 32p incorporation into PA caused by sound stimulation by about 22%. In order to understand the effect of neomycin on PA formation, another approach was carried out using different systems.

EFFECT OF NEOMYCIN ON MECHANOSENSITIVITY OF LATERAL LINE ORGAN OF MUDPUPPY

To investigate the effect of neomycin on mechanosensitivity, neomycin was applied to the lateral line organ of mudpuppy and nerve discharges were recorded by a suction electrode enclosing one of the neuromasts of a stitch in that organ (Shiozawa, 1979). The degree of synchronization between 8 Hz mechanical vibrations and afferent nerve discharges was examined, and expressed as the probability of afferent responses being elicited in phase with the stimulation.

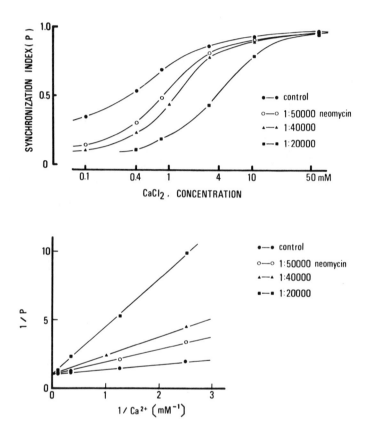

Fig. 2 Upper: Semi logarithmic plot of degree of synchronization vs concentration of Ca^{2+} in the presence of various
concentrations of neomycin.
       Lower: Double reciprocal plot of the data shown in
upper part of figure.

In order to determine the effect of neomycin, various
concentrations of neomycin were applied to the neuromast.
When a solution of neomycin at a concentration of 1:10,000

was used, afferent synchronization was completely suppressed 5 min after application. Afferent synchronization was remarkably improved by application of 20 mM $CaCl_2$ solution, to 5 times greater synchronization than in distilled water. The data obtained in the various concentrations of neomycin are charted in the upper part of Fig. 2. The result of a double reciprocal plot analysis, shown in the lower part of Fig. 2, is a family of straight lines which cross the ordinate at a common intercept. From these data it can be concluded that neomycin is a competitive inhibitor of Ca^{2+} on the receptor membrane. Wersäll and Flock (1964) reported that streptomycin reversibly reduced the microphonic output from the lateral line organ and suggested that this antibiotic primarily affected the surface membrane of the receptor cells. The suppressive effect of neomycin on the mechanosensitivity supports this conclusion.

TRANSPORT OF Ca FROM AQUEOUS PHASE TO ORGANIC PHASE BY PA

The site of aminoglycoside action is generally accepted to be the cell membrane, but the mechanism is uncertain. This is primarily due to the lack of knowledge regarding the specific structural and functional modifications of the receptor membrane. It thus seemed that if PA could bind Ca, and this binding activity were suppressed by neomycin, PA might be a key molecule in neomycin suppression of mechanosensitivity.

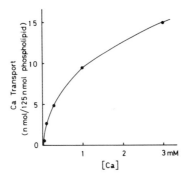

Fig. 3 Dependence of ^{45}Ca transport by PA on Ca^{2+} concentration.

Recently, Schacht et al. (1977) proposed, on the basis of in vivo and in vitro experiments (Orsulakova 1976; Stockhorst 1977), that the polyphosphoinositides, such as phosphatidylinositol phosphate (DPI) and phosphatidylinositol bisphosphate (TPI), serve as in vivo receptors for aminoglycosides. They further proposed that the transduction of acoustic energy into a generator potential involves regulation of the phosphorylation-dephosphorylation reaction in the receptor membrane. As shown in the autoradiograph pattern in Fig. 1, the phosphorylation of phosphatidylinositol (PI) to DPI and to TPI was far less than that of DG to PA. Recently, Green (1980) and Tyson (1976) proposed that PA can function as a calcium ionophore. Experiments reported in this section were undertaken to determine if interaction might occur between PA and Ca, and whether their binding properties might be affected by neomycin. Calcium binding assays were similar to those described by Feinstein (1964). Three milliliters of $CHCl_3$-CH_3OH (2:1 v/v) including 0.5 mg PA and 1 ml Ringer's solution (116 mM NaCl, 2.5 mM KCl, 2 mM Tris-HCl buffer pH adjusted to 7.4, with various concentrations of $CaCl_2$ containing 1 μCi of ^{45}Ca) was added to the 15 ml capped culture tubes. The tubes were vortexed thoroughly for 2 min and the mixture was centrifuged at 500xg for 5 min at room temperature. Then, 1 ml of a lower organic phase was taken from the tube and dissolved in Aquasol II for submission to scintillation count.

The results are shown in Fig. 3. Other phospholipids, phosphatidylethanolamine (PE) and phosphatidylcholine (PC), were tested for comparison, but they did not transport ^{45}Ca from the aqueous phase to the organic phase (data not shown).

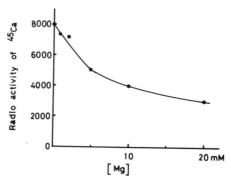

Fig. 4 Suppression of ^{45}Ca transport by Mg^{2+}. Concentration of Ca was 1 μM.

The effect of Mg^{2+} on ^{45}Ca transport was investigated using
the same methods described above and it was found that Mg^{2+}
reduced ^{45}Ca binding to PA to some extent (Fig. 4).   The
results indicate that PA can selectively bind Ca^{2+} in ordin-
ary physiological buffer solution.

The effect of neomycin on ^{45}Ca transport was also tested.
Neomycin solution at a concentration of 1:100 was diluted
100 times by the Ringer's solution (final concentration, 1:
10,000) containing ^{45}Ca, and ^{45}Ca binding activity to PA was
then measured.   The radioactivity of the organic phase was
reduced to about 74% of the control.

PRODUCTION OF MONOCLONAL ANTIBODY OF PA

In order to further understand the role of PA in mechanical transduction, we investigated whether monoclonal antibody to PA and neomycin recognize the same PA binding site
or not. Monoclonal antibody of PA was obtained as a byproduct
in the production of monoclonal antibody of TPI. Monoclonal
antibodies to TPI were raised in mice immunized with a hapten
suspension of TPI, lecithin and cholesterol (1:4:30 by wt)

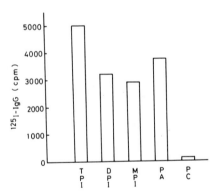

Fig. 5   Cross reactivity of monoclonal antibody secreted from
hybridoma cell line MCB-3 with several kinds of phospholipid.
Abbreviations: TPI (phosphatidylinositol bisphosphate), DPI
(phosphatidylinositol phosphate), MPI (phosphatidylinositol),
PA (phosphatidic acid), PC (phosphatidylcholine).

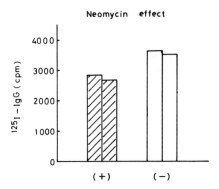

Fig. 6 Suppression of PA and MCB-3 monoclonal antibody reactivity by neomycin at a dilution 1:10,000.

in 0.1% methylated bovin serum albumin(BSA). The hapten suspension was administered every two days for three weeks with each mouse receiving a total of 3 mg TPI (Greenberg, 1979).

The reactivity between monoclonal antibody and antigen was assessed by radioimmuno-assay using ^{125}I labelled anti-mouse IgG. After screening, the antibody secreted from an MCB-3 cell line was selected and its cross reactivity with TPI, DPI, PI, PA and PC was tested. The results are shown in Fig. 5. The reactivity of this monoclonal antibody to PA was as large as to TPI. When neomycin solution at a dilution of 1:10,000 was applied to the antigen-antibody reaction medium, the reactivity between the monoclonal antibody and PA was reduced to 76% of the control value as is shown in Fig. 6. Neomycin interferes with both the PA-monoclonal antibody reaction and with transportation of ^{45}Ca by PA to the same degree.

CONCLUDING REMARKS

As described above, neomycin reduced PA formation in the hair cell of guinea pig about 22%, Ca transport 26%, and PA-PA antibody reaction 26%. Physiological synchronization data revealed that neomycin interacts competitively with Ca in the receptor membrane.

According to Schacht (1976), the polyphosphoinositides, DPI and TPI, may be sites of calcium binding in the membrane. Hydrolytic attack of the phosphomonoesterase will split off the monoester phosphate to liberate calcium and thereby change membrane permeability to cations. Evidence for such a mechanism stems from experiments on erythrocytes which demonstrated correlation between polyphosphoinositide metabolism and calcium binding (Buckley and Hawthorne, 1974).

This hypothesis can be applied to our proposed model without modification, since PA formation is regulated by DG kinase as well as by PI kinase in DPI and TPI formation. As is seen in Fig. 1, increase or decrease of ^{32}P incorporation into PA, DPI, and TPI occurred in phase. The established characteristics of PA as a Ca ionophore make the hypothesis more acceptable.

REFERENCES

Buckley JT and Hawthorne JN (1974). Erythrocyte membrane polyphosphoinositide metabolism and the regulation of calcium binding. J Biol Chem 247:7218

Feinstein MB (1964). Reaction of local anesthetics with phospholipids. J Gen Physiol 48:357

Green DE, Mitchell F and Blondin GA (1980). Phospholipids as the molecular instruments of ion and solute transport in biological membrane. Proc Natl Acad Sci 77:257

Greenberg AJ, Trevor AJ, Johnson DA and Loh HH (1979). Immunological studies of phospholipids; production of antibodies to triphosphoinositides. Molec Immun 16:193

Hawkins JEJr (1970). Biochemical aspects of ototoxicity. In Paparella MM (ed): "Biochemical mechanisms in hearing and deafness" Thomas, Springfield, P323

Katsuki Y (1982). "Receptive mechanisms of sound in ear" Cambridge University Press

Orsulakova A, Stockhorst E and Schacht J (1976). Effect of neomycin on phospholipid labelling in guinea pig inner ear tissue in vivo and in vitro. J Neurochem 26:285

Schacht J (1976). Biochemistry of neomycin ototoxicity. J Acoust Soc Am 59:940

Schacht J, Lodhi S and Weiner ND (1977). Effect of neomycin on polyphosphoinositides in inner ear tissues and monolayer films. In Miller MW and Shamoo AE (ed): "Membrane toxicity" Plenum Press P191

Shiozawa K and Yanagisawa K (1979). Effect of neomycin on the lateral line organ of the mudpuppy. Proc Japan Acad 55B:374

Stockhorst E and Schacht J (1977). Radioactive labelling of phospholipids and proteins by cochlea perfusion in the guinea pig and the effect of neomycin. Act Otolaryngol 83:401

Tyson CA, Vande Zande H and Green DE (1976). Phospholipids as ionophores. J Biol Chem 251:1326

Wersäll J and Flock A (1964). Suppression and restoration of the microphonic output from the lateral line organ after local application of streptomycin. Life Sci 3:1151

Yanagisawa K, Yoshioka T, Inoue H, Hayashi H and Tanaka Y (1982). Potassium effects on phosphatidylinositol phosphorylation in cochlea tissues. Neuroscience Letts suppl 9:s113

Yoshioka T, Inoue H and Hotta Y (1982). Phosphatidylinositol metabolism in photo receptor membrane of visually defective mutant in Drosophila melanogaster. Neuroscience Letts suppl 9:s124

The Physiology of Excitable Cells, pages 515–521

DIRECTIONAL SENSITIVITY AND SPONTANEOUS DISCHARGES OF
SENSORY NEURONS IN CRAYFISH ANTENNA

Shiko Chichibu, M.D.

Department of Physiology
Kinki University School of Medicine
Sayama, Osaka, Japan 589

Sensory hairs on the inner flagellum of the crayfish
(Procambarus clarkii) are mechanosensory organs, serving
mainly as detectors of water currents (Chichibu, Suzuki
1978). These hairs are innervated by two sensory neurons
located at the base of the hair socket. The response
characteristics of these two neurons are complementary,
both in preferred directions and in response types
(Laverack 1964; Chichibu 1972; Wiese 1976). These characteristics are very similar to those of sensory hairs found
in other arthropods (Chapman, Smith 1963; Thurm 1965;
Gnatzy 1980). The relationship between the hair types and
their receptive field properties was examined, based on
the character of impulse train responses to mechanical
stimuli.

The morphology of the hairs was examined with a scanning
electron microscope (JEM-100C, JEOL, Japan). Preparations
were fixed with 2% glutaraldehyde and 1% osmium tetroxide
(Chichibu, Wada, Komiya 1978). Electrophysiological activities were recorded extracellularly on the antennal nerve
bundle with a pair of stainless steel needle electrodes.
Impulse frequencies in response to mechanical stimuli were
determined by averaging twenty successive responses (Chichibu
1978a). The mechanical stimuli were applied to single hairs
by driving a fine glass capillary mounted on a piezo-electric
plate made of lead titan-zirconate.

There are three hair types on the inner flagellum,
namely L, S and F. Of these three the L-type was less
restricted in movement at its base. The hair socket was

wider than the other two, and there was a larger elastic
fold surrounding the base. The S- and F-type hairs had
narrower sockets, and the elastic folds were poorly
developed. The longer axis and the direction of hair
inclination were parallel to the axis of the flagellum.

The innervating neurons were classified into three
groups, namely A, V and D units, according to the degrees
of the adaptation (Chichibu 1978a). The L-type hairs were
innervated in most cases by a combination of A and D units,
and in a smaller fraction of cases, by V and D units.

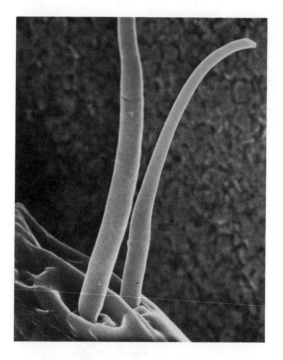

Fig. 1. Two L-type sensory hairs. Articulations were
observed on the hair-shafts. The hair tip on the right
upper corner was not broken, but was forming a 'Cyclopean-
head' structure (Chichibu, Wada, Komiya 1978). Note the
asymmetry of the hair bulb and the socket. Magnification:
x 780.

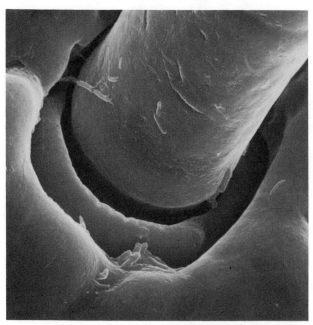

Fig. 2.   Scanning EM view of the base of an L-type hair.
The longitudinal axis of the flagellum was oriented toward
the upper right. Magnification: x 3000.

    Mechanical stimuli applied to a single L-type hair from
various directions evoked responses in two units, but with
different adaptation rates. The receptive fields of single
units were obtained with 50 ms mechanical step stimuli
applied to single hairs. The examples shown in Fig. 3 were
obtained by calculating the peak instantaneous impulse frequency of responses. These A units belonging to L-type hairs
showed the greatest impulse increases to stimuli coinciding
with the direction of the flagellar axis. On the other hand,
the responses to stimuli in the direction perpendicular to
the flagellar axis were small, even less than the maximum
spontaneous discharge level. Thus the movability of a hair
is dependent on the shape and size of the hair socket, and
on the elasticity of the elastic fold around the hair-base.

    In addition to differences in external morphology,
sensory axons of the three groups differed in their conduction velocity (Tsukada, Sata, Chichibu 1969), and response
frequency characteristics (Tani, Chichibu, Sato 1974).

	Axis ratio	Gain	Spont. disch. level	Gain ratio	Contr. ratio
A-unit	420+630 / 60+60	8.75	85	6.18	0.70
	380+650 / 50+60	9.36	80	6.44	0.69
	380+480 / 70+60	6.62	110	3.91	0.59
	600+400 / 70+50	8.33	90	5.56	0.67
	350+680 / 60+60	8.58	100	5.15	0.60
	mean	8.34	93	5.45	0.65
		±0.92	±11	±0.89	±0.05
V-unit	240+90 / 55+55	3.00	130	1.27	0.42
	85+190 / 50+60	2.50	120	1.15	0.46
	230+90 / 70+60	2.46	90	1.78	0.72
	250+100 / 60+50	3.18	90	1.94	0.61
	80+200 / 50+70	2.33	85	1.65	0.71
	mean	2.69	103	1.56	0.58
		±0.33	±18	±0.30	±0.12
D-unit	200+50 / 70+70	1.78	85	1.47	0.82
	150+35 / 45+45	2.06	55	1.68	0.82
	180+40 / 70+80	1.47	60	1.83	1.25
	130+40 / 70+60	1.31	70	1.21	0.93
	170+50 / 60+80	1.83	80	1.38	0.88
	mean	1.69	70	1.51	0.94
		±0.27	±11	±0.22	±0.16

Table 1. Indices representing the receptive field characteristics of 5 mechanoreceptive units of each type. The axis
ratio represents the summed maximum responses (imp/sec) to
50 msec step stimuli in either direction along the axis of
the flagellum divided by the sum of responses to movement
from side to side with respect to that axis. Other parameters are explained in the text.

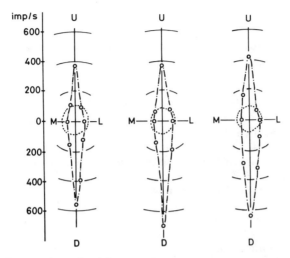

Fig. 3. Receptive fields of three A-units. U, D, M and L denote movement up, down, medially, and laterally. Ordinate: the peak instantaneous response frequency to 50 ms step mechanical stimuli. The dotted lines indicate the maximum spontaneous levels in each unit.

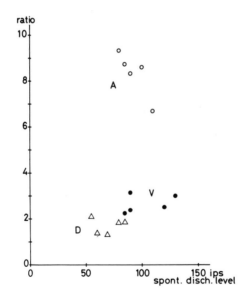

Fig. 4. Gain (ordinate) versus spontaneous discharge level (abscissa) in each unit type.

To characterize the receptive fields of each type of unit, three indices were calculated, based on the peak instantaneous impulse frequencies in response to step mechanical stimuli. These were as follows: the axis ratio (long axis/short axis), the gain ratio (long axis/spontaneous discharge frequency) and the contrast ratio (short axis/spontaneous discharge frequency), respectively. The gain, which was calculated from the peak impulse frequency divided by the spontaneous discharge frequency, was different for the three unit types. The largest values were obtained in the A-units while the smallest values were found in the D-units. As adaptation was disregarded in obtaining this index, the gain reflects the maximum signalling capacity for transient stimuli. On the other hand, the gain ratio and the contrast ratio indicate the shape of the receptive field.

As the measurements of maximum spontaneous discharge level were obtained without locking the hair tip in one position with the glass capillary, spontaneous movements of the hair caused by disturbances in the surrounding medium were not suppressed. Thus the spontaneous discharge level was influenced by hair movement both longitudinal and perpendicular to the flagellar axis. The smaller figures in the contrast ratio of each unit-type reflected the values when hair movement was restricted to the perpendicular direction (Chichibu 1978b; Chichibu, Kitagawa 1982).

Sensory neurons innervating setae on the internal flagellum of sensory hairs also differ in their adaptation rate, and these divergences in response characteristics can be related to efficiency of detection and analysis of water currents in the environment.

ACKNOWLEDGMENTS

The author is indebted to C. Kitagawa for assistance in preparing the manuscript.

REFERENCES

Chapman KM, Smith RS (1963). A linear transfer function underlying impulse frequency modulation in a cockroach mechanoreceptor. Nature 197:699.

Chichibu S (1972). Response characteristics of the tactile
hairs on the crayfish antenna to mechanical stimulation.
J Physiol Soc Jap 34:599.
Chichibu S (1978a). Activities of the velocity-sensitive
mechanoreceptor in the crayfish. Acta med Kinki Univ
3:167.
Chichibu S (1978b). Response patterns and the direction
sensitivities of the crayfish setae units. Acta med
Kinki Univ 3:177.
Chichibu S, Kitagawa C (1982). Spontaneous discharge levels
in crayfish mechanoreceptive neurons. Neurosci Lett Suppl
9:S105.
Chichibu S, Suzuki K (1978). Rheotaxic behaviours and the
function of the crayfish antennal mechanoreceptors. Acta
med Kinki Univ 3:241.
Chichibu S, Wada T, Komiya H (1978). The terminal outgrowths
of the crayfish mechanosensory hairs. Acta med Kinki Univ
3:209.
Gnatzy W (1980). Ultrastructure and mechanical properties
of an insect mechanoreceptor: Stimulus-transmitting
structures and sensory apparatus of the cercal filiform
hairs of Gryllus. Cell Tiss Res 213:441.
Laverack MS (1964). The antennular sense organs of Panulirus
argus. Comp Biochem Physiol 13:301.
Tani Y, Chichibu S, Sato R (1974). Frequency response characteristics of the mechanoreceptors on the crayfish antenna. Abst 13th Ann Meeting Jap Soc MEBE 2E-92:544.
Thurm U (1965). An insect mechanoreceptor. Part I: Fine
structure and adequate stimulus. Cold Spring Harbor Symp
Quant Biol 30:75.
Tsukada M, Sato R, Chichibu S (1969). Relationship between
sensory stimuli and the impulse discharge patterns. Abst
Jap U for Soc Electr Eng 3444.
Wiese K (1976). Mechanoreceptors for near-field water dis-
placements in crayfish. J Neurophysiol 39:816.

The Physiology of Excitable Cells, pages 523–534
© 1983 Alan R. Liss, Inc., 150 Fifth Avenue, New York, NY 10011

TRANSMITTER SUBSTANCE OF BARNACLE PHOTORECEPTOR

Hiroyuki Koike, M.D., Ph.D.

Department of Neurophysiology
Tokyo Metropolitan Institute for Neurosciences
Tokyo, 183 JAPAN

Light produces a large sustained membrane depolarization in barnacle photoreceptors (Gwilliam 1963, 1965;
Brown, Hagiwara, Koike, Meech 1970). The depolarizing
receptor potential spreads electrotonically towards the
presynaptic axon terminal several millimeters away from the
cell body (Shaw 1972; Ozawa, Hagiwara, Nicolaysen, Stuart
1976; Ozawa, Hagiwara, Nicolaysen 1977; Hudspeth, Stuart
1977). The receptor potential itself activates Ca-channels
at the terminal (Ross, Stuart 1974), and acts to release
the synaptic transmitter. The synaptic action of the photoreceptor cell is likely to be inhibitory. The postsynaptic
second order sensory cells hyperpolarize with membrane
conductance increase in the light (Stuart, Oertel 1978).

In this paper the transmitter substance of the photoreceptor cell will be discussed. Many sensory cells of
arthropods are cholinergic (Barker, Herbert, Hildebrand,
Kravitz 1972). Acetylcholine (ACh) is also suggested to be
an inhibitory transmitter substance of the photoreceptor
cell of dragonfly (Klingman, Chappell 1978) and of a
mollusk (Heldman, Grossman, Jerussi, Alkon 1979). However,
in the barnacle photoreceptor cell, the transmitter substance is not ACh, but is likely to be gamma-amminobutyric
acid (GABA) (Koike, Tsuda 1974, 1979, 1980).

PHARMACOLOGICAL EFFECTS

The photoreceptor axon terminates in the supra-
esophageal ganglion, in which at least two synapses are

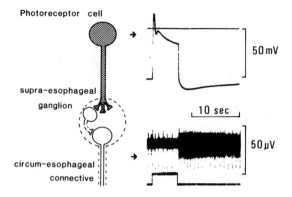

Fig. 1. A schematic picture of the synaptic connections of the photoreceptor (left) and light responses in the barnacle. During a step of light, which is indicated as an upward deflexion of the right bottom trace, the photoreceptor produced an initial overshooting and sustained depolarization, with a small hump but no spike potentials (upper trace). Upon termination of the light pulse, it hyperpolarized more than the previous resting membrane potential level, due to the action of an electrogenic sodium pump (Koike, Brown, Hagiwara 1971). The spike activity recorded from circum-esophageal connective, shown at lower right, increased suddenly upon termination of the light pulse and lasted for minutes thereafter. There was some suppression of the activity during the light pulse and also some activity which had no relationship to photoreception.

affected (Fig. 1). The spike activity of the circum-esophageal connective monitored by a suction electrode is suppressed during light and increased upon termination of the light pulse (Fig. 1). The increased activity upon termination of light (off-response) may correlate with the behavioral "shadow reflex" (Darwin 1854), which is the only known function of photoreception in the barnacle.

When ACh at a concentration of 10^{-5} M was applied in the perfusing solution, the spike activity in the connective persisted even during the light. The off-response was slightly suppressed and masked by the background actvity (Fig. 2). A cholinergic antagonist, d-tubocurarine, also effectively suppressed the off-response. This suggests that cholinergic synaptic transmission might be involved

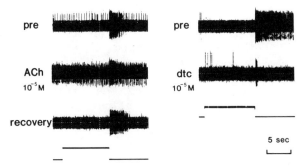

Fig. 2.   Spike activity in the circum-esophageal connective in response to light pulse (bottom traces). Middle traces show responses during the bath applications of acetylcholine (ACh)   and   d-tubocurarine   (dtc)   at   the   concentrations indicated.

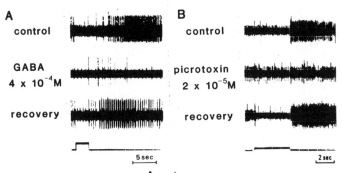

Fig. 3.   Effects of bath applications of gamma-aminobutyric acid (GABA) and picrotoxin on light responses in the circum-esophageal connective.

somewhere in the the path from the photoreceptor to the output connective of the supra-esophageal ganglion.

Among other agonists and antagonists of transmitter candidates tested, gamma-aminobutyric acid (GABA) and related substances showed similar effects (Fig. 3). GABA, although at relatively high concentration in the bathing solution, suppressed the off-response reversibly (Fig. 3A). A GABA antagonist, picrotoxin, also blocked the off-response with moderate increase of background activity (Fig. 3B).

CELLULAR SYNTHESIS ABILITIES FOR ACH AND GABA.

To examine ACh synthesis ability of the photoreceptor cell, a precursor of ACh, radioactive choline (^{3}H-choline: Amersham TRK.179) was injected into the photoreceptor cell body. The injection was performed by intracellular penetration with a double-barrelled capillary; one barrel for recording membrane potential and the other for pressure injection of the ^{3}H-choline solution (Koike, Eisenstadt, Schwartz 1972; Koike, Tsuda 1980). Injecting pressure was controlled by an electric circuit to avoid cell damage due to too rapid injection (Koike 1979).

Following the injection of ^{3}H-choline, the photoreceptor was cultured for 2 hours, and the cell and surrounding tissues were homogenized in ice-cold acetone-formic acid (85:15 in volume) solution. The homogenates were paper electrophoresed to analyze the synthesis of ACh in the cell (Koike et al 1972). The result shown in Fig. 4 revealed that not even a trace of ACh was formed in the photoreceptor cell of the barnacle.

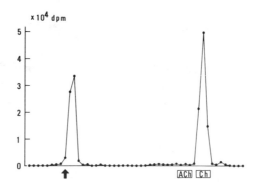

Fig. 4. Results of high voltage paper electrophoresis of the homogenate of a photoreceptor cell cultured for 2 hours following the injection of ^{3}H-choline. The radioactivity on the paper was measured by liquid scintillation counter. Started at the position indicated by the arrow, the movements of internally added cold choline (Ch) and acetylcholine (ACh) were visualized by iodine vapor on the paper. The location of each is indicated below the abscissa. The radioactivity near the origin was phospholipids. (from Koike, Tsuda 1979)

ACh is known to be synthesized primarily in the pre-synaptic terminal of the cholinergic neuron. However, the synthesizing enzyme, choline acetyltransferase, is synthesized in the cholinergic neuron's cell body and axonally transported towards the presynaptic terminal. Therefore, synthetic enzymes for ACh would be expected even in the cell body, if the cell is cholinergic (Koike et al 1972; Koike, Tsuda 1979). On the contrary, if the cell has no enzyme activity for the synthesis of a particular transmitter candidate, this constitutes strong evidence that the substance is not a transmitter of the cell. The present result, therefore, indicates that the transmitter substance of the photoreceptor cell of the barnacle is not ACh.

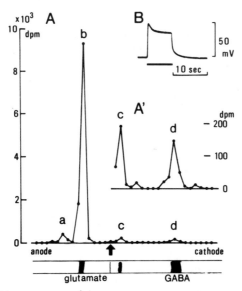

Fig. 5. A: Results of paper electrophoresis of a photoreceptor 2 hours following the injection of ^{3}H-glutamate. Starting at the arrow, internally added cold glutamate and gamma-aminobutyric acid (GABA) moved in opposite directions toward the anode and cathode, and were visualized by ninhydrin reaction (lower figures). Another ninhydrin-reacted substance coinciding with radioactivity marked c was a mixture of several endogenous amino acids, mainly glutamine. A': Part of A (cathodal direction only) with magnified ordinate scale. Peaks c and d were identical to c and d' in A. B: Receptor potential just after the injection. (from Koike, Tsuda 1980)

Another candidate for the transmitter of the photo-receptor cell was GABA, and we injected ^{3}H-glutamate (Amersham TRK.445) into the cell to test the enzyme activity for the synthesis of GABA. Fig. 5 shows a typical result of paper electrophoresis from the photoreceptor cell cultured for 2 hours following the ^{3}H-glutamate injection. Besides the high radioactivity peak (marked b in the figure) of injected ^{3}H-glutamate, several other radio-activity peaks were seen. They were identified as aspartate (a), glutamine (c) and GABA (d) from their positions on the paper in reference to the positions of the internal markers (Koike, Tsuda 1980). Insert A' illustrates more clearly the formation of GABA with higher ordinate scale.

Fig. 6 shows examples of electrophoresis results from several different preparations following similar injection of ^{3}H-glutamate. ^{3}H-GABA was formed in the barnacle photo-receptor cell (A) and in a neuron of the abdominal ganglion of the spiny lobster (D), but it was not in a giant cholin-ergic neuron of Aplysia (B) or another neuron of the spiny lobster (C). Thus the neurons can be divided into those

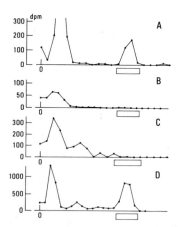

Fig. 6: Examples of paper electrophoresis results (cathodal directions only; origin, 0, and positions of GABA, rectangles, are marked) from (A) photoreceptor cell of the barnacle, (B) a cholinergic neuron (R2) of the abdominal ganglion of Aplysia, (C) and (D) neurons in the abdominal ganglion of spiny lobster, each injected with ^{3}H-glutamate. Ordinate scales were normalized by the injected amounts in each neuron.

that form GABA in the cell and those that do not. The latter neurons' transmitter should not be GABA, but the former neurons' transmitter substance could possibly be GABA.

The amount of conversion to GABA from injected glutamate was not high. A possible reason for the small conversion ratio even in a lobster neuron might be that the injected radioactive glutamate was diluted by endogenous glutamate existing at high concentration in the cell (Otsuka, Kravitz, Potter 1967), so that the radioactive GABA detected represented only a fraction of newly synthesized GABA in the cell (see Discussion of Koike, Tsuda 1979).

RELEASE OF RADIOACTIVE GABA FROM PHOTORECEPTOR

The photoreceptor and supra-esophageal ganglion were immersed in sea water containing 10 μCurie of ^{3}H-GABA and tetrodotoxin (TTX 10^{-6}M) and given repeated light pulses. The photoreceptor axon terminal was expected to be loaded by ^{3}H-GABA, if the transmitter of the receptor cell was GABA. TTX added in the bath did not affect the receptor potential, which probably regulates transmitter release from the photoreceptor axon terminals. Therefore, TTX was expected to be helpful for the somewhat selective loading of ^{3}H-GABA into the photoreceptor and for the possible reduction of background ^{3}H-GABA, unrelated to the light stimuli, contaminating the perfusate. After 30 to 60 minutes loading of ^{3}H-GABA, the preparation was perfused with a solution containing cold GABA (10^{-4}M) and TTX. The perfusates were collected every 2 minutes and the radioactivity in the solution was measured by liquid scintillation counter.

After 50 minutes wash-out in complete darkness in the experiment shown in Fig. 7, a bright light was applied to the photoreceptor preparation for 10 minutes. The radioactivity collected during the light was plotted by open circles. Applications of the bright light were repeated three times with 16 minute intervals of complete darkness in between. The radioactivity appearing in the perfusate tended to decrease with time, and it seemed slightly higher during the light than the expected values from before and following the dark periods.

Similar results of the increase in release during the light were obtained from three other experiments (Koike 1978). The increase of the release was not so high but seemed to be significant. Moreover the increase in release corresponding to the light period was seen only when the preparation was GABA-loaded. Fig. 8 shows an example of non-significant change of the release by the light when the photoreceptor was loaded by 10 μ Curie of ^{14}C-choline.

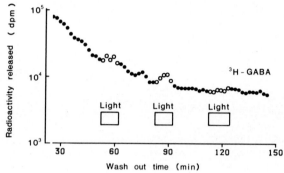

Fig. 7. A semilog plot of the time course of the appearance of radioactivity in the solution perfusing a photoreceptor-supraesophageal preparation which had been loaded by ^{3}H-GABA. The perfusate was collected every 2 minutes and the radioactivity involved was measured by a liquid scintillation counter, and expressed as disintegrations per minute (dpm). The radioactivity of the perfusate collected during dark is indicated by filled circles and that during light by open circles.

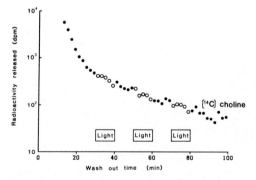

Fig. 8. The appearance of radioactivity in the perfusate of a photoreceptor preparation which had been loaded by ^{14}C-choline. Perfusate contains 10^{-5}M choline.

SHADOW   RESPONSE   RELEASE   OF   RADIOACTIVITY: POSSIBLE
INVOLVEMENT OF CHOLINERGIC TRANSMISSION

From actions of bath applied ACh and d-tubocurarine
shown previously, cholinergic synaptic transmission has
been suggested for some synapses of the photoreceptor.
However, ACh had already been ruled out as a transmitter of
the photoreceptor. Another possible site for cholinergic
transmission is the synapse from the second order sensory
interneuron (I cell) to the third order neurons (see
Fig. 1). Since the I cell is depolarized upon termination
of light (Stuart, Oertel 1978), the transmitter might be
released upon termination of light or during dark, but stop
being released during light.

In an experiment shown in Fig. 9, following ^{14}C-choline
loading of the photoreceptor-supraesophageal ganglion

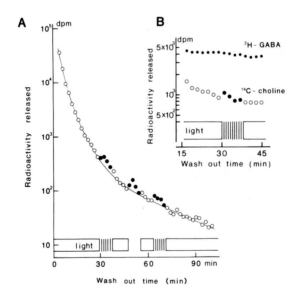

Fig. 9. The release of radioactivity in the perfusate of
the photoreceptor in response to turning off of the light
or shadowing the preparation. Duration of light is indicated by squares, and the perpendicular bars represent periods
when short light pulses were repeated with longer dark
periods.
A: Loaded by ^{14}C-choline.   B: Loaded by ^{14}C-choline and
^{3}H-GABA.

preparation, initial wash-out was performed under bright light with the perfusate containing cold choline and TTX. Since I cell is also a non-spiking neuron (Ozawa et al 1976; Stuart, Oertel 1978), TTX may be expected to reduce the background without affecting transmitter release from the I cell. As expected, the radioactivity was increased significantly when the light was off, and also when the short light pulses were repeated (periods indicated by perpendicular bars) instead of continuous darkness. Increase of radioactivity was not observed when radioactive substances other than choline were loaded. In Fig. 9 B, ^{3}H-GABA and ^{14}C-choline were loaded in a preparation, and we observed an increase of release of ^{14}C-labeled substance, but no change of ^{3}H-labeled during repeated light pulses. Loaded ^{3}H-GABA would possibly be released completely during the early wash-out period under light. It seems likely that the second order sensory neuron (I cell) is cholinergic.

GABA AS A TRANSMITTER SUBSTANCE OF BARNACLE PHOTORECEPTOR

There are several pieces of evidence suggesting that GABA is the transmitter substance of the photoreceptor cell of the barnacle (this paper; Koike 1978; Koike, Tsuda 1974, 1979, 1980):
(a) The photoreceptor cell has enzyme activity in the cell for the synthesis of GABA from glutamate. The enzyme activity is sufficiently high to consider the cell's transmitter to be GABA, from an estimate of endogenous glutamate.
(b) The newly synthesized ^{3}H-GABA, as well as intracellularly injected ^{3}H-GABA, moved down the photoreceptor axon towards the axon terminal.
(c) Electron-microscope autoradiography revealed the transported ^{3}H-GABA in the axon terminal in the dark, but it faded away in the light.
(d) A GABA analogue, ^{3}H-di-aminobutyric acid, which is known to be selectively taken up into the GABA-ergic nerve terminal (Kelly, Dick 1975), was selectively taken up into the photoreceptor axon terminal.
(e) Finally, we observed the release of radioactive GABA, apparently from the photoreceptor, in response to light.

The action of picrotoxin on the off-response was reported earlier (Millechia, Gwilliam 1972), and we have confirmed their result. Bath applied GABA suppressed the off-response, with overall suppression of spike activity;

but picrotoxin has a tendency to enhance the background activity, possibly because of the block of inhibitory action of GABA at the synapse. In contrast, bath applied ACh increased spike activity even during light, and d-tubocurarine tended to suppress it generally. Every evidence presented supports quite well the conclusion that the inhibitory synaptic action of the photoreceptor on the I cell is mediated by GABA. Upon termination of light, the I cell depolarizes by disinhibition of GABA action, and releases the excitatory transmitter, ACh.

REFERENCES

Barker DL, Herbert E, Hildebrand JG, Kravitz EA (1972). Acetylcholine and lobster sensory neurones. J Physiol 226:205.

Brown HM, Hagiwara S, Koike H, Meech RM (1970). Membrane properties of a barnacle photoreceptor examined by the voltage clamp technique. J Physiol 208:385.

Darwin C (1854). "Monograph On Subclass Cirripedia", London: The Ray Society, p 88.

Gwilliam GF (1963). The mechanism of the shadow reflex in cirripedia. I. Electrical activity in the supra-esophageal ganglion and ocellar nerve. Biol Bull mar biol Lab Woods Hole 125:470.

Gwilliam GF (1965). The mechanism of the shadow reflex in cirripedia. II. Photoreceptor cell response, second-order responses, and motor cell output. Biol Bull mar biol Lab Woods Hole 129:244.

Heldman E, Grossman Y, Jerussi TP, Alkon DL (1979). Cholinergic features of photoreceptor synapses in Hermissenda. J Neurophsiol 42:153.

Hudspeth AJ, Stuart AE (1977). Morphology and responses to light of the somata, axons, and terminal regions of individual photoreceptors of the giant barnacle. J Physiol 272:1.

Kelly JS, Dick F (1975). Differential labelling of glial cells and GABA-inhibitory interneurons and nerve terminals following the micro-injection of ß-^{3}H-alanine, ^{3}H-DABA and ^{3}H-GABA into single folia of the cerebellum. Cold Spring Harb Symp quant Biol 40:93.

Klingman A, Chappel RL (1978). Feed back synaptic integration in the dragonfly ocellar retina. J gen Physiol 71:157.

Koike H (1978). Light induced GABA release from photoreceptor cell of barnacle. J Physiol Soc Japan 40:234.

Koike H (1979). A new device for controlled intracellular injection using pneumatic pressure. Integrative Control Functions of the Brain 2:39.

Koike H, Brown HM, Hagiwara S (1971). Hyperpolarization of a barnacle photoreceptor membrane following illumination. J gen Physiol 57:723.

Koike H, Eisenstadt M, Schwartz JH (1972). Axonal transport of newly synthesized acetylcholine in an identified neuron of Aplysia. Brain Res 37:152.

Koike H, Tsuda K (1974). Transmitter identification of crustacean sensory cells by means of intracellular injection of labelled precursors. Proc Int Union Physiol Sci 26:152

Koike H, Tsuda K (1979). Intracellular acetylcholine synthesis and GABA synthesis in some crustacean neurons. in Otsuka M, Hall Z (ed): "Neurobiology of Chemical Transmission", New York: John Wiley & Sons, p 65.

Koike H, Tsuda K (1980). Cellular synthesis and axonal transport of gamma-aminobutyric acid in a photoreceptor cell of the barnacle. J Physiol 305:125.

Millechia R, Gwilliam GF (1972). Photoreception in a barnacle: Electrophysiology of the shadow reflex pathway in Balanus cariosus. Science N Y 177:438.

Otsuka M, Kravitz EA, Potter DD (1967). Physiological and chemical architecture of a lobster ganglion with particular reference to gamma-aminobutyrate and glutamate. J Neurophysiol 30:725.

Ozawa S, Hagiwara S, Nicolaysen K (1977). Neural organization of shadow reflex in a giant barnacle, Balanus nubilus. J Neurophysiol 40:982.

Ozawa S, Hagiwara S, Nicolaysen K, Stuart AE (1976). Signal transmission from photoreceptors to ganglion cells in the visual system of the giant barnacle. Cold Spring Harb Symp quant Biol 15:563.

Ross WN, Stuart AE (1974). Voltage sensitive calcium channels in the presynaptic terminals of a decrementally conducting photoreceptor. J Physiol 274:173.

Shaw SR (1972). Decremental conduction of the visual signal in barnacle lateral eye. J Physiol 220:145.

Stuart AE, Oertel D (1978). Neuronal properties underlying processing of visual information in the barnacle. Nature London 275:287.

The Physiology of Excitable Cells, pages 535-547
© 1983 Alan R. Liss, Inc.. 150 Fifth Avenue, New York, NY 10011

DOPAMINERGIC NEURONS REGULATE ELECTRICAL AND DYE COUPLING
BETWEEN HORIZONTAL CELLS OF THE FISH RETINA

Koroku Negishi, Tsunenobu Teranishi and
Satoru Kato

Department of Neurophysiology, Neuroinformation
Research Institute, University of Kanazawa
School of Medicine, Kanazawa, Ishikawa 920, Japan

Interplexiform cells with processes extending towards
both the outer (OPL) and inner plexiform layers (IPL) are
found in many vertebrate retinas (Boycott et al 1975).
Histofluorescence studies have revealed that dopamine (DA)-
containing cells are generally present at the innermost
level of the inner nuclear layer (INL) (Ehinger 1976). However, DA-containing interplexiform cells have been found
only in the retina of the Cebus monkey and of various species of teleost fishes (Dowling, Ehinger 1975, 1978; Negishi
et al 1981c). The processes of the DA cells of the goldfish
and Cebus monkey retinas have reciprocal synapses between
amacrine cells in the IPL and are presynaptic to horizontal
and bipolar cells in the INL (Dowling, Ehinger 1975). Therefore, the DA cells provide a centrifugal pathway from the
IPL to the OPL (Dowling, Ehinger 1975, 1978).

Horizontal cells (the second order neurons) in a given
cellular layer of the fish retina have been shown to be
coupled electrically via gap junctions (Naka, Rushton 1967;
Kaneko 1971). The amplitude of the S-potentials evoked by
a small spot of light decrements with distance from the
illuminated area. The half decay distance varies in
different horizontal cell layers (Negishi, Sutija 1969;
Teranishi et al 1982) and between different spectral
responses even in a given cellular layer (Teranishi et al
1982), suggesting that the lateral spread of S-potentials is
not based simply on electrical coupling. A feedback input
from the DA interplexiform cells to external horizontal
cells may regulate synaptically the spatial properties of S-
potentials (Negishi, Drujan 1979a, b).

The present paper summarizes our recent fluorescence microscopic observations on the DA interplexiform cells in the teleost retina (Negishi et al 1980, 1981a-c, 1982a, b), and also studies on the effects of DA on the membrane potential of external horizontal cells in isolated fish retinas (Negishi, Drujan, 1979a-c; Laufer et al 1981; Teranishi et al 1983).

METHODS

A histochemical technique modified from the "Faglu" water-stable fluophores method (Furness et al 1977; Nakamura 1979) was used for cryosectioned (15 μm thickness) and flat-mounted preparations of the retina of mullet (Mugil cephalus) and carp (Cyprinus carpio). Following a modified method from the original technique (Ehinger, Nordenfelt 1977), DA cells were destroyed in carp retinas by treatment with 6-hydroxydopamine (6-OHDA) (DA cell deprived retina). In these cases, the eyes were intravitreally injected with 6-OHDA (20 μg) on 2 successive days 10-15 days before electrophysiological experiments. For details of the technique, refer to our previous papers (Negishi et al 1980, 1981a-c, 1982b).

In the electrophysiological studies, isolated carp retinas exposed to humidified oxygen have been used. A small amount (approximately 5 μl) of Ringer solution containing DA (2 mM) and pargyline (2 mM) or one of DA-related compounds was applied to the residual vitreous fluid beneath the isolated retinas 30 min before the start of the electrophysiological experiment (Teranishi et al 1983). This manner of drug application was intended to give a constant concentration in the retina for the 30-90 min during which the measurements were made. The wet weight of the preparation (retina plus residual vitreous fluid) was 180 mg on the average. Therefore, if the applied DA was assumed to diffuse homogenously throughout the entire preparation, its concentration was approximately 10 μM. After electrophysiological experiments, the retinas were processed with the histofluorescence method to examine horizontal cells intracellularly marked with Lucifer Yellow (LY) and monoamine-accumulating cells in flatmounts and cryosections. In this series, 108 isolated retinas were used; 72 normal and 36 DA cell-deprived retinas. Although 3-4 recording sites were marked with LY in each preparation, approximately 30% of marked points corresponded well to the electrophysiological data. In all, photopic L-type S-potentials were recorded from 104 somata and 44 axon terminals of external horizontal cells.

RESULTS

Fluorescence photomicrographs of cryosections of the mullet retina, which had been treated with intravitreal injection of noradrenaline (NA; 20 μg) 2 hr prior to enucleation to enhance cellular fluorescence, are shown in Fig. 1. Typically a DA interplexiform cell seen in a radial section of the central region of the retina has a cell body in the innermost level of the INL and sends processes towards both the OPL and IPL. These processes form dense fiber networks in the distal part of the INL surrounding the external horizontal cells (marked with "a" in Fig. 1B) and in the IPL ("c" in Fig. 1B) as seen in tangential sections through the levels marked with "a" (Fig. 1A) and "c" (Fig. 1C).

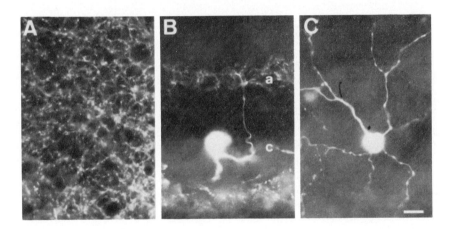

Fig. 1. Fluorescence photomicrographs taken from cryosections of the central region of the mullet retina. A and C, tangential; B, radial; scale, 10 μm. Modified from Negishi et al (1982a).

In the electrophysiological experiments, the effects

on photopic L-type S-potentials of displacing the 621-nm
light spot (0.5 mm in dia.) were examined. The amplitude
of the S-potential was usually maximal when the spot was
above the recording point (0 mm) (mean ± S.D. = 7.8 ± 0.5 mV),
and decreased exponentially as the spot was moved away from
this point (Fig. 2A). The data are plotted in Fig. 4.

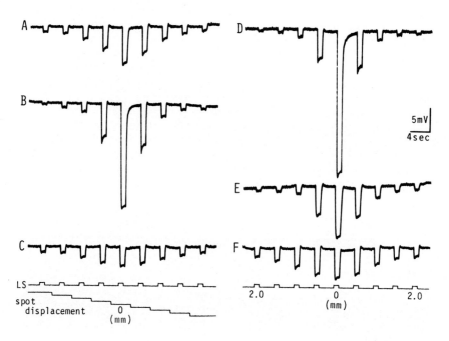

Fig. 2. Changes in S-potential amplitudes with spot displacement. The isolated retina was stimulated by a spot of
light beam from a 500 W Xenon arc lamp. The stimulus was
given with a 1-sec duration at 4-sec intervals from below
the preparation. The beam had a set of representative monochromatic interference filters (halfband width of 10-15 nm)
calibrated so as to deliver, at the retinal surface, 1×10^{13}
quanta/cm^2/sec at 621 nm. The interference filters used
consisted of 423, 459, 517, 575, 621 and 674 nm. A set of
neutral filters was used to attenuate the light intensity
in steps of 0.5 log in a range of 4.0-0 log units. A
flashing spot (0.5 mm dia.) of red (621 nm; 0.5) light was
displaced in steps of 0.5 mm along a straight 4-mm line
passing over the recording point (at 0 mm); the bottom
tracings indicate the light pulse (LS) and its stepwise

displacement. A, recorded from a normal retina; B, from a
retina treated with dopamine (see below); and C, from a
retina from which dopamine (DA) cells had been deprived by
intravitreal injection of 6-hydroxydopamine 15 days prior
to enucleation (DA cell-deprived retina); D, from a DA cell-
deprived retina treated with DA; E, from a DA cell-deprived
retina treated with dibutyryl-cAMP; F, from a normal retina
treated with haloperidol and subsequent DA. Amplitude (5
mV) and time (4 sec) scales are indicated in D. From
Teranishi et al (1983).

When LY was injected into a horizontal cell, usually 5-6
cell bodies with an axon terminal (faintly in a few flatmounts) were seen to be fluorescent in normal retinas (Fig.
3A), as observed by others (Stewart 1978; Kaneko, Stuart
1980). These cell bodies were found to lie in the external
horizontal cell layer when examined in radial cryosections
of the retina (not illustrated). The correspondence of S-
potential recordings to marked spots was established in 41
sites (cell bodies) explored in 34 normal retinas.

The application of a small amount (about 5 µl) of DA
(2 mM)-containing Ringer solution to the residual vitreous
fluid beneath the isolated retina 30 min before recording
increased the amplitude of the S-potential recorded just
under the light spot (22.0 ± 4.0 mV) (Fig. 2B). The average
increase was about 2.8 fold. When the light spot was displaced in steps, the amplitude of the S-potential decayed
sharply (Fig. 4). In retinas so treated, the injected LY
was found in the flatmount to be restricted to the one cell
from which the recording was made (Fig. 3B). This response
was found in 13 cell bodies investigated in 10 retinas.

In retinas deprived of DA cells the S-potentials
recorded at the central point were small (about 60% of
normal) in amplitude (4.7 ± 0.7 mV) (Fig. 2C). Examination
of retinas in which the dye was injected into a recorded
cell consistently showed numerous fluorescent horizontal
cell bodies (more than 10 cells) (Fig. 3C). This finding
was observed in 18 recording sites (cell bodies) in 12 DA
cell-deprived retinas.

The amplitude of the S-potentials in DA cell-deprived
retinas (n=8) was increased at centered spots (0 mm) by
application of DA but it was sharply reduced with spot

displacement. In two of the 10 cells examined the central response was larger than that found in the DA-treated normal retinas; the record from one of them is illustrated in Fig. 2D. The LY injected into the DA treated cells in the DA cell-deprived retinas was restricted to one horizontal cell (Fig. 3D).

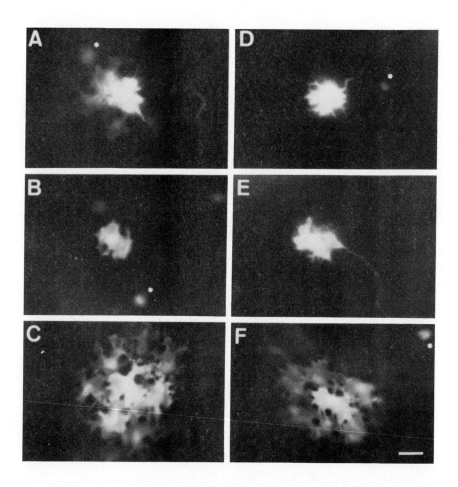

Fig. 3. Fluorescence photomicrographs showing horizontal cells intracellularly marked with Lucifer yellow (LY) and dopamine (DA)-accumulating cells. A fluorescent dye, LY, was ionophoretically injected with a 4-Hz sinusoidal current of ±10 nA for 15 sec into individual horizontal cells, from

which photopic L-type S-potentials (ascertained with their spectral response curve) were recorded with a glass micro-electrode filled with a 4% LY solution, having a resistance of about 200 MΩ after beveling the tip. The concentration of LY, the injection time and the electrode resistance were roughly constant throughout the experiments, and so the amount of dye injected into each cell was supposed to be about the same. At the end of the electrophysiological experiments, the preparations were kept in the dark for 30 min, fixed in a formaldehyde-containing (FGS) solution over-night, and then prepared as flatmounts. After the LY-marked cells were examined in half-dried flatmounts under a fluorescence microscope (Nikon EF), the preparations were softened again in the FGS solution, and then subjected to cryosectioning to 15 μm thickness (not illustrated). The retinal sections were dried over Drierite in a desiccator for several hours, sealed with Entellan and again examined under the fluorescence microscope. In this manner, intracellularly marked horizontal cells as well as DA-accumulating cells, if they existed, were seen simul-taneously in both flatmounts and sections of the same pre-parations. A, a normal retina; B, a normal retina treated with DA; C, a DA cell-deprived retina; D, a DA cell-deprived retina treated with DA; E, a DA cell-deprived retina treated with dibutyryl-cAMP; and F, a normal retina treated with haloperidol and subsequent DA. The fluorescence intensity of LY-marked cells varied to some extent depending upon the background fluorescence of retinal tissue. It should be pointed out that an endogenously DA-containing cell (marked with *; out of focus) is seen faintly in A, while one or a few DA-accumulating cells (*) are visible in B, D and F. The cell (*) in D belongs to the class of indoleamine-accumulating amacrine cells (Ehinger, Florén 1976; Ehinger, Holmgren 1979), which also took up applied DA. No fluo-rescent cells other than those marked with LY are seen in C and E, because DA was not applied to these DA cell-deprived preparations. The bar in F indicates 25 μm. From Teranishi et al (1983).

In some experiments the photopic L-type S-potentials were recorded from and the dye was injected into the axon terminals of horizontal cells. In those cases a similar but much less consistent correlation was observed between the size and spatial decrement of the S-potentials and the dye diffusion area. This occurred in both normal and DA cell-deprived retinas.

It is known that in the retina DA stimulates adenylate cyclase, the enzyme which synthesizes cAMP from ATP (Brown, Makman 1972; Watling, Dowling 1981). The effects of DA on horizontal cells, observed electrophysiologically in the fish retina, are assumed to be mediated by the activation of DA-sensitive adenylate cyclase in the horizontal cell membrane. If this is the case, dibutyryl-cAMP should produce the same effects as DA. A small amount (about 5 μl) of Ringer containing dibutyryl-cAMP (10 mM) and a phospho-diesterase inhibitor, 3-isobutyl-1-methylxanthine (1 mM), was applied to DA cell-deprived retinas in the same manner as was DA. The amplitude of the S-potential at 0 mm was larger than those seen in DA cell-deprived and in normal retinas (Fig. 2E). The injected LY was found restricted to one horizontal cell (n=7 marked sites) in such prepara-tions (Fig. 3E).

A dopaminergic blocker, haloperidol (10 mM), was simi-larly applied to 8 normal retinas in combination with DA. Haloperidol was added 45 min prior to the application of DA, because the former was assumed to act more slowly on the horizontal cells than did the latter. The blocker tended to prevent the appearance of the DA effect on S-potentials (Fig. 2F). The reduction in S-potential amplitude with spot displacement seen in 15 sites resembled that shown in the retinas deprived of DA cells (Fig. 2C). The injected LY was found to diffuse through many horizontal cell bodies (Fig. 3F).

DISCUSSION

The effects of some putative neurotransmitters on the membrane potential of horizontal cells in the fish retina have been explored extensively (Murakami et al 1972; Sugawara, Negishi 1973; Wu, Dowling 1979; Negishi, Drujan 1979a-c; Ishida, Fain 1981). Among them, L-glutamate and GABA were most effective; the former markedly depolarized while the latter hyperpolarized horizontal cells, and the two agents diminished or abolished both center and surround responses at the same time. Therefore, the effects of DA on external horizontal cells, shown in the present study, are unique among those of the putative neurotransmitters examined thus far.

The spatial decrements of the S-potentials recorded under three conditions are shown in Fig. 4A and the normalized values are plotted in Fig. 4B. DA enhanced the sensitivity in the center of the horizontal cell receptive field and the lack of DA had the opposite effect. Moreover, a comparison of the normalized spatial decay curves of the S-potentials (Fig. 4B), suggests that the spatial properties of S-potentials appear to be regulated by DA; DA reduced the receptive field and the lack of DA had the opposite effect.

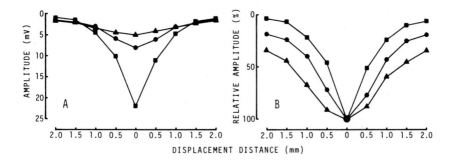

DISPLACEMENT DISTANCE (mm)

Fig. 4. Curves showing distribution of S-potential amplitudes with spot displacement in different preparations. The absolute (A) and normalized (B) amplitude values are averaged (n=5) and plotted against spot displacement distance (mm). The standard deviation is less than 0.5, 4.0 and 0.7 mV in graph A for the points in the normal retina (●), normal retina treated with DA (■) and DA cell-deprived retina (▲), respectively. Taken in part from Teranishi et al (1983).

Furthermore, in the present experiments, the amplitude of the S-potentials increased markedly following intravitreal application of DA to DA cell-deprived preparations, indicating that the postsynaptic site of the horizontal cell membrane remained intact after neurotoxic destruction of DA cells.

All of the apparent effects of DA observed in the present study can be explained on the basis of an uncoupling action of DA at gap junctions between horizontal cells. Recently, in the teleost retina DA has been reported to reduce markedly the receptive field of horizontal cells and at the same time to increase their membrane and coupling resistances (Laufer, Sales 1981). The absence of DA from the retina may reduce the membrane and coupling resistances, resulting in the reduced amplitude of S-potentials, enlarged receptive field and wider dye-coupling area.

A clear-cut correlation between electrical and dye coupling has been demonstrated with external horizontal cells in the turtle retina (Piccolino et al 1982), where GABA is assumed to participate in regulation of the spatial properties of the S-potentials. In the turtle retina, the DA cells are of the amacrine cell-type, and so differ from the interplexiform cell-type in the carp retina. However, DA in the carp and GABA in the turtle appear to play a common role in regulation of the S-potential in the external horizontal cells. However, our recent experiments (Negishi et al 1983) showed that in the carp retina the GABA system is involved also in the spatial regulation, because a GABA antagonist, bicuculline, produced similar effects to those of DA on external horizontal cells. The effects of bicuculline were prevented by the presence of haloperidol and not observed in retinas deprived of DA cells. Therefore, bicuculline appears to act on the horizontal cells through the dopaminergic system in the carp retina. Further, it is of interest to note another difference between the two tissues, although the experimental conditions were different in both studies. In the carp retina the dye injected into a horizontal cell body diffuses mainly to neighboring cell bodies, whereas in the turtle retina it seems to spread from the cell body to the axon terminal, or vice versa. Therefore, in the carp the gap junctions at the horizontal cell body level may participate predominantly in electrical coupling while in the turtle those between axon terminals may function similarly.

SUMMARY

Horizontal cells in a given cellular layer of the fish retina are known to be electrically coupled via gap junctions, so that intracellularly injected dye normally diffuses to several neighboring cells. Dopamine (DA) application altered the spatial properties of photopic L-type S-potentials, by increasing the amplitude of the center responses to spots and decreasing that of surround responses to annuli. The present study on the carp retina showed that

DA also restricts intracellular Lucifer Yellow (LY) to single injected horizontal cells. This effect, like that of DA on the S-potentials, was antagonized by a dopaminergic blocker, haloperidol. Prior destruction of interplexiform cells with 6-hydroxydopamine (6-OHDA) reduced the amplitude of the S-potentials to centered light spots and increased the lateral spread of these potentials; the injected dye diffused extensively to numerous neighboring cells. After 6-OHDA treatment, however, DA application had the usual effect on the S-potential; it increased the amplitude at the centered spot, narrowed the receptive field and restricted LY to the injected cell. These effects were closely mimicked by dibutyryl-cAMP. The results suggest that dopaminergic interplexiform cells in the fish retina normally function to regulate the spatial properties of S-potentials in external horizontal cells, probably by acting on their junctional resistance via a DA-receptor mediated mechanism.

REFERENCES

Boycott BB, Dowling JE, Fisher SK, Kolb H, Laties AM (1975). Interplexiform cells of the mammalian retina and their comparison with catecholamine containing cells. Proc R Soc Lond Biol B 207:7.

Brown JH, Makman MH (1972). Stimulation by dopamine of adenylate cyclase in retinal homogenates and of adenosine 3',5'-cyclic monophosphate formation in intact retina. Proc Natl Acad Sci USA 69:539.

Dowling JE, Ehinger B (1975). Synaptic organization of the amine-containing interplexiform cells of the goldfish and Cebus monkey retinas. Science 188:270.

Dowling JE, Ehinger B (1978). The interplexiform cell system. I. Synapses of the dopaminergic neurons of the goldfish retina. Proc Roy Soc Lond Biol B 201:7.

Ehinger B (1976). Biogenic monoamines as transmitters in the retina. In Bonting SL (ed): "Transmitters in the Visual Process:, London: Pergamon Press, p145.

Ehinger B, Florèn I (1976). Indoleamine-accumulating neurons in the retinas of rabbit, cat, goldfish. Cell Tiss Res 175:37.

Ehinger B, Holmgren I (1979). Electron microscopy of the indoleamine accumulating neurons in the retina of the rabbit. Cell Tiss Res 197:175.

Ehinger B, Nordenfelt L (1977). Destruction of retinal

dopamine-containing neurons in rabbit and goldfish. Exp Eye Res 24:179.

Furness JF, Costa M, Willson AL (1977). Waterstable fluorophores, produced by reaction with aldehyde solutions, for the histochemical localization of catechol and indolethyl amines. Histochemistry 52:159.

Ishida AT, Fain GL (1981). D-Aspartate potentiates the effects of L-glutamate on horizontal cells in goldfish retina. Proc Natl Acad Sci USA 78:5890.

Kaneko A (1971). Electrical connections between horizontal cells in the dogfish retina. J Physiol (London) 213:95.

Kaneko A, Stuart AE (1980). Coupling between horizontal cells in the carp retina examined by diffusion of Lucifer yellow. Biol Bull 159:486.

Laufer M, Negishi K, Drujan BD (1981). Pharmacological manipulation of spatial properties of S-potentials. Vision Res 21:1657.

Laufer M, Salas R (1981). Intercellular coupling and retinal horizontal cell receptive field. Neurosci Lett 7:S339.

Murakami M, Ohtsu K, Ohtsuka T (1972). Effects of chemicals on receptors and horizontal cells in the retina. J Physiol (London) 227:899.

Naka K-I, Rushton WAH (1967). The generation and spread of S-potentials in fish (Cyprinidae). J Physiol (London) 192:437.

Nakamura T (1979). Application of the Faglu method (Furness et al) for the histochemical demonstration of catecholamines to the cryostat section method. Acta histochem cytochem 12:182.

Negishi K, Drujan BD (1979a). Reciprocal changes in center and surround S-potentials of the fish retina in response to dopamine. Neurochem Res 4:313.

Negishi K, Drujan BD (1979b). Effects of catecholamines and related compounds on horizontal cells in the fish retina. J Neurosci Res 4:311.

Negishi K, Drujan BD (1979c). Effects of some amino acids on horizontal cells in the fish retina. J Neurosci Res 4:351.

Negishi K, Drujan BD, Laufer M (1980). Spatial distribution of catecholaminergic cells in the fish retina. J. Neurosci Res 5:621.

Negishi K, Kato S, Teranishi T (1982a). Density ratio of dopaminergic cells to other cells in the amacrine cell layer of the fish retina. Acta histochem cytochem 14:317.

Negishi K, Sutija V (1969). Lateral spread of light-induced

potentials along different cell layers in the teleost retina. Vision Res 9:881.

Negishi K, Teranishi T, Kato S (1981a). Similarity in spatial distribution between dopaminergic cells and indoleamine-accumulating cells of carp retina. Acta histochem cytochem 14:449.

Negishi K, Teranishi T, Kato S (1981b). Density of retinal dopaminergic cells and indoleamine-accumulating cells in different-sized carp. Acta histochem cytochem 14:596.

Negishi K, Teranishi T, Kato S (1981c). 5,7-Dihydroxy-tryptamine destroys indoleamine-accumulating cell bodies in carp retina. Acta histochem cytochem 14:654.

Negishi K, Teranishi T, Kato S (1982b). Neurotoxic destruction of dopaminergic cells in the carp retina revealed by a histofluorescence study. Acta histochem cytochem 15:768.

Negishi K, Teranishi T, Kato S (1983). Regulating effects of dopamine on electrical and dye couplings of external horizontal cells in carp retina. Abstracts of the 6th Ann Meeting of the Jpn Neurosci Soc, Kyoto, January 18-19, 1983. Neurosci Letts (in press).

Piccolino M, Neyton J, Witkovsky P, Gerschenfeld HM (1982). γ-Aminobutyric acid antagonists decrease junctional communication between L-horizontal cells of the retina. Proc Natl Acad Sci USA 79:3671.

Stewart WW (1978). Functional connections between cells as revealed by dye-coupling with a highly fluorescent naphthalimide tracer. Cell 14:441.

Sugawara K, Negishi K (1973). Effects of some amino acids on light-induced responses in the isolated carp retina. Vision Res 13:2479.

Teranishi T, Kato S, Negishi K (1982). Lateral spread of S-potential components in the carp retina. Exp Eye Res 34:389.

Teranishi T, Negishi K, Kato S (1983). Dopamine modulates S-potential amplitude and dye-coupling between external horizontal cells in the carp retina. Nature (in press).

Watling KT, Dowling JE (1981). Dopaminergic mechanisms in the teleost retina. I. Dopamine-sensitive adenylate cyclase in homogenates of carp retina; effects of agonists, antagonists and ergots. J Neurochem 36:559.

Wu SM, Dowling JE (1978). L-aspartate: Evidence for a role in cone photoreceptor synaptic transmission in the carp retina. Proc Natl Acad Sci USA 75:5205.

K. Negishi; S. Chichibu; Y. Saito

The Physiology of Excitable Cells, pages 549–556

INTRACELLULAR RECORDING WITH HORSERADISH-PEROXIDASE
ELECTRODES REVEALS DISTINCT FUNCTIONAL ROLES OF RETINAL
PLEXIFORM LAYERS

P.L. Marchiafava, R. Weiler* and E. Strettoi

Istituto di Neurofisiologia del CNR, Pisa,
Italy and *Zoologisches Institut der
Universität, München, FRG

The following is a brief account of our study on the
functional architecture of the inner plexiform layer in the
turtle retina.

The significance of the retinal structure is that it
contains all cellular inputs converging onto ganglion cells,
which in turn are known to forward the visual message to the
brain.

The work consisted in finding the detailed morphology
of intracellularly recorded retinal cells and analyzing possible functional relationships with the aid of the structural data. For this, we have used microelectrodes filled
with a solution of horseradish peroxidase (see Marchiafava,
Weiler 1982 for a complete description of methods).

As an aid in following this report, Figure 1 shows a
schema of the retina, as it may be seen in a radial section.

The retinal cell types are represented here to illustrate the path followed by the visual response: from
photoreceptors (rod & cones) to bipolar cells and to ganglion cells. Horizontal and amacrine cells are known to
provide the lateral spread of visual signals. In this
schema, however, the actual connections between the various
cellular subtypes are arranged following the results to be
shown below. It is evident that the inner plexiform layer
stands as a large portion of the retina, including the terminal arborizations of bipolar and amacrine cells, and the
dendrites of ganglion cells.

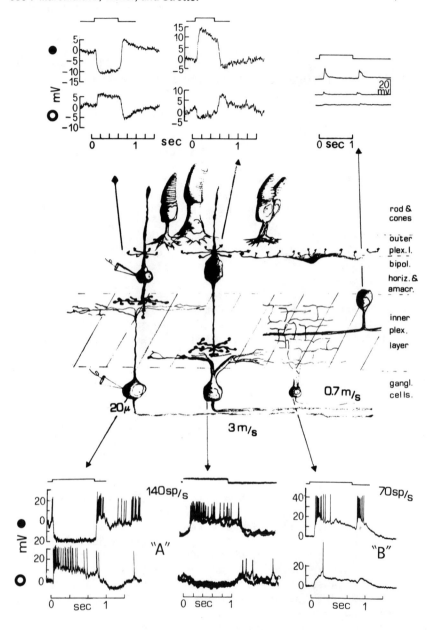

Fig. 1. Schematic drawing of the turtle retina, as seen in a transverse section. Arrows relate the different retinal cell types to their photo-responses, recorded intracellularly.

In our study, we recorded photoresponses intracellularly from ganglion cells that were subsequently identified histologically by the presence of the axon. Photoresponses were also recorded from the other types of cells known to impinge onto ganglion cells, that is the bipolar and amacrine cells.

Connectivity between a given input and the recipient cell was thought to exist whenever the time course of their photoresponses coincided, and whenever the structural organization of the two cells would support such a conclusion.

The responses of the ganglion cells to illumination of a small circle or an annulus, centered on the cell's receptive field, are shown at the bottom of Figure 1.

The ganglion cell photoresponses are of two types, distinguished by their time course: in one case they are sustained, but of opposite polarity depending on the nature of the stimulus; in the other case the responses are represented by depolarizing "on" and "off" transients to both central and peripheral illumination. These latter cells fired at lower frequency (below 70 spikes/sec) than the sustained cells (up to 140 spikes/sec) and had a lower conduction velocity (0.7 vs. 1-3 m/sec; Marchiafava, Weiler 1980).

When the recorded ganglion cells were examined histologically, following the histochemical peroxidase reaction with diaminobenzidine (see Marchiafava, Weiler 1980 for reference), we found that the "sustained ganglion cells" were structurally quite different than the "transient ganglion cells." The sustained cells had a large cell body, a thick axon and dendrites that were monostratified at either the outermost or the innermost portion of the inner plexiform layer. In contrast, the transient ganglion cells were radially diffused across the whole inner plexiform layer (Marchiafava, Weiler 1980).

To understand which inputs could account for the different types of ganglion cell photoresponses, intracellular responses to the same form of illumination were recorded from bipolar and amacrine cells. Bipolar cell responses (Fig. 1, upper records) were similar to those previously recorded from other animal species (Werblin, Dowling 1969; Kaneko 1970; Richter, Simon 1975). They consisted of a sustained response which varied in polarity according to the form of illumination.

Considering the structure of these cells, also injected with peroxidase, it was interesting to note that the bipolar cells which hyperpolarized to central illumination had processes stratified at the outermost portion of the inner plexiform layer, i.e., in the same layer where ganglion cells producing homologous photoresponses were distributed. Accordingly, bipolar cells showing depolarizing responses to central illumination are distributed close to the dendrites of centre-depolarizing ganglion cells (Marchiafava, Weiler 1980).

The good matching of the photoresponses produced by bipolar and ganglion cells with processes intermingled in the same level of the inner plexiform layer suggests direct transmission only between homologous cell types. This conclusion is supported by two types of evidence: (1) Artificial polarization of a bipolar cell produces a membrane potential change of the same sign in a nearby ganglion cell (Marchiafava 1983); (2) The sustained ganglion cell responses all showed similar reversal potentials, suggesting that a single input transmitter produced photoresponses to all forms of illumination (Fig. 2) (Marchiafava, Weiler 1980).

In contrast to the sustained ganglion cells, those producing transient photoresponses showed two distinct reversal potentials, depending on the form of illumination used (Fig. 2). The additional, more negative, reversal potential may be attributed to the amacrine input (see Marchiafava, Weiler 1980 for reference).

In conclusion, two structurally different classes of ganglion cells have been found, characterized by different photoresponses (sustained and transient) determined by two different types of input synaptic organizations: represented in one case by bipolar cells alone, in the other by the addition of amacrine cells.

It is interesting to note that the sustained ganglion cells, because they receive input from bipolar cells alone, become the isolated output of the outer plexiform layer in those cases where bipolar cells receive visual signals resulting from complex interactions among photoreceptors, horizontal and bipolar cells themselves. The transient ganglion cells, on the other hand, because they receive the additional amacrine input, represent the output of the inner plexiform layer.

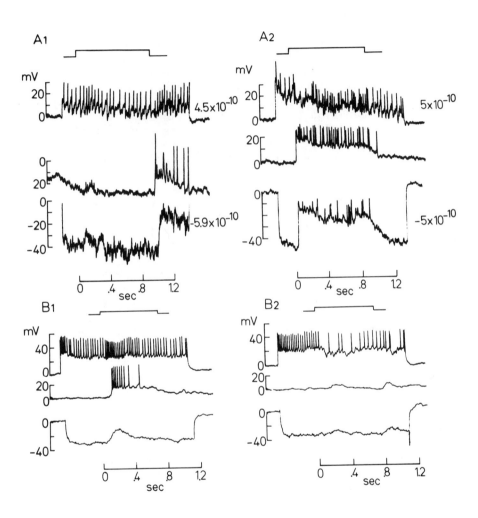

Fig. 2. Direct current injection (indicated to the right of record, amp.) across sustained (A) and transient ganglion cells (B) reveals reversal of photoresponse to central (left) and peripheral illumination (right). The middle records in A and B are the responses in absence of current injection.

It became apparent, therefore, that we had possibly found an appropriate means for exploring, independently, the functional properties of each of the two plexiform layers. First, we measured spectral sensitivity of ganglion cells to determine which retinal plexiform layers are involved in color discrimination.

It was found that all sustained ganglion cells showed either single or double color opponency (Fig. 3), with peak sensitivity at various wavelengths, depending on the cell. Transient cells, instead, all showed peak at about 646 nm (Marchiafava, Wagner 1981). These results indicate that color discrimination is a retinal function entirely dependent on interactions among photoreceptors and horizontal cells, from which the amacrine cells are excluded. Indeed, all amacrine cells recorded were most sensitive at about 646 nm (Marchifava, Torre 1978).

Another important visual function tested was directional selectivity which was found to be an exclusive property of the transient ganglion cells. Moreover, in the turtle retina, as in other animals studied, directional selectivity consisted of a preferred and a null response to a visual stimulus moving along a specifically oriented axis of the receptive field (Marchiafava 1979). We have found that the null response, recorded intercellularly for the first time in this study, resulted from a critically timed inhibitory input, whose reversal potential could be clearly attributed to amacrine cells.

In conclusion, the present results indicate the existence of two classes of ganglion cells which are distinguishable by their structural and functional characteristics. The new aspect of the present classification is that the two classes of ganglion cells are characterized by two different organizations of inputs. In one case (Type A cells), the input originates from bipolar cells alone while the other class of cells (Type B) instead receives mixed inputs from bipolar and amacrine cells.

As an attractive consequence, important retinal functions like spectral sensitivity and directional selectivity may now be interpreted as the outcome of discrete and identified retinal circuits, independently converging on either of the two classes of ganglion cells.

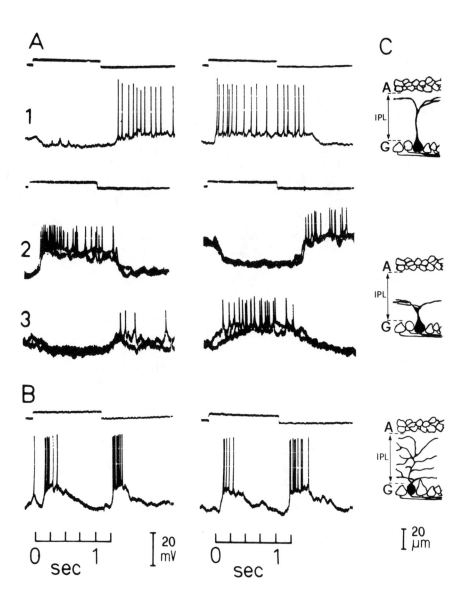

Fig. 3. Sustained (A) and transient ganglion cell responses (B) to monochromatic illumination of central (left)
and peripheral areas (right). Wavelengths are at about
640 nm, except in A3, which is 489 (B, left) and 434 nm
(B, right).

Acknowledgement

The authors thank Mr. G. Bottaro for organizing and typing the manuscript in the present form.

References

Kaneko A (1970). Physiological and morphological identification of horizontal, bipolar and amacrine cells in the goldfish retina. J Physiol Lond 207:623.

Marchiafava PL (1979). The responses of retinal ganglion cells to sationary and moving visual stimuli. Vision Research 19:1203.

Marchiafava PL (1983). The organization of inputs establishes two functional and morphologically identifiable classes of ganglion cells in the retina of the turtle. Vision Research (in press).

Marchiafava PL, Torre V (1978). The responses of amacrine cells to light and intracellularly applied currents. J Physiol Lond 276:83.

Marchiafava PL, Wagner HG (1981). Interactions leading to color opponency in ganglion cells of turtle retina. Proc R Soc Lond B 211:261.

Marchiafava PL, Weiler R (1980). Intracellular analysis and structural correlates of the organization of inputs to ganglion cells in the retina of the turtle. Proc R Soc Lond B 208:103.

Marchiafava PL, Weiler R (1982). The photoresponses of structurally identified amacrine cells in the turtle retina. Proc R Soc Lond B 214:403.

Richter A, Simon EJ (1975). Properties of centre-hyperpolarizing red-sensitive bipolar cells in the turtle retina. J Physiol Lond 248:317.

Werblin FS, Dowling JE (1969). Organization of the retina of the mudpuppy, *Necturus maculosus*. II Intracellular recording. J Neurophysiol 32:339.

The Physiology of Excitable Cells, pages 557–567
© 1983 Alan R. Liss, Inc., 150 Fifth Avenue, New York, NY 10011

VISUAL CORTICAL PLASTICITY STUDIED IN A BRAIN SLICE
PREPARATION

K. Toyama and Y. Komatsu

Department of Physiology, Kyoto Prefectural
School of Medicine and Department of Physiology
School of Medicine, Nagoya University, Japan

Visual cortical plasticity has been studied extensively
during the last two decades (Hubel, Wiesel 1963, 1970, 1977;
Wiesel, Hubel 1963, 1965; Hirsch, Spinelli 1970; Blakemore,
Cooper 1970; Blakemore, Van Sluyters 1975). It has been
established that neuronal responsiveness in the visual
cortex is highly modifiable by visual experience. The
modifiability is found only during the early infancy of
the animal (Wiesel, Hubel 1963, 1965a, 1965b; Hubel, Wiesel
1970; Blakemore, Van Sluyters 1975). Once the neuronal
responsiveness is modified, the modified state persists
throughout the life of the animal. It is postulated that
neuronal networks in the visual cortex are still immature
in infancy and that their structure can be modified by
visual inputs within certain limitations (Wiesel, Hubel
1963, 1965a, 1965b; Hubel, Wiesel 1970; Blakemore, Van
Sluyters 1975).

In contrast to its remarkable plasticity during infancy,
the visual cortex of the adult animal is characterized by an
orderly, fixed pattern of organization. As schematically
illustrated in Fig. 1, cells responding predominately to the
inputs through the right eye are clustered into a slab perpendicular to the cortical surface (R columns) and those
responding to the left eye into another slab (L columns).
Orthogonal to the slabs of ocular dominance are smaller slabs
which contain cells with the same preference for the orientation of light stimuli. The orientation slabs are arranged
in a well-ordered sequence, each selectively responsive to
stimuli of slightly different orientation, from vertical
through horizontal and back to vertical (Hubel, Wiesel 1977).

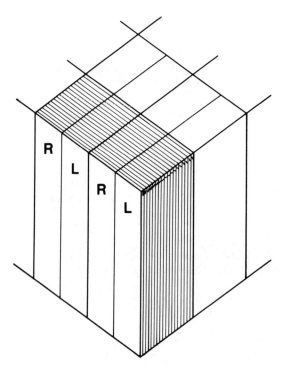

Fig. 1. Three-dimensional scheme of columnar structures in the visual cortex. Two features of cortical organization, ocular dominance and orientation columns, are represented along the two coordinates orthogonal to each other. The columns denoted by L and R are those for the inputs through the left and right eyes; those denoted by vertical and horizontal bars represent the columns for the vertical and horizontal orientation, respectively (Hubel, Wiesel 1977).

Finally there is a tangential pattern of organization (Toyama, Matsunami, Ohno, Tokashiki 1974). Cells of the same efferent connectivity are arranged in the same lamina; 1) cells whose axons project to the contralateral visual cortex or to the ipsilateral visual association cortex are located in layer III (E1 in Fig. 2A), 2) those projecting to the superior colliculus in layer V (E2) and 3) those to the lateral geniculate nucleus (LGN) in layer VI (E3). Tangential organization is also found in afferent connectivity. Only cells in layers III-V (E1, E2, I1 in Fig. 2B) receive excitatory inputs from LGN. Cells in layers II and VI (I′ and E3) receive the inputs indirectly through an

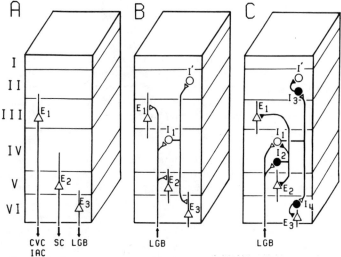

Fig. 2. Laminar structures in cat's visual cortex. A.
efferent organization. E1: efferent cell projecting to the
contralateral visual cortex (CVC) or the ipsilateral association cortex (IAC), E2: corticotectal cell projecting to
the superior colliculus (SC), E3: cortigogeniculate cells to
the lateral geniculate nucleus (LGN). B and C: organization
of geniculate inputs, B for excitation, C for inhibition.
I1: interneuron mediating disynaptic excitation of cells in
layers II (I´) and VI (E3), I2: interneuron mediating disynaptic inhibition of cells in layers III-V (E1, E2, I1),
I3 and I4: interneurons mediating trisynaptic inhibition of
cells in layers II and VI (I´ and E3) (Toyama et al. 1974).

excitatory interneuron in layer IV (I1). Inhibition is exerted on cells in layers III-V (E1, E2 and I1) through an
inhibitory interneuron in layer IV (I2 in Fig. 2C), which
receives excitation from LGN. On cells in layers II and VI
(I´ and E3), inhibition is exerted through additional inhibitory interneurons (I3 and I4). These perpendicular
and tangential organizations in the visual cortex have been
demonstrated either functionally as clusters of cells with
similar responsiveness to light stimuli (Hubel, Wiesel 1977)
or structually as clusters of cells of similar input-output
connections (Toyama et al. 1974).

     The reconciliation of this rigid pattern of organization
in the adult animal with the remarkable plasticity in neuronal responsiveness in the infant animal has been a basic

problem in developmental neurobiology. One approach has
been to assume that only certain features of cortical organization can be modified by visual experience. This assumes
that the modification of photic responsiveness by visual
experience involves plastic change at a specific subset of
synapses in the visual cortex. This approach is plausible
because a population of cells in the newborn animal display
mature photic responsiveness, while responsiveness is immature in most neurons (Blakemore, Van Sluyters 1975).

Approaches to the visual cortical plasticity have been
devoted so far to investigation of the modification of neuronal responsiveness by experimental manipulation of the
visual experience of the animal. The modification of neuronal responsiveness is probably a consequence of structural
modification of neuronal networks by the visual inputs experienced by the animal during its critical period. All
of these inputs may not be known or under experimental
control.

Another approach is the reductive one of keeping experimental conditions as simple as possible. We have
attempted to take this approach. Using a slice preparation
of striate cortex from infant kittens, we have studied the
modification of synaptic transmission after electrical
stimulation of the white matter. Synaptic transmission
was studied in two ways: 1) by extracellular recording
analysis of the delays in orthodromic activation of cells.
and 2) by current-source density (CD) analysis of the field
potentials (FPs) evoked by stimulation of the white matter.
The former technique provides information concerning the
conduction velocities of orthodromic impulses, central
delays in orthodromic activation, and the relationship
between the modes of orthodromic activation and cellular
locations. Most of the FPs analyzed by the second technique
can be ascribed to excitatory synapses in the visual cortex.
We found that modification of synaptic transmission occurred
in two restricted zones in the visual cortex, in the granular and supragranular layers. Modification was stronger
and more long-lasting in the supragranular layer than in
the granular layer. This restricted localization of synaptic plasticity is consistent with the view that development
of the visual cortical networks proceeds by modification of
immature structures by visual inputs along the frame-works
of the mature structures.

Kittens (7-49 days old) were decapitated, and transverse slices (0.5-0.7 mm thick) of striate cortex were dissected and incubated in Krebs-Ringer solution saturated with a mixture of O_2 and CO_2 (95:5) at 33°C for one hour. After the incubation, the slice was placed in a recording chamber perfused with the same solution as that used for the incubation. The temperature of the perfusing solution was maintained at 33°C. A glass microelectrode (ml in Fig. 3A) was inserted into the visual cortex (VC) for recording extracellular field potentials and impulses from individual cortical cells. Another microelectrode (m2) was placed in the white matter (WM) to monitor afferent volleys produced by WM stimulation. Two pairs of bipolar stimulating electrodes (s1 and s2), each of which were composed of teflon-coated platinum wires (interpolar distance, 0.15 mm), were placed in WM at about 1 mm separation. Pulse stimuli (0.5-3 mA, 0.1 ms) were supplied to these stimulating electrodes from constant current stimulators (internal resistance, 8 Mohm). The s1 electrode was used for both conditioning and test stimulation of WM. Paired stimulation with s1 and s2 was used for studying neuronal connections of cortical cells within the WM. Conduction velocities were determined by measuring the difference in latency of orthodromic impulses evoked by s1 and s2 stimulation and knowing the distance between s1 and s2. The central delays (time spent for orthodromic activation of cortical cells after arrival of impulses at the synapses) were determined by extrapolating the latency-distance relationship in orthodromic activation with s1 and s2 to zero distance. For conditioning, repetitive pulses at 2 Hz were given for one hour. Test stimulation was repeated throughout the entire period of the recording session attt 0.05 Hz. The 0.05 Hz stimulation was infrequent enough to avoid cumulative effects such as depression or potentiation (Tsumoto, Suda 1979).

Events in individual cortical cells were compared between conditioned and unconditioned slices which had been dissected from the same hemisphere of the same kittens. Seven to sixteen cells were sampled extracellularly from each slice during a recording session which lasted 3-7 hours. Ninety-three and 77 cells were collected from 8 pairs of unconditioned and conditioned slices, respectively. All conditioned slices exhibited FPs potentiated to a level more than twice the control. Recordings in the conditioned slices started 2 hours after the onset of conditioning

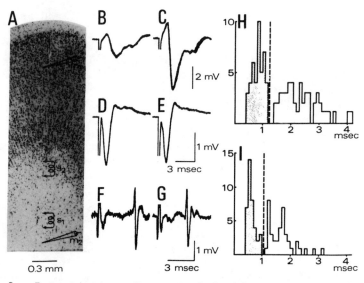

Fig. 3. Potentiation of cortical field potentials and shortening of latencies of cortical units. A illustrates arrangements
of stimulating and recording electrodes in a histological section of a cortical slice (Nissl staining). B and C: cortical
field potentials evoked by test stimulation before (B) and 2
hours after (C) the onset of conditioning stimulation at 2 Hz.
D and E: volleys in WM evoked by test stimulation before and
after conditioning. F and G: impulses extracellularly recorded
from cortical cells in response to test stimulation with S1
and S2, sampled from an unconditioned (F) and conditioned (C)
slice. H and I: histograms of central delays for unconditioned
(H) and conditioned (I) slices. Hatched columns represent
monosynaptic and blank columns polysynaptic values. Dotted
lines indicate the critical delay discriminating between mono-
and polysynaptic cells.

Figs. 3B-E illustrate FPs in the visual cortical slice and afferent volleys in the WM before and 2 hours after the onset of
conditioning stimulation. The FPs were 2 times larger after
conditioning stimulation (C) than those before conditioning
stimulation (B), while the afferent volleys in WM remained
essentially unchanged (cf. D and E). the potentiation of the
FPs started immediately after the end of the conditioning
stimulation, reached a maximum within a few hours, and lasted
as long as the slice preparation remained intact (observed up
to 15 hours). The potentiation of FPs was dependent upon the
age of the kittens used; it occurred in more than 80% (30/36)

of slices sampled from kittens at ages 21-23 days, about 50%
(3/6 and 4/10) at 14-20 and 35-41 days, and none (0/2 and
0/2) of those at 7-13 and 42-49 days.

stimulation. The sampled cells were located in a region
0.3-0.8 mm from the cortical surface, where a large current
source increase for FPs was revealed by current source-
density analysis (see below). In each cell the conduction
velocity and the central delays were determined for orthodromic impulses evoked by s1 and s2 stimulation (Figs. 3F
and G). Values of central delays in the unconditioned
slices comprised two peaks at 0.8 and 2 ms, having a clear
separation at 1.2 ms (dotted line in Fig. 3H). These two
peaks probably represent monosynaptic (hatched columns)
and polysynaptic activation (blank columns). Central
delays in the conditioned slices were also distributed
bimodally (Fig. 3I). However the central delays in the
conditioned slices were significantly shorter than those
in control slices for both monosynaptic (mean delay, 0.61 $\pm$
0.15 ms in conditioned slices vs. 0.80 $\pm$ 0.21 ms in unconditioned slices) and polysynatpic activation (1.71 $\pm$ 0.46
vs. 2.35 $\pm$ 0.67 ms). Both of these differences are statis-
tically significant (P < 0.001). It is concluded that the
FP potentiation may involve both mono- and polysynaptic
transmission from cortical afferents to cortical cells.
By contrast, there was no significant differences in conduction velocity between impulses in afferent fibers of
the WM in conditioned (0.74 + 0.31 m/sec) and unconditioned
(0.70 + 0.32 m/sec) slices (p - 0.4). The slow conduction
velocities of afferent impulses are consistent with the
finding that the visual afferents remain unmyelinated until
32 days after the birth (Grafstein 1963; LeVay, Stryker
1978).

The locations of the potentiated synapses were determined by CD analysis of FPs (Nicolson, Freeman 1975;
Mitzdorf, Singer 1978). Precision of the analysis depends
upon the following three conditions: 1) precision in
placement of the recording microelectrode, 2) reproducibility of the FPs evoked by test stimulation in amplitude
as well as in the time course, and 3) constancy of the
electrical resistance in the visual cortex along the track
of the recording. These requirements are favourably satisfied in the slice preparation. Placement of the recording
microelectrode can be made with great preciscion using a

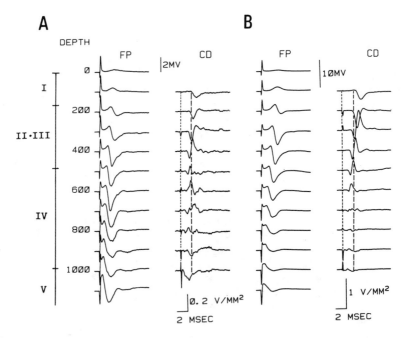

Fig. 4. Current source-density analysis of cortical field potentials. A and B: cortical field potentials. FP and current source-densities (CD) evoked by test stimulation. A: before conditioning. B: after conditioning. Dotted lines in traces of CD represent the onsets of test stimulation and interrupted lines the border value to discriminate mono- and polysynaptic current-sinks.

3-dimensional manipulator under direct visual control. Fluctuations of the FPs are also minimized in the slice preparation, in which the visual cortical circuitry is isolated from the rest of the brain. Constancy of the electrical resistance is also relatively assured in the brain slice which is surrounded by a large volume of Ringer solution and has no flowing circulatory system.

The CD analysis was made of the FPs evoked by test stimulation before and after conditioning stimulation of the WM. Before conditioning stimulation, the FPs were largest in the granular layer, and smaller in the supra- and infragranular layers (traces FP in Fig. 4A). CD analysis demonstrated prominent current-sinks in the granular and supragranular layers, which probably represent excitatory post-synaptic potentials, and action currents generated in the local area of the visual cortex. The current-sinks were divided into mono- and polysynaptic components by taking 2.2 ms as the border (interrupted line in traces CD in Fig. 4A) which is a sum of 1.2 ms for a central delay (see above) and 1 ms for conduction of orthodromic impulses over a distance of 0.7 mm from WM to the center of the granular layer at the mean conduction velocity of 0.7 m/s (see below).

Monosynaptic current-sinks were most prominent in the granular layer at 500 and 600 μm from the cortical surface, while polysynaptic sinks were largest in the supragranular layer at 300 and 400 μm. Eight hours after conditioning stimulation, the FPs increased markedly, i.e., the FP at 400 μm was 5 times as large as that of controls (cf. traces FP in Fig. 4B with those in Fig. 4A). There was corresponding enhancement of CD in two restricted zones: a six-fold increase in the polysynaptic current-sinks in the supragranular layer (cf. traces CD at 300 μm in Fig. 4B and at 400 μm in Fig. 4A) and a three-fold increase in the monosynaptic current-sinks in the granular layer (cf. traces at 600 μm in Figs. 4A and B).

The time course of the enhancement was studied by CD analysis of the FPs evoked by test stimulation at different time intervals after conditioning stimulation. It was demonstrated that monosynaptic currents were reduced slightly during the conditioning stimulation, increased rather rapidly after conditioning stimulation (within 2 hours), and declined gradually (for about 10 hours). Polysynaptic currents were depressed strongly during conditioning stimulation, and increased gradually for 10 hours after conditioning stimulation.

The results indicate that conditioning activation of WM fibers causes a long-term enhancement of synaptic transmission in the visual cortex. The enhancement was greatest for intracortical transmission in the supragranular layer II, moderate for geniculo-cortical transmission in the granular layer and negligible for intracortical transmission in the infragranular layers. Thus synaptic plasticity that can be demonstrated by electrical activation of the white matter exists in a slice preparation of the visual cortex of the infant kitten. Since electrical stimulation activates all neural elements contained in the white matter, localization of the synaptic plasticity in the visual pathway remains unclear. However, the fact that monosynaptic currents were confined to the granular layer, which receives terminations of geniculate axons predominantly suggests that the geniculostriate synapses are involved in the enhancement of the synaptic transmission. In support of this view we could demonstrate similar enhancement of FPs in the visual cortex after conditioning stimulation of the lateral geniculate nucleus or the optic nerve.

In summary, synaptic circuitry that is modifiable by activity of the geniculate inputs exists in restricted zones of the visual cortex, with a great deal of plasticity in the supragranular layer, a smaller amount in the granular layer and probably no plasticity in the remaining layers. The restricted localization of the synaptic plasticity is consistent with the view that development of the immature structures in the visual cortical networks proceeds within constraints set by the mature structures. It remains to be investigated whether the synaptic plasticity demonstrated in the visual cortex by electrical stimulation of the visual pathways is the same as the modification of responsiveness by altered visual experience. The possibility that the same form of plasticity is involved is supported by the finding that the critical period for synaptic plasticity we have studied roughly agrees with that for the altered visual experience.

This work was supported by Grant-in-Aid for Scientific Research (No. 557029) and that for Special Project Research (No. 410808) of the Ministry of Education, Science and Culture of Japan.

REFERENCES

Blakemore C, Cooper GF (1970). Development of the brain depends on the visual environment. Nature 228:477.
Blakemore C, Van Sluyters RC (1975). Innate and environmental factors in the development of the kitten visual cortex. J Physiol 248:663.
Ferster D, LeVay S (1978). The axonal arborizations of lateral geniculate neurones in the striate cortex of the cat. J comp Neurol 182:923.

Gilbert CD, Wiesel TN (1979). Morphology and intracortical projections of functionally characterized neurones in the cat visual cortex. Nature 280:120.

Grafstein B (1963). Postnatal development of the transcallosal evoked response in the cerebral cortex of the cat. J Neurophysiol 26:79.

Hirsch HVB, Spinelli SN (1970). Visual experience modifies distribution of horizontally oriented receptive fields in cats. Science 168:869.

Hubel DH, Wiesel TN (1963). Receptive fields of cells in striate cortex of very young, visually inexperienced kittens. J Neurophysiol 26:994.

Hubel DH, Wiesel TN (1970). The period of susceptibility to the physiological effects of unilateral eye closure in kittens. J Physiol 206:419.

Hubel DH, Wiesel TN (1977). Functional architecture of macaque monkey cortex. Proc Roy Soc B 198:1.

LeVay,S, Stryker MP (1978). The development of ocular dominance columns in the cat. In Ferrendelli TA (ed): "Aspects of Developmental Neurobiology," Society for Neuroscience Symposia Vol IV.

Mitzdorf V, Singer W (1978). Prominent excitatory pathway in cat visual cortex (A17 and A18): A current source-density analysis of electrically evoked potentials. Exp Brain Res 33:371.

Nicolson C, Freeman JA (1975). Theory of current source-density analysis and determination of conductivity tensor for anuran cerebellum. J Neurophysiol 38:356.

Tsumoto T, Suda K (1979). Cross-depression: an electrophysiological manifestation of binocular competition in the developing visual cortex. Brain Res 168:190.

Toyama K, Matsunami K, Ohno T, Tokashiki A (1974). An intracellular study of neuronal organization in the visual cortex. Brain Res 21:45.

Wiesel TN, Hubel DH (1963). Single-cell responses in striate cortex of kittens deprived of vision in one eye. J Neurophysiol 26:1003.

Wiesel TN, Hubel DH (1965). Comparison of the effects of unilateral and bilateral eye closure on cortical unit responses in kittens. J Neurophysiol 28:1029.

Wiesel TN, Hubel DH (1965). Extent of recovery from the effects of visual deprivation in kittens. J Neurophysiol 28:1060.

C. Edwards; K. Toyama; H. Sakata

The Physiology of Excitable Cells, pages 569–586
© 1983 Alan R. Liss, Inc., 150 Fifth Avenue, New York, NY 10011

ENDOGENOUS CHEMICAL SUBSTANCES AND CONTROL OF FOOD INTAKE

Yutaka Oomura, Nobuaki Shimizu & Toshiie Sakata

Department of Physiology, Faculty of Medicine,
Kyushu University 60, Fukuoka 812, Japan

Introduction
    Mankind is beset with two diametrically opposite problems. Large parts of the world's population exist in various states of malnutrition ranging from improper diet or insufficient food to outright starvation. At the other extreme are those who have more food available than they can consume, and in many cases appetites which either exceed their nutritional requirements or are directed toward unhealthy diets. It is ironical that man, having taken such great strides in controlling the world in which he lives, has lost control of his own appetite. This situation makes it important for us to determine the nature and functions of appetite so that, having lost natural control of it, we can consciously exert some degree of control over our food intake.
    Early study of feeding control led to a prediction of glucose responsive neurons in the hypothalamus (Mayer, 1955), and proof of their existence and functions. Extensive investigation of these neurons, others which are closely associated with them, and their central and peripheral counterparts produced hypothetical identification of the feeding and satiation centers in the hypothalamus, the lateral hypothalamic area (LHA) and the ventromedial hypothalamic nucleus (VMH). The interrelations of these centers with each other and with other sites in the body, as well as their effects on behavior have all been verified and greatly clarified by lesion and stimulation experiments.

Control Theory

Theories which could explain control of feeding and satiation by effects of endogenous factors on neural centers have been propounded. Mayer's glucostat theory (1955) involves the metabolite, glucose. The subject of glucose is important for its association with nutrition, feeding behavior (Fig.1), obesity, and anorexia. LHA and VMH neurons respond to glucose in different ways and by different mechanisms. Glucoresponsive neurons in the VMH have specific glucoreceptor sites, the existence and specificity of which have been demonstrated (Oomura, 1980).

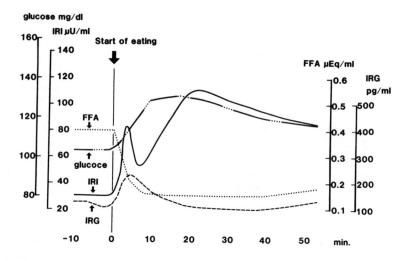

Fig. 1 Periprandial variation of two metabolites and two hormones. At the very beginning of meal, free fatty acid (FFA) decreases rapidly while insulin (IRI) and glucagon (IRG) rise fairly rapidly and glucose level rises gradually. A short time later, as glucose continues to rise, insulin drops transiently and glucagon returns gradually to its baseline level. Insulin again rises and its level then continues to run parallel to the glucose level, while FFA and glucagon remain at their intermeal levels   (Strubbe, Steffens, 1975).

The activity of about 25% of VMH glucoreceptor (GR) neurons is facilitated by glucose and by glucose plus insulin. Poligalitol, a glucose analogue (1-deoxy-D-glucose), has no effect on GR neurons when applied alone, but it does suppress a certain type of observed glucose effect (Oomura, 1980). The activity of most GR neurons increases in the presence of glucose and usually returns quickly to the preapplication level when glucose is stopped. The increased activity of some of these neurons persists after the glucose application has stopped. The reason for this continued activity, and its appearance in a few neurons but not in most, is not known, but it is believed to be related in some way to metabolism within the cell. Poligalitol appears to suppress this long term activity. It may be that these few anomalous neurons in the VMH are not glucoreceptors per se, but are simply neurons with some other function, and which happen to accept glucose which then supplies energy to increase their activity. The answer to this question requires further investigation.

The anomers of methyl-glucose (MG), α-MG and β-MG, affect GR neurons differently (Oomura, 1980). This discrimination indicates binding sites on the GR neurons, since such discrimination is not a metabolic function (Niki et al., 1974). Phlorizin, which depolarizes and then blocks the glucose effect and suppresses the electrical activity of pancreatic β-cells (Dean et al., 1975), acts similarly when applied to VMH GR neurons. It does not affect non-glucoreceptor neurons. Both phlorizin and glucose affect GR neurons in a dose related manner which indicates that each receptor site responds to two molecules of either agent (Oomura, 1980). These results are compatible with concepts of receptor sites, but do not prove their existence. After preparation and verification, fresh rat VMH antibody, applied to a VMH GR neuron, produced a large transient increase followed by irreversible cessation of activity. Old antibody, probably somewhat deteriorated, produced more moderate results, which were reversible. The presence of receptor sites on GR neurons of the VMH was thus proved (Oomura, 1980).

The GR neurons in the VMH, and their characteristics, meet the criterion of the glucostat theory which requires the existence of such neurons. It does not, however, indicate that these neurons exclusively control short or long term feeding behavior, nor does it prove that any given feeding episode is initiated or terminated by any effect of glucose. Other substances which will be discussed might contribute to initiation or termination of feeding, or both.

Certain unique neurons in the LHA respond to glucose by hyperpolarizing. LHA glucose-sensitive (GS) neurons seem to differ from other cells of the body in having Na^{+}

pump capacity substantially greater than that of the other cells, and glucose furnishes the excess energy needed to drive this extra pump capacity. The Na$^+$ pump of the VMH GR neuron is normal. Ingestion of glucose by either LHA or VMH neurons produces ATP which drives their Na$^+$ pumps. In the VMH these increases are commensurate with normal functions of neurons, so spontaneous activity increases and appropriate signals are transmitted. In the LHA GS neuron, however, since the Na$^+$ pump capacity greatly exceeds that of other cells the neuron is quickly driven into hyperpolarization by the Na$^+$ pump so that neuron activity decreases or stops (Oomura et al., 1974). Thus, neuronal signals which might otherwise induce feeding are suppressed by glucose. About 25% of LHA neurons are glucose-sensitive.

These glucose effects in the hypothalamus were demonstrated 20 years ago, but in light of more recent data (Niijima, 1981), those earlier results might well have been due, at least in part, to peripheral glucose receptors. Precise placement of glucose applications as small as 0.1 pmole within the hypothalamus by electrophoresis (Oomura, 1980) has proved that there are neurons in the hypothalamus that respond directly to glucose. There is no doubt that these chemoresponsive neurons are involved in feeding behavior, and partly so through their responses to glucose. One should, however, avoid the conclusion that feeding behavior is controlled solely by either these neurons or by glucose. Feeding is a complicated set of behaviors from motivation to consumation and the materials and networks (Oomura, 1980), both neural and humoral, which control feeding, are comparably intricate.

Kennedy's lipostat theory suggests that feeding behavior depends on monitoring the body's adipose tissue level (Kennedy, 1953). There are neurons which respond to FFA as well as to other materials in the VMH and LHA. Monitoring of FFA requires that it be related to the amount of adipose tissue in the body. Obese subjects have high FFA levels (Nisbett, 1972). FFA in the blood diminishes during lipogenesis and increases, at least transiently, during lipolysis. Insulin enhances glucose uptake by fat cells and affects lipogenesis.

FFA has been shown to affect glucose responsive neurons in the VMH and LHA oppositely from the effects of glucose (Oomura, 1980). FFA is not normally used by brain cells, so its modification of the glucose response is apparently not due to direct effects on or in the neurons and has been presumed to be due to the blocking of glucose uptake by FFA. FFA level in the blood falls after the termination of eating and then rises slowly until the beginning of the next meal (Fig.1) (Strubbe, Steffens, 1975). It appears that if there is lipostatic control of body weight, it must be in cooperation with other factors.

Triglyceride breakdown during lipolysis yields fatty acids and glycerol. During metabolism, excess carbohydrate is converted to triglyceride and glycerol-P is a byproduct in one step of this process. Davis et al. (1981) applied glycerol subcutaneously, and intraventricularly (i.c.v.) in mice. The results were qualitatively similar but were more dramatic for the i.c.v. injection. In both cases, there was significant weight loss which was maintained for as long as either injections or infusion of glycerol were maintained. Glucose injected i.c.v. had less than half of the glycerol effect on food intake, but not body weight; it was almost equivalent to glycerol in reducing body weight. It was concluded that blood glycerol may be important in controlling body weight and that this is probably mediated through the central nervous system, since effective i.c.v. doses were far below the threshold levels of subcutaneous doses. Also, glycerol was effective in only ventricle III, which is immediately adjacent to the VMH, and not in the lateral ventricle. Glycerol has not yet been applied directly to neurons, and until such applications are investigated in both the LHA and the VMH, it is inappropriate to speculate on either the effects of glycerol, or any mechanism which might possibly be related to such application.

I suggest, at this point, another feeding control theory. The effects of many metabolites, hormones and peptides on certain LHA and VMH neurons are well known. ATP is necessary for the energy required to operate the Na^+ pump or the Ca^{2+} pump of any membrane. Normally, at some relative activity level of VMH and the LHA neurons, feeding will be controlled. If there is an increase in a metabolite which produces ATP in either GR or GS neurons, a decrease in feeding behavior will follow, and vice versa. Whatever could adjust set point has not yet been determined, but putative candidates are available. Anything which changes the activity of the ATPase in a cell will change its pump activity, or it may be possible to change the number of carriers (if such exist) which are available to transport Na^+.

The above speculation considers conditions within a

neuron.

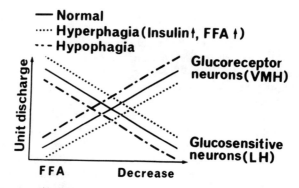

Fig. 2   Schematic diagram of one concept of set point
control of feeding behavior and body weight. Abscissa:
level of metabolites(s) (e.g. FFA) or other effective
agent(s) in the environment of hypothalamic neurons.
Ordinate: activity rate of neurons in the lateral area, and
ventromedial nucleus of the hypothalamus. The reciprocal
effects which are produced by the effective agent(s) will
cause the activity levels of the two neuron types to change
in opposite directions. At some point in the relative
activity levels, the effects of the two centers will be
equal. Note that the set point need not necessarily be the
point of equal activity, although that point can be used
for explanation of the phenomenon. On one side of the
point of equality, feeding will be induced and on the other
side it will be suppressed (Oomura, 1981).

It might also be possible to adjust the set point by
variation of extracellular factors. In the normal hunger
state, the activity of the GS neuron is high and the
activity of the GR neuron is low.   Referring to Fig. 2,
hunger is indicated by the left side of the graph, and
movement toward the right approaches satiation.   At the
left, the magnitude or extent of the hunger feeling is
indicated by the vertical distance between the lines of the
GR and GS neuron graphs, the greater the hunger feeling the
greater   the   vertical   distance.   As   one   approaches
satiation, i.e. toward the   right   of   the figure,   the
frequencies of the GR and the GS neurons will be equal at
some point. We can call this point the set point, and for
a normal animal it will be at the intersection of the two
solid lines.   In hyperphagic animals the FFA and insulin

levels are higher than normal (Nisbett, 1972), so neuronal activity in the VMH is lower and activity in the LHA is higher than normal. This condition is indicated by the dotted lines in the figure. This condition would move the intersection to the right, so in order to get the same neuron activity as that of a normal animal, i.e. to get a normal feeling of satiety, the FFA concentration must be decreased. Thus, the set point of a hyperphagic animal is located at the right of the normal set point, so to achieve a normal feeling of satiation the animal must eat more to increase the glucose and insulin concentrations which in turn reduce the FFA concentration. Otherwise, a feeling of satiation cannot be realized. If VMH activity is higher than normal and LHA activity is lower, the intersection moves to the left as indicated by the two broken lines. In this situation an increase in the level of FFA is needed to create a normal hunger feeling. The term "set point" is very easily used and the word is sometimes convenient to explain experimental data while its true properties are still unknown. This subject requires more future study, but recently found properties of insulin and its receptors in the brain could provide an answer.

Putative Set Points

The role of insulin in the body's general utilization of glucose is well known, but important details of its characteristics are less familiar. Shortly after describing GS neurons in the rat hypothalamus, the modulatory effects which insulin exerts on glucose sensitivity were reported (Oomura, 1973). Activity of GR neurons in the VMH is facilitated by simultaneous application of insulin and glucose, but it is slightly inhibited by insulin alone. Activity of GS neurons in the LHA is facilitated by insulin in a dose-dependent manner (Oomura, 1980).

Insulin binding sites in the brain have been identified (Havrankova et al., 1979; Oomura, Kita, 1981) although measurements of insulin in the blood plasma and brain of rats have provided a wide range of absolute values (Baskin, Dorsa, 1982; Havrankova et al., 1979; Oomura, Kita, 1981). Effects of a meal on blood insulin concentration are shown in Fig. 1. Possible reasons for these discrepancies cannot be discussed in the space available here, but all investigators appear to agree on two points: plasma insulin level varies widely from lean to obese animals and from one strain to another of lean animals; brain insulin content, although it varies with either the strain or the investigator, appears to be independent of plasma level and highly stable at whatever level it may be within each individual (Baskin, Dorsa, 1982; Havrankova et al., 1979; Oomura, Kita, 1981). Specific binding sites of insulin in the brain are most abundant in the hypothalamus and olfactory bulb, and are also unaffected by peripheral insulin concentration. The stability plus the known direct and modulatory effects of insulin might justify speculation that constant brain insulin could provide the set point to which other materials such as glucose, for instance, are compared in order to control energy input and defend body weight. Such comparisons could be quickly carried neurally to all concerned centers and organs.

A high incidence of glucagon-like immunoreactivity (GLI) was detected in the rat hypothalamus, and little was found in the pons and olfactory bulb. Immunoreactive pancreatic glucagon (IRG) was not found in any brain region studied (Lorén et al., 1979; Tager et al., 1980). Effects of a meal on IRG concentration are shown in Fig. 1.

GLI was demonstrated in some nuclei of the rat hypothalamus by indirect immunofluorescence. In the periventricular region, paraventricular nucleus, supraoptic nucleus, anterior hypothalamus, LHA, dorsomedial hypothalamic nucleus (DMH) and VMH, anti-GLI reacted with nerve fibers, but not with cell bodies (Lorén et al., 1979) although such cells had been previously reported in the paraventricular nucleus (Tager et al., 1980).

Neuronal activity of LHA, DMH and VMH neurons was recorded extracullularly during electrophoretic micro application of pancreatic glucagon (Oomura, 1981; Oomura et al., 1982). The suppression of hypothalamic neuron activity by glucagon was significant compared to its effect on cortical neurons. Glucagon suppressed the activity of GS neurons in the LHA, and the few effects which appeared in the VMH were also suppression of GR neurons. The inhibitory effect of glucagon in the LHA (the only place this was tested) was blocked by ouabain and enhanced by glucose at low concentration, so glucagon apparently regulates cyclic AMP and, consequently, Na^+-K^+ pump activity.

When glucagon was injected into ventricle III the reaction was rapid, transient decrease followed by increase of blood glucose. The time course of this action agrees with the initial effect being due to the closer nucleus, the VMH, and the final effect being due to the predominant activity of the farther nucleus, the LHA. Thus, slight suppression in the VMH produces pancreatic insulin through splanchnic disinhibition and this insulin secretion is finally reduced because of decreased vagal activity through the predominant suppression of glucagon by the LHA. Pancreatic glucagon may suppress LHA GS neurons hematogenically to contribute to the termination of feeding (Langhans et al., 1982), and GLI originating in the hypothalamus might act as an inhibitory neurotransmitter or neuromodulator.

Chemicals which affect the hypothalamus

The GS and GR neurons in the LHA and VMH are not glucose specific in their sensitivities. Many other materials either affect these neurons directly or modulate the effects of other factors.

There seems to be a general agreement that a large part of feeding activity of an animal is mediated in the brain and at peripheral sites. The problem still remaining, of course, is resolution of the mechanism of this mediation. It is apparent that there are no known effects or relations of either glucose or FFA which, by themselves, solve all of the problems of weight control.

Figure 3 summarizes the effects of various chemicals on LHA and VMH neurons when applied iontophoretically. It is interesting (and important) to note that those substances which match glucose in its LHA effects also suppress short term feeding behavior. In the long term the comparison is not so reliable, but this could be due, in part at least, to toxic effects, since some of the materials (for instance, 2-DG) are not physiological in the way they are used (Shiraishi et al., 1982). Those which have effects opposite to glucose also tend to have short term effects which are opposite and long term effects which may or may not be so.

FFA, insulin, 2-DG, and 3-OMG have already been described as facilitators of LHA activity and suppressors of VMH activity (Oomura, 1980). Glucose analogues probably enter a neuron by the same channels that are used by glucose. Once inside the cell they interfere with glucose metabolism, but cannot themselves be metabolized to produce energy.

TRHA, which is found in the urine of anorexia nervosa patients, has a Gly residue in place of the Pro residue of TRH (Reichelt et al, 1978). This material was reported to depress body weight when injected i.p. into mice. TRHA suppressed 80% of LHA GS neurons (n=14). TRH, on the other hand, stimulated 41% and inhibited 6% of VMH GR neurons (n=17), while it had no effect on GS neurons (Ishibashi et al., 1979). Neither TRHA nor TRH affected non-glucose responding neurons in either the VMH or the LHA. Feeding was inhibited by i.c.v. application of TRH (Vijayan, McCann, 1977).

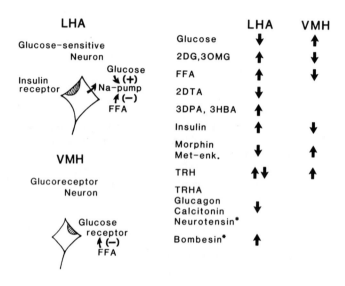

Fig. 3 Effect of various chemicals on rat LHA GS and VMH GR neurons. Left, mechanism of effect of glucose, free fatty acids (FFA) and insulin on both neuron types. LHA: electrogenic Na-pump is facilitated by glucose but slowed by FFA which inhibits uptake of glucose into the neuron. This neuron membrane has insulin receptor sites. VMH: GR neuron has glucoreceptor sites. FFA inhibits firing of this neuron by interposing between glucose molecules and the receptor site, hence inhibiting binding and/or uptake of glucose by the neuron (Oomura, 1980). Right, 2-DG, 2-deoxy-D-glucose; 3-OMG;, 3-0-methyl-D-glucose, 2-DTA, 2-deoxytetronic acid; 3-DPA, 3-deoxypentonic acid; 3-HBA, 3-hydroxy-butyric acid; Met-enk, methionin-enkephalin; TRH, thyrotropin releasing hormone; TRHA, TRH analog found in anorexia nervosa patient (pGlu-His-Gly); CCK, cholecysto-kinin. Neurotensin*, no direct effect on neuron, but inhibites excitatory effect of insulin or 2-DG; Bombesin*, no direct effect on GS neuron but facilitates effect insulin or 2-DG (Shiraishi et al., 1980) (Oomura, 1982).

Morphine and met-enkephalin were shown to have the same effects as glucose in the LHA and VMH of naive rats (Ono et al., 1980). The effects of morphine and met-enkephalin, but not the effects of glucose, were blocked by naloxone, a morphine antagonist. Opioids which have been reported to depress food intake under various acute conditions are found in the hypothalamus, but some results are difficult to evaluate because of addiction and withdrawal effects (King et al., 1979). If data is collected for 24 or more hours, β-endorphin facilitates food intake and naloxone blocks this effect (Grandison, Guidotti, 1977).

Another material which more or less mimics the action of glucose in the LHA is calcitonin. One known function of calcitonin is facilitation of Ca deposits in bone. The blood level of calcitonin increases after eating, and applications of calcitonin i.p. or i.c.v. were reported to suppress food intake in rats (Freed et al., 1979). Its inhibitory effects on GS neurons in the LHA were 65% for a small number (n=14) of neurons tested, but it did affect 14% of the non-GS neurons (n=21) while 86% of these were not affected. Ouabain blocks suppression by glucose but not that by calcitonin, and phenoxybenzamine blocks noradrenalin (NA) but not calcitonin. Calcitonin hyperpolarizes neurons and reduces membrane conductance (Oomura et al., 1982).

Bombesin, found in the visceral region and brain of mammals, is a potent gastric acid releasing factor, possibly through its effect on gastrin secretion. When electroosmotically applied by itself, it had no effect on LHA neurons, but it modulated the effects which other agents induced on LHA GS neurons. Bombesin enhanced both insulin induced and 2-DG induced activity of 70% of the LHA GS neurons tested (n=17) and had no effect on 86% of non-GS neurons (n=42) (Shiraishi et al., 1980). Neurotensin, which has been localized in various parts of the brain and gut, was also ineffective in the LHA when used alone, but like bombesin, it modulated the effects of other agents. Neurotensin suppressed activity which was induced by either insulin or 2-DG in 79% (n=14) of LHA GS neurons, and was ineffective in 74% (n=35) of non-GS neurons (Shiraishi et al., 1980). Bombesin and neurotensin depress feeding (Gibbs et al., 1979; Hoebel, 1982). This might be partly explained by the hyperglycemia, hyperglucagonemia and hypoinsulinemia observed after i.c.v. injection (Brown et al., 1979). Gastrointestinal peptide gastrin was ineffective on 100% of LHA GS neurons (n=40) and 94% (n=17) of VMH GR neurons (Shiraishi et al., 1982).

Newly Identified Organic Acids

Since some metabolites, hormones and endogenous peptides are known to be closely related to feeding and satiety a search was conducted for other, unknown factors by gas chromatograph-mass spectrometry-computer analysis of the blood of rats which were starved for up to 132 hr. Nine blood factors were tentatively identified as feeding related from their variations during the starvation period (Fig.4). Peripheral and central administration of some of these substances have been reported to affect feeding behavior (Novin, Vanderweele, 1977), and electrophoretic microapplication of some in the LHA have been shown to affect feeding-related peripheral processes such as gastric secretion (Shiraishi et al., 1980, 1982). Not every glucoresponsive neuron is affected by all of these materials, but where tests have been made, results indicate that those which could affect food intake might do so through glucose responsive neurons.

Two short organic acids, 2-deoxytetronate (2-DTA), $CH_2(OH) \cdot CH(OH) \cdot CH_2 \cdot COOH$, and 3-deoxypentonate (3-DPA), $CH_2(OH) \cdot CH(OH) \cdot CH_2 \cdot CH(OH) \cdot COOH$, were selected from among these nine materials for further testing. Serum levels of these two acids increased until normal feeding time of the second day and then decreased to a minimum at the beginning of daytime of the third day. At 132 hr of starvation, 2-DTA was about 2.5 times and 3-DPA was about 2 times the levels which existed before deprivation (Fig.4) (Oomura, 1981; Oomura et al., 1982).

Food intake was induced in rat at a rate well above control by 3-DPA within 20 min after ventricle III infusion was started, and this lasted for up to 6 hr. Food consumption was suppressed by 2-DTA for 24 hr. The suppression of feeding by 2-DTA was apparently not due to general depression, since motor activity remained unchanged during the test period. Within the 2 hr period after 3-DPA infusion, food intake was induced with latencies of 8 to 25 min in 9 of 15 rats tested. Feeding was not induced by 2-DTA in any of 17 rats tested. Control infusions of saline or mannitol had no effect. In the 1 day period following infusion, 2-DTA decreased food intake to 65% of control (p<0.01, Student's t test). When 2-DTA or artificial cerebrospinal fluid was injected i.c.v. and food made available after 72 hr of food deprivation, food intake by the 2-DTA treated

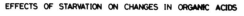

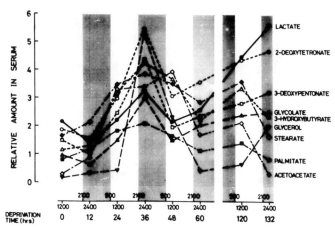

Fig. 4 Relative levels of nine blood serum factors at various times during 132 hr of food deprivation of rats. Glycerol, which has been reported to depress body weight when administered i.p. over an extended period (Davis et al., 1981), had the largest relative changes of all materials shown. Free fatty acids (palmitate, stearate) have been previously tested for their effects on feeding behavior (Adair et al., 1968), and on hypothalamic neurons (Oomura, 1980). The two substances, 2-deoxytetronate and 3-deoxypentonate, see text. Abcissas: lower-deprivation time in hr; upper-actual time of each day. Ordinate: relative amounts of blood serum concentration of each substance. Shaded parts: dark periods, normal feeding times (Oomura et al., 1982).

animals was half that of the controls. This depression persisted for up to 2 hr. After 2 hr the controls and treated animals both appeared to be sated, but the amounts of food consumed in that period were significantly different. Increase of food intake after application of 3-DPA depended on the time of application. If it was infused just prior to normal eating time, increased food intake was not significant, but if it was infused at the beginning of a normal eating hiatus, food intake increased significantly.

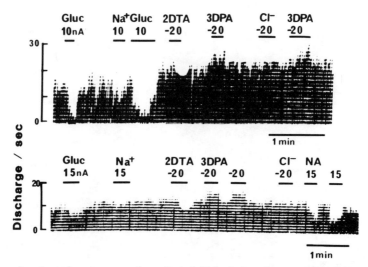

Fig. 5 Specimen records of inhibition of rat LHA GS neurons by 2-DTA and facilitation by 3-DPA. Inhibition by glucose, and no effect by Na$^+$ or Cl$^-$, verifing not osmosensitive or current sensitive. Lower, typical GS neurons, inhibited by noradrenaline (NA). Two different neurons (Oomura et al., 1982).

Simultaneous changes of rat single LHA neuron activity and feeding behavior were recorded following 2-DTA and 3-DPA application. When food was presented at 1800 hr after 24 hr of food deprivation, neuronal activity and food intake both increased for a substantial time. Infusion of 2.5 μmole 2-DTA decreased the neuronal activity and stopped food consumption after about 8 min and for as long as 4 hr. Infusion of 3-DPA at 1400 hr provoked food intake and prandial drinking which was accompanied by increased neuronal activity.

When these organic acids were applied iontophoretically to LHA GS neurons, 2-DTA depressed activity in a dose-related manner, and 3-DPA increased activity (Fig.5). In tests on 150 LHA neurons, the activity of GS neurons was significantly inhibited by 2-DTA and facilitated by 3-DPA (p<0.01) while fewer than 7% of non-GS neurons and no cortical neurons were affected (Table 1). The high correlation between 2-DTA and 3-DPA modification of GS neuron activity and feeding behavior strongly suggests that these two acids might mediate feeding behavior through hypothalamic neurons.

The only apparent reason for the opposite effects induced by these relatively simple and nearly identical molecules is the absence of an OH group from the second (2-DTA) or third (3-DPA) carbon of their respective chains. This conclusion is supported by the effects which are induced by 3-deoxytetronate, $CH_2(OH) \cdot CH_2 \cdot CH(OH) \cdot COOH$. This 4 carbon acid, which is similar to 3-DPA in structure, had similar effects on feeding behavior.

Table I

Effects of 2-DTA and 3-DPA on
LHA Neurones in Anesthetized Rats.

	2-DTA			3-DPA		
	↑	→	↓	↑	→	↓
Glucose-sensitive neuron	3	18 (38)	17*	12*	8 (24)	4
Non glucose-sensitive neuron	1	57 (61)	3	1	23 (27)	3

↑, facilitated; →, no effect; ↓, inhibited
* significant (p<0.01)

Note, 15 GS neurons were each tested for response to both 2-DTA and 3-DPA (Oomura et al., 1982).

Summary

There are neurons in the LHA and the VMH which respond to blood borne glucose in ways which could control feeding behavior and possibly weight, if timing and other factors were appropriate. Some of these other factors include metabolites other than glucose, such as FFA, 2-DTA, 3-DPA, glycerol, etc. They also include hormones such as insulin, TRH, glucagon, calcitonin and others, plus peptides such as enkephaline-like substances, neurotensin, bombesin, etc. Analogues of some of these substances, such as 2-DG, 3-OMG, poligalitol, and TRHA also affect LHA neurons. The problem which remains is twofold: First, the factor or factors which determine the set point must be determined. Second, it must be shown that proper timing exists for the rise and

fall of the blood levels of these substances to terminate
feeding behavior by appropriately affecting LHA and/or VMH
chemosensitive neurons. When the answers to both phases of
this problem have been ascertained, we should be able to
not only affect cures of extreme feeding related problems
such as obesity and anorexia nervosa, but greatly improve
our general health by giving attention to factors which
affect our more subtle eating habits. Despite the great
amount of illuminating results, the important goal of
understanding the feeding process, although it comes ever
nearer, has yet to be reached.

Acknowledgements
      I thank Prof. A. Simpson for help in preparation of
this manuscript. The work was partly supported by
Grant-in-Aid 587035, 00587035 and 57440085 from the
Japanese Ministry of Education, Science and Culture.

References
Adair ER, Miller NE, Booth DA (1968). Effects of
   continuous intravenous infusion of nutritive substances
   on consummatory behavior in rats. Communications in
   Behavioral Biology 2:25.
Brown M, Taché Y, Fisher D (1979). Central nervous system
   action of bombesin: Mechanism to induce hyperglycemia.
   Endocrinology 105:660.
Baskin DG, Dorsa DM (1982). Brain insulin concentrations
   in hyperinsulinemic fatty Zacker rats are not elevated.
   Abstr Soc for Neurosci 8:273.
Davis JD, Wirthshafter D, Asin DE, Brief K (1981).
   Sustained intracerebroventricular infusion of brain fuels
   reduces body weight and food intake in rats. Science
   212:81.
Dean PM, Matthews EK, Sakamoto Y (1975). Pancreatic islet
   cells: Effects of monosaccharides, glycolytic inter-
   mediates and metabolic inhibitors on membrane potential
   and electrical activity. J Physiol 246:459.
Freed WJ, Perlow MJ, Wyatt RJ (1979). Calcitonin:
   Inhibitory effect on eating in rats. Science 206:850.
Gibbs J, Fauser DJ, Rowe EA, Rolls BJ, Rolls ET, Maddison
   SP (1979). Bombesin suppresses feeding in rats. Nature
   (London) 282:208.
Grandison L, Guidotti A (1977). Stimulation of food intake
   by muscimol and beta endorphin. Neuropharmacology
   16:533.
Havrankova J, Roth J, Brownstein MT (1979). Concentrations

of insulin and of insulin receptors in the brain are independent of peripheral insulin levels: Studies of obese and streptozotocin-treated rodents. J Clin Invest 64:636.

Hoebel BG, Hernandez L, McLean S, Stanley BG, Aulissi Ed F, Glimcher P, Margolin D (1982). Catecholamines, enkephalin and neurotensin in feeding and reward. In Hoebel BG, Novin D (eds): "The Neural Basis of Feeding and Reward", Brunswick, Maine: Haer Institute for Electrophysiological Research, p 465.

Ishibashi S, Oomura Y, Okajima T (1979). Facilitatory and inhibitory effects of TRH on lateral hypothalamic and ventromedial neurons. Physiol Behav 22:785.

Kennedy GC (1953). The role of depot fat in the hypothalamic control of food intake in the rat. Proc Roy Soc B 140:578.

King MG, Kastin AJ, Olson RD, Coy DH (1979). Systematic administration of Met-enkephalin (D-Ala^2)-beta-endorphin: Effects on eating, drinking and activity measures in rat. Pharmacol Biochem Behav 11:407.

Langhans W, Zieger V, Scharrer E, Geary N (1982). Stimulation of feeding in rats by intraperitoneal injection of antibodies to glucagon. Science 218:894.

Lorén I, Alumets J, Hakanson R, Sundley F, Thorell J (1979). Gut-type glucagon immunopreactivity in nerves of the rat brain. Histochemistry 61:335.

Mayer J (1955). Regulation of energy intake and the body weight: The glucostatic theory and the lipostatic hypothesis. Ann N Y Acad Sci 63:15.

Niijima A (1981). Visceral afferents and metabolic function. Diabetologia 20:325.

Nisbett RE (1972). Hunger, obesity, and the ventromedial hypothalamus. Psychol Rev 79:433.

Novin D, Vanderweele DA, Rezek M (1973). Infusion of 2-deoxy-d-glucose into the hepatic-portal system causes eating: Evidence for peripheral glucoreceptors. Science 181:858.

Ono T, Oomura Y, Nishino H, Sasaki K, Muramoto K, Yano I (1980). Morphine and enkephalin effects on hypothalamic glucose responsive neurons. Brain Res 185:208.

Oomura Y (1973). Central mechanisms of feeding. In Kotani M (ed): "Advances in Biophysics", Tokyo: Univ. Tokyo Press, p 65.

Oomura Y, Ooyama H, Sugimori M, Nakamura T, Yamada Y (1974). Glucose inhibition of the glucose-sensitive neurons in the lateral hypothalamus. Nature (London)

247:284.

Oomura Y (1980). Input-output organization in the hypothalamus relating to food intake behavior. In Morgane PJ, Panksepp J (eds): "Handbook of the Hypothalamus", Vol 2, Physiology of the Hypothalamus, New York, Basel: Marcel Dekker Inc., p 557.

Oomura Y (1981). Chemosensitive neurons in the hypothalamus related to food intake behavior. Japan J Pharmacol 31, Suppl:1.

Oomura Y, Kita H (1981). Insulin acting as a modulator of feeding through the hypothalamus. Diabetologia 20:290.

Oomura Y, Shimizu N, Miyahara S, Hattori K (1982). Chemosensitive neurons in the hypothalamus: Do they relate to feeding behavior? In Hoebel B, Novin D (eds): "The Neural Basis of Feeding and Reward", Brunswick, Maine: Haer Institute Electrophysiological Research, p 551.

Reichelt KL, Foss I, Trygslag O, Edminson PD, Johanson JH, Bøler JB (1978). Humoral control of appetite-II. Purification and characterization of an anorexogenic peptide from human urine. Neuroscience 3:1207.

Shiraishi T, Inoue A, Yanaihara N (1980). Neurotensin and bombesin effects on LHA-gastrosecretory relations. Brain Res Bull 5, Suppl 4:133.

Shiraishi T, Tsutsui K, Sakata T, Simpson A (1982). 2-deoxy-d-glucose effects on hypothalamic neurons and on short and long term feeding. In Hoebel BG, Novin D (eds): "The Neural Basis of Feeding and Reward", Brunswick, Maine: Haer Institute for Electrophysiological Research, p 373.

Strubbe JH, Steffens AB (1975). Rapid insulin release after ingestion of a meal in the unanesthetized rat. Am J Physiol 229:1019.

Tager H, Hohenboken M, Markese J, Dinerstein RJ (1980). Identification and localization of glucagon-related peptides in rat brain. Proc Nat'l Acad Sci, U S A 77:6229.

Vijayan E, McCann SM (1977). Suppression of feeding and drinking activity in rats following intraventricular injection of thyrotropin releasing hormone (TRH). Endocrinology 100:1727.

The Physiology of Excitable Cells, pages 587–595

NEUROETHOLOGICAL ROLE OF DYNAMIC TRAITS OF EXCITABLE CELLS:
A PROPOSAL FOR THE PHYSIOLOGICAL BASIS OF
SLOTHFULNESS IN THE SLOTH

Theodore Holmes Bullock

Neurobiol. Unit, Scripps Inst. of Oceanography,
Dept. of Neurosciences, School of Medicine,
Univ. of Calif., San Diego, La Jolla, CA 92093

INTRODUCTION

The question I pose in the context of this book is whether the dynamic properties of excitable cells, amply illustrated elsewhere in this volume and in the corpus of work of Susumu Hagiwara, can be of direct neuroethological significance at the level of gross features of vertebrate behavior, for example in causing the slothfulness of the sloth (Bradypus, Edentata), or whether such behavioral anomalies must be explained in terms of system properties, that is emergent properties of constellations of cells. The answer I will come to is affirmative, in the form of a proposition that, for a class of behavioral differences between species, such as the obligate sluggishness of the sloth, cellular properties of large numbers of central neurons do play a large role, but that other mechanisms at several levels of integration also play their roles, for example the balance of modulator substances in local brain regions that exert pervading influences.

The more general question that subsumes this one is: what special features of the brain of an animal with behavior notably different from that of other related species are causally related to the behavioral differences? This is a core problem for neuroethology and comparative neurology; few differences in the anatomy, physiolgy or chemistry of the brain of different species are known to be relevant to their differences in behavior. Yet differences in behavior are surely the principal achievement of evolution, and the corresponding differences between brains

surely go beyond the traditionally studied differences in
size, packing density, lamination and hodology.

STRATEGY

In order to facilitate progress on the relation of
brain to behavior, I am here advocating the comparison of
taxa. We can not be satisfied with finding the common
denominators of nerve impulses, or synaptic transmission or
of the neural basis of posture, or locomotion or motivation.
We should recognize and put to use, sooner rather than
later, the facts that evolution has produced a vast
diversity of animals and that this diversity involves
behavioral adaptation as much as skin, teeth and kidneys.
Just as the geneticists got a handle on the mechanisms of
inheritance by looking at alleles, perhaps we can make
progress by studying the bases of behavioral differences
between species or higher taxa as well as between
pathological and developmental states.

Selection of the cases to be studied is important.
Behavioral differences between taxa can be ill suited for
our purposes if they are too small or subtle, at too high a
level - e.g. habitat preferences, or at too low a level -
e.g. based on differences in the sensory or the motor
apparatus. Ideally, we would like simple and gross
behavioral differences entirely or largely due to midlevel
central neural factors.

A SPECIFIC CASE

In this light, the slothfulness of sloths forced itself
on my attention during an expedition from the U.C.L.A.
Brain Research Institute to Belem in Brazil in 1964, mainly
to study electric fish - an expedition shared with Susumu
Hagiwara, Lawrence Kruger, Horst Schwassmann, Hans Lissmann,
Thomas Szabo and Shiko Chichibu. I observed three toed
sloths in the park of the Museu Goeldi. This is the slower
of the two kinds of sloth and in spite of the occasionally
surprising swipe of its arm - or rather including this
movement, it is obligatorily, inescapably slow, unlike a
lazy cat or even the marsupial koala bear which on occasion
jumps from limb to limb.

The upside down habit had been studied, but the only paper I could find on the apparent laziness (Britton and Kline 1939) was entitled "On deslothing the sloth" and argued that correcting for the normally low body temperature and hypothyroid metabolism and providing enough of a stimulus would convert a sloth into an animal with a normal mammalian time frame. After some preliminary experiments and computations I could not accept this. Per Enger from Oslo joined me and we undertook the first step, a study of how much could be attributed to the muscles, peripheral nervous system and lowest spinal levels. We found (Enger and Bullock 1965) that the muscles of <u>Bradypus</u> are exceedingly slow at any temperature, especially the distal muscles. Nerve conduction is slow, axons are small, monosynaptic reflex central time (dorsal to ventral root time) is at least twice that in a rabbit. But these contribute very little to the overall slowness. Neuromuscular delay (measured by Alan Grinnell, intracellularly) is not importantly elongated.

However, the sloth is not simply a frustrated animal with muscles acting as a bottleneck but with a brain as fast as a cat's. As a comparative physiologist I supposed there must be concomittant specializations in the central nervous system beyond axon conduction velocity and synaptic transmission time. As occasion has permitted I have studied a series of both two toed and three toed sloths, looking for such specializations - not to explain the ecological or evolutionary significance or adaptive value of slowness, but its physiological basis. After a number of experiments in Brazil, in Los Angeles and in Panama (on an expedition with Alan Grinnell), I called for help from my friends Robert B. Livingston, Donald B. Lindsley, James F. Toole and Robert Galambos. A paper on the reflexology has appeared (Toole and Bullock 1973), but the main body of results is still in preparation. The present paper is essentially speculative and although based on those results does not undertake to summarize them in detail.

It appears that motor neurons in the cord slowly attain a low maximum firing rate so that although minimal reaction time to pin prick or electric shock is about 150 ms, essentially the same as in the human, the rate of rise and the time to peak of response are slow. The minimal latency of muscle action potentials to tapping a tendon or a muscle with a reflex hammer is 20-25 ms, not materially different from our own. But the recruitment of tension, due both to muscle properties and to motor neuron acceleration of firing rate, singly and as a population, is so slow that sloths can never jump or even leave the substratum. The minimal latency of mechanical response to motor cortex stimulation is 50-60 ms; most of that is in the muscle, after the muscle action potential.

We know little about the physiology of the motor system. Incidentally, the cerebellum is as large as that of a cat! This is just one of many comparisons between species that violate the classical claims based on a few comparisons such as the small size in frogs and large size in birds. Nystagmus beats are slow, typically 1-3 per second. Tetanus fusion frequency in various muscles is from a maximum of 25 down to less than 5 per second; in a cat or rabbit these numbers are from 85 down to 70. Motor convulsions are largely tonic with some slow clonus.

The EEG power spectrum looks like that of a cat, both during the alert state and in slow wave sleep. The EEG during seizures is not appreciably different from that in familiar mammals.

Evoked potentials to sensory stimuli such as clicks and flashes are significantly later and recover more slowly than those in a cat; the difference is probably less than a factor of two and amounts to a small fraction of the response latency to a visual or auditory stimulus. Values in this domain are quite variable among specimens and even in the same individual depending on the stimulus conditions, regime, status of the animal and unknown factors. The generalization is based on a good many individuals, recorded both under barbiturate and with implanted electrodes in the alert state.

Each property measured either looks much like that in a cat, or is slower but not enough to explain the conspicuous slothfulness.

Looking "deeper" we had a flurry of excitement after noticing that the striatum is large; the caudate nucleus is huge. Could the sloth be excessively inhibited? In several animals we implanted two pairs of fine cannulas through which we injected small doses of novocaine (<0.01 - 0.05 ml) on one or both sides. Effects wore off in 10-20 minutes and

we could repeat the local anesthetic or inject a saline
control. The visible effects of depressing the caudate in
this way are mainly in the direction of enhanced activity.
Spontaneous movements increase, from slightly to remarkably.
The head may turn and peer irregularly or may bob at about
1/s or quiver at 5-7/s or shudder intermittently; the
eyelids may move at ca. 2/s; the eyes may move randomly or
show nystagmus which can be vertical or oblique. The
mandible may show chewing movements; the arms may quiver
intermittently or grope; the tail may be alternately
elevated and depressed. Sometimes there is a suggestion of
increased locomotion and agonistic behavior. With
unilateral injection the signs are mainly contralateral.
They develop within 3 min and subside by 20 min. Novocaine
in the caudate does release signs of increased motility but
they are not deslothed or speeded up. They provide at most
a partial explanation, limited to the laziness aspect but
not the sluggish aspect of the characteristic behavior. The
large caudate, we learned, is just as much a feature of
edentate relatives of Bradypus, such as anteaters and
armadillos, including species that are not at all slothful
but rapid runners and modest jumpers.

To test the hypothesis that the sloth has a
parasympathomimetic syndrome permanently and little or no
sympathetic arousability, we did two things. One was to
stimulate the cervical sympathetic trunk and measure
pupillary dilation. From a light adapted pin hole start,
such stimulation causes an increase in area of the pupil of
8 fold with a half time of 10-15 s. In the cat we could get
an increase of 15 fold with a half time of 2 s. The other
manipulation was to inject a large dose of adrenalin to look
for acceleration of heart rate. The rate increased by 19%
to a 10 ug/kg rapid injection, compared to 22% in a cat;
the onset of the acceleration was slightly quicker in the
cat. A 36% increase in heart rate can be caused by
attempting to open the claws of a resting Bradypus - which
is quite arousing. We can reject the hypothesis of a
permanently placid animal; emergency sympathetic mechanisms
are not lacking though some of the effects are slow.

SPECULATIONS

Clearly, the goal of accounting for the apparently
simple anomaly of behavior in this animal is not at hand and

not likely to be achieved with one more crucial experiment. It may be heuristic to attempt to list the possibilities, as I see them today, even if the main value proves tomorrow to be improving a stupid formulation or highlighting others I have overlooked. I have already rejected three possible explanations.

(1) One is that the basic differences between sloths and less sluggish mammals are a low thyroid, low body temperature and lack of effective stimuli. I do not believe these are the main distinctions from familiar species. Britton and Kline (1939) got a 38% increase in speed of locomotion, in meters/s, when the sloth was under the influence of the most effective of the drugs tried, prostigmin; and an average of 42% with the optimum increase in body temperature – from a rate of 0.36 to 0.80 km/hr! The kinds of startle behavior, muscle twitch, and evoked potential measurements we have made would be somewhat altered by these factors but would still appear conspicuously slothful compared to a rabbit, cat, anteater or armadillo.

(2) The second possibility already rejected was that, since Enger and I found sloths to have slow muscles, that might be enough; the muscles are a bottleneck and ensure slow movement. I am betting there are additional, essential, parallel, central specializations, even though all the rate functions so far measured, in spite of being substantially slower than standard species, do not seem to add up to an adequate accounting.

(3) The third possibility already rejected was that sloths are so completely parasympathomimetic as to be incapable of arousal. We found a good sympathetic response, although it was slow to build up.

(4) A fourth possibility has not been tested. This is that some balance of transmitters or modulators, perhaps a deficiency in dopamine or an excess of GABA in some strategic region, most likely limbic, is the important deviation from familiar species. Some anomalies or pathologic states within familiar species, affecting behavior, mood or movements, have been attributed to such deficiencies. I have no basis for placing a high or a low probability that this factor is playing a significant role in the sloth. It might be well worth investigation.

(5) A different sort of balance, at least conceptually, is represented by the classical concept of "half centers", such as those controlling expiration and inspiration, heart rate acceleration and deceleration, hunger and satiety. Like differences in vagal tone, i.e. the tonic activity of inhibitory centers on the heart rate, there might be differences in the activity of limbic structures that cause movements, autonomic output and emotions in the domain of temperament. In the sloth these might lean toward low levels of fear and escape as well as pursuit and prey capture. Such a balance of tone in a half center may or may not involve a difference in size or number of cells, processes, synapses or quantity of the corresponding transmitters. It does predict that we should be able to desloth a sloth by the right lesion or stimulation or drug, at least within the limitations of its downstream specializations, such as slow interneurons, motor neurons and muscles.

(6) A sixth possibility is that circuit dynamics will turn out to be the main specialization. The time constants of the connectivities involved, therefore the system properties, may be the most important determinants of the slow movements. This is in principle distinguishable from the next alternative and as such I am not betting on it to be important.

(7) The last possibility I will mention, and one I would emphasize, by asserting the proposition in the introduction to this essay, is that cellular properties are important, at least as bottlenecks, all along the pathways. By this I mean the same kinds of properties that are familiar contrasts between small as opposed to large sensory neurons, interneurons and motor neurons, as pointed out by Katsuki et al.(1951,1954), Bullock (1953) and Henneman (1957, see also Henneman et al. 1965). These include the tendency for small cells to be longer in latency, lower in maximum firing frequency and more sustained or tonic during maintained stimulation. Other kinds of "personality" differences are known, for instance among identified cells of Aplysia – in respect to tendency to burst, to afterdischarge or to rebound, to abrupt starts versus gradual acceleration of firing rate, to autoinhibition and the like. Just as the muscles of Bradypus are different in respect to cellular properties of contraction rate, time to peak tension, and relaxation time, compared to homologous

muscles of a cat, I am betting the neurons will also be different, from the sensory right through to the motor neurons, and that concatenations of such cells will go a long way to account for the jungle mysteries we started with.

CONCLUDING COMMENTS

While favoring the last possibility, I do not exclude some of the others, as indicated in each item, and I explicitly adhere to the pluralistic view that co-adaptations exist at each of several levels. I do not expect the higher level specializations by themselves will explain the sloth's characteristic behavior. Parenthetically, the cell properties listed above as most significant are themselves still mysteries, for the most part; we can not explain most of them in terms of membrane or intracellular parameters as yet, though doubtless this is well on its way already, as many chapters in this book suggest.

I have given a bit of detail about an obviously incomplete investigation to make a point that seems to me of rather general interest in the domain of putting the physiology of excitable cells into the intact, functioning animal. The point is this: for all the seeming sophistication of instrumentation and measurement in neurophysiology, we are seriously deficient in thinking up the measurable variables relevant to some superficially simple traits of behavior by which animals differ widely, whether taxa or pathologic or developmental states. We are in need of more ideas as to what quantitative assessments are more than trivially correlated with such traits. I see it as an intellectual challenge, one demanding a good ethological understanding of the species under comparison, plus a very practical, operational approach to explanation in brain terms, and a large dose of uninhibited imagination. In spite of my several experimental forays, a series of some three dozen animals, and collaborators much more perspicacious than I, it is my opinion that the main secrets of the central specialization for slothfulness are still unknown.

This may not trouble you! After all, who but sloth lovers would care if we can not understand an exotic edentate? I think the dilemma is general, however. We do not know what we need to know about differences between brains to understand behavior in the variety of ways by which one vertebrate species differs from others. We do not have a list of measurements of the physiology, anatomy or chemistry of the brain which, taken together would adequately characterize the brain in respect to its performance. We can not even account for most of the central time it takes for us, or for a cat to do something!

LIST OF REFERENCES

Britton SW, Kline RF (1939) On deslothing the sloth. Science 90: 16-17.

Bullock TH (1953) Comparative aspects of some biological transducers. Fed Proc 12: 666-672.

Enger PS, Bullock TH (1965) Physiological basis of slothfulness in the sloth. Hvalradets Skrifter 48:143-160.

Katsuki Y, Yoshino S, Chen J (1951) Neural mechanism of the lateral-line organ of fish (Fundamental neural mechanism of sensory organs in general). Jap J Physiol 1: 264-268.

Katsuki Y, Chen J, Takeda H (1954) Fundamental neural mechanism of the sense organ. Bull Tokyo Med Dent Univ 1: 21-31.

Henneman E (1957). Relation between size of neurons and their susceptibility to discharge. Science 126: 1345-1346.

Henneman E, Somjen G, Carpenter DO (1965). Functional significance of cell size in spinal motoneurons. J Neurophysiol 28: 560-580.

Toole JF, Bullock TH (1973) Neuromuscular responses of sloths. J Comp Neurol 149: 259-270.

T.H. Bullock; D. Junge; R. Gruener

Index